Lecture Notes in Computational Science and Engineering

79

For further volumes:
http://www.springer.com/series/3527

Michael Griebel · Marc Alexander Schweitzer
Editors

Meshfree Methods for Partial Differential Equations V

Springer

Editors
Michael Griebel
Universität Bonn
Institut für Numerische Simulation
Wegelerstr. 6
53115 Bonn
Germany
griebel@ins.uni-bonn.de

Marc Alexander Schweitzer
Universität Stuttgart
Institut für Parallele und Verteilte Systeme
Universitätsstr. 38
70569 Stuttgart
Germany
marc.alexander.schweitzer@ipvs.uni-stuttgart.de

ISSN 1439-7358
ISBN 978-3-642-26583-9   ISBN 978-3-642-16229-9 (eBook)
DOI 10.1007/978-3-642-16229-9
Springer Heidelberg Dordrecht London New York

Library of Congress Control Number: 2010937847

Mathematics Subject Classification (2010): 65N99, 64M99, 65M12, 65Y99

Cover design: deblik, Berlin

Printed on acid-free paper

Springer is part of Springer Science+Business Media (www.springer.com)

Preface

The Fifth International Workshop on *Meshfree Methods for Partial Differential Equations* was held from August 17 to August 19, 2009 in Bonn, Germany. One of the major goals of this workshop series is to bring together European, American and Asian researchers working in this exciting field of interdisciplinary research on a regular basis. To this end Ivo Babuška, Ted Belytschko, Michael Griebel, Antonio Huerta, Wing Kam Liu, and Harry Yserentant invited scientist from all over the world to Bonn to strengthen the mathematical understanding and analysis of meshfree discretizations and to promote the exchange of ideas on their implementation and application.

The workshop was again hosted by the Institut für Numerische Simulation at the Rheinische Friedrich-Wilhelms-Universität Bonn with the financial support of the Sonderforschungsbereich 611 *Singular Phenomena and Scaling in Mathematical Models* which is kindly acknowledged. Moreover we would like to thank Christian Rieger who carried most of the load as local organizer of this workshop.

This volume of LNCSE now comprises selected contributions of attendees of the workshop. Their content ranges from applied mathematics to physics and engineering which clearly indicates the maturaty meshfree methods have reached in recent years. They are becoming more and more mainstream in many areas of applications due to their flexiblity and wide applicability.

Bonn,
July, 2010

Michael Griebel
Marc Alexander Schweitzer

Contents

Global-local Petrov-Galerkin formulations in the Meshless Finite Difference Method
Slawomir Milewski, Janusz Orkisz 1

Treatment of general domains in two space dimensions in a Partition of Unity Method
Marc Alexander Schweitzer, Maharavo Randrianarivony 27

Sampling Inequalities and Support Vector Machines for Galerkin Type Data
Christian Rieger 51

Meshfree Vectorial Interpolation Based on the Generalized Stokes Problem
Csaba Gáspár 65

Pressure XFEM for two-phase incompressible flows with application to 3D droplet problems
Sven Gross 81

Special-relativistic Smoothed Particle Hydrodynamics: a benchmark suite
Stephan Rosswog 89

An exact particle method for scalar conservation laws and its application to stiff reaction kinetics
Yossi Farjoun, Benjamin Seibold 105

Application of Smoothed Particle Hydrodynamics to Structure Formation in Chemical Engineering
Franz Keller, Ulrich Nieken 125

Numerical validation of a constraints-based multiscale simulation method for solids
Konstantin Fackeldey, Dorian Krause, Rolf Krause 141

Coupling of the Navier-Stokes and the Boltzmann equations with a meshfree particle and kinetic particle methods for a micro cavity
Sudarshan Tiwari, Axel Klar 155

Accuracy and Robustness of Kinetic Meshfree Method
Konark Arora, Suresh M. Deshpande 173

Kinetic meshless methods for unsteady moving boundaries
V. Ramesh, S. Vivek, S. M. Deshpande 189

Efficient cloud refinement for kinetic meshless methods
M. Somasekhar, S. Vivek, K. S. Malagi, V. Ramesh, S. M. Deshpande .. 207

Fast exact evaluation of particle interaction vectors in the finite volume particle method
Nathan J. Quinlan, Ruairi M. Nestor 219

Parallel summation of symmetric inter-particle forces in smoothed particle hydrodynamics
Johannes Willkomm, H. Martin Bücker 235

Meshfree Wavelet-Galerkin Method for Steady-State Analysis of Nonlinear Microwave Circuits
Alla Brunner 249

Global-local Petrov-Galerkin formulations in the Meshless Finite Difference Method

Slawomir Milewski[1] and Janusz Orkisz[2]

[1] Institute for Computational Civil Engineering, Cracow University of Technology, Warszawska St.24, Cracow, Poland slawek@L5.pk.edu.pl
[2] Institute for Computational Civil Engineering, Cracow University of Technology, Warszawska St.24, Cracow, Poland plorkisz@cyf-kr.edu.pl

Summary. The paper presents the recent developments in both the Local Petrov-Galerkin (LPG) formulations of the boundary value problems of mechanics, and the Meshless Finite Difference Method MFDM of numerical analysis. The MLPG formulations use the well-known concept of the Petrov-Galerkin weak approach, where the test function may be different from the trial function. The support of such test function is limited to chosen subdomains, usually of regular shape, rather than to the whole domain. This significantly simplifies the numerical integration. MLPG discretization is performed here for the first time ever, in combination with the MFDM, the oldest and possibly the most developed meshless method. It is based on arbitrarily irregular clouds of nodes and moving weighted least squares approximation (MWLS), using here additional Higher Order correction terms. These Higher Order terms, originated from the Taylor series expansion, are considered in order to raise the local approximation rank in the most efficient manner, as well as to estimate both the a-posteriori solution and residual errors. Some new concepts of development of the original MLPG formulations are proposed as well. Several benchmark problems are analysed. Results of preliminary tests are very encouraging.

Key words: Meshless Local Petrov Galerkin, Meshless Finite Difference Method, Higher order approximation

1 Introduction

The MFDM ([13, 14]) is one of the basic discrete solution approaches to analysis of the boundary value problems of mechanics. It belongs to the wide group of methods called nowadays the Meshless Methods (MM, [3,13,14,27]). The MM are contemporary tools for analysis of boundary value problems. In the meshless methods, approximation of a sought function is described in terms of nodes, rather than by means of any imposed structure like elements, regular meshes etc. Therefore, the MFDM, using arbitrarily irregular clouds

M. Griebel, M.A. Schweitzer (eds.), *Meshfree Methods for Partial Differential Equations V*, Lecture Notes in Computational Science and Engineering 79,
DOI 10.1007/978-3-642-16229-9_1, © Springer-Verlag Berlin Heidelberg 2011

of nodes, and the Moving Weighted Least Squares (MWLS, [11, 14]) approximation falls into the category of the MM, being in fact the oldest [14] and possibly the most developed one of them. The bases, and the recent state of the art in the research on the MFDM, as well as several possible directions of its development are briefly presented in [14, 25].

The MFDM may deal with boundary value problems posed in every formulation [14], where the differential operator value at each required point may be replaced by a relevant difference operator involving a combination of searched unknowns of the method. Using difference operators and an appropriate approach, like the collocation, Petrov-Galerkin variational principle, and functional minimisation, simultaneous MFDM equations may be generated for any boundary value problem analysed.

In the recent years, in many applications of mechanics, the Local Petrov-Galerkin (LPG) formulations [2] have gained popularity. They use the old concept of the Petrov-Galerkin approach, in which the test function (v) may be different from the trial function (u) but is limited to subdomains, rather than to the whole domain at once.

The objective of this paper is to present brief outlines of the actual development of the MFDM, especially in combination with the MLPG formulations. A presentation of some results of the current research on the Higher Order approximation (HOA) [19–26] in the MFDM will be also given, with special emphasis laid upon the MFD discretization of the MLPG formulations. Such solution approach is performed here for the first time ever. Both the original MLPG formulations, and a new one (e.g. the MLPG7) proposed here, are tested for the MFDM discretization and compared with the standard formulations of boundary value problem. A variety of benchark problems was analysed. Convergence and precision of solutions were compared for various formulations. Fast and high quality results were obtained when using a combined MLPG/MFDM solution approach in the benchmark boundary value problems.

2 Boundary value problem formulations

In the MFDM, any boundary value problem formulation that involves function and its derivatives may be used [14]. Boundary value problem posed in the local formulation is understood as one or a set of differential equations and appropriate boundary conditions, satisfied at every point P of the domain $\Omega \subset \Re^n$, and its boundary $\partial\Omega$.

$$\begin{cases} Lu = f \\ Gu = g \end{cases} \quad u = u(P) \quad \begin{cases} P \in \Omega \\ P \in \partial\Omega \end{cases} \tag{1}$$

Also the global formulations may be analysed by the MFDM nowadays. These may be posed either in the form of optimisation of the functional

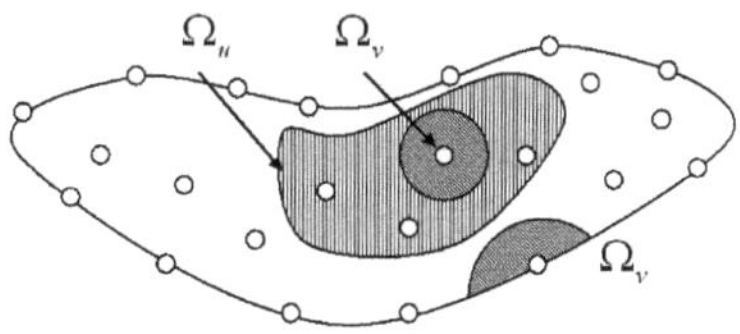

Fig. 1. Concept of the MLPG formulations

$$I(u) = \frac{1}{2}b(u,u) - l(u) \tag{2}$$

satisfying the boundary conditions from (1), or as variational principles (e.g. the principle of virtual work)

$$b(u,v) = l(v) \quad \text{for} \quad v \in V_{adm} \tag{3}$$

with or without constraints.

The global approach involves integration over the domain Ω. Mixed, global / local formulations may be also considered. Those may be given in form of a functional or the variational principle with local constraints. In the present work specially investigated are the global-local Petrov-Galerkin formulations ([2, 26]) when the functional or variational principle is only locally satisfied.

3 Meshless local Petrov-Galerkin formulations

These group of formulations uses the old Petrov-Galerkin concept in which the test function (defined on subdomain Ω_v) may be different from the trial function (defined on subdomain Ω_u). The test function may be defined in such a way that it is different from zero in a subdomain and is equal to zero elsewhere in the domain considered. Thus in the MLPG, the domain is already discretized using arbitrarily scattered nodes, without any structure imposed on them. This type of discrete formulations seems to be natural for the meshless methods. One may prescribe to each node the local subdomain, with simple regular shape which stands for the support of the test function (Fig.1), where it is different from zero.

Atluri [2] proposed a classification of MLPG formulations. They all differ from each other in the form of the test function and they are consequently named from the MLPG1 to MLPG6. This classification is presented in the Fig.2, together with the equivalent formulation from mathematics and some examples of applications.

NAME	FORM OF TEST FUNCTION (v)	FORMULATION EQUIVALENT TO	EXEMPLARY APPLICATIONS
MLPG1	WEIGHT FUNCTION (w) IN THE MWLS APPROXIMATION	-	VARIOUS MESHLESS METHODS
MLPG2	DIRAC DELTA FUNCTION $\delta(x-\bar{x})$	LOCAL FORMULATION	POINT COLLOCATION METHOD
MLPG3	RESIDUAL ERROR FUNCTION $r=b(\bar{u},v)-l(v)$	FUNCTIONAL MINIMISATION	ENERGY METHODS
MLPG4	FUNDAMENTAL SOLUTION $\hat{u} \rightarrow b(\hat{u},v)=\delta(x-\bar{x})$	SOMIGLIANA FORMULATION	BOUNDARY ELEMENT METHOD
MLPG5	HEAVISIDE STEP FUNCTION $H(x-\bar{x})$	-	VARIOUS MESHLESS METHODS
MLPG6	TRIAL FUNCTION $u(x)$	STANDARD GALERKIN APPROACH	FINITE ELEMENT METHOD

Fig. 2. Classification of the MLPG formulations, proposed by Atluri

4 Meshless local Petrov-Galerkin 5 (MLPG5) formulation

In this paper specially investigated is the MLPG5 formulation, in which the test function is given in the form of the Heaviside (step) function [2,26]. In such a way, the test function has a constant value over the subsequent subdomains prescribed to the each node (Fig.3). It should be stressed that the **MLPG5** name may be decoded, as follows:

- Meshless – because the domain is discretized using cloud of arbitrarily distributed nodes,
- Local – because the variational principle (3) is only locally satisfied,

$$\begin{array}{l} b(u,v) = l(v) \\ v \in \Omega_i \quad , \quad i = 1,...,N \end{array} \tag{4}$$

- Petrov-Galerkin – because the approximation of the test function differs from the approximation of the trial function,
- **5** (see Fig.2) – because the test function is the Heaviside function.

$$x_i \quad \longrightarrow \quad \Omega_i : \quad v(x,y) = \begin{cases} 1, & (x,y) \in \Omega_i \\ 0, & (x,y) \notin \Omega_i \end{cases} , \quad i = 1,2,....,N \tag{5}$$

Such discrete formulation (MLGP5) has numerous advantages. Among them one may mention the following:

- in practice the test function is defined on a local subdomain, where it has a constant value,
- in variational principle all terms containing derivatives of the test function vanish,

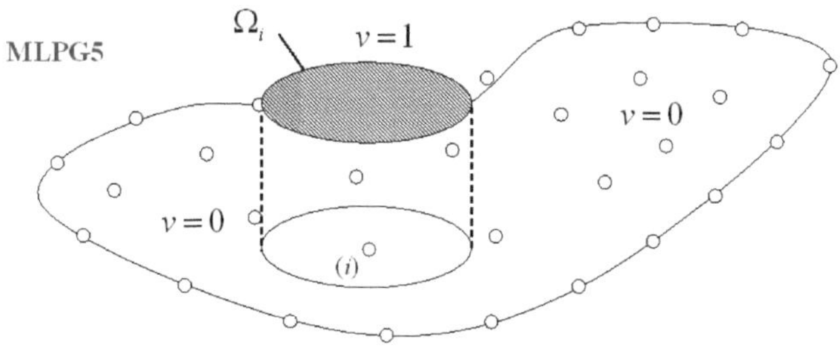

Fig. 3. Concept of the MLPG5 formulation

- only trial function needs to be approximated using its nodal values from the neighbourhood,
- the support of the trial function may be different from the test function subdomain,
- integration is limited to simple subdomains only,
- in some cases integration may be reduced only to the subdomain boundary.

In the original Atluri's concept of the MLPG5 [2], the domain has to be covered with simple subdomains prescribed to each node (circles, rectangles, etc), which may overlap but have to cover the whole domain. It was shown ([2]), how to determine the shape and size of such subdomains which produce stable and unique solution. The final results, however, may strongly depend on the way of choosing these subdomains.

Moreover, the method (MLPG5) can not be applied to those boundary value problem formulations where differentiation is prescribed to the test function only, e.g. the second order non-symmetric variational principle, commonly applied in the partition of unity methods [27].

All these drawbacks may be removed when using the MLPG formulation together with the Meshless Finite Difference Method (MFDM) solution approach. We propose here, for the first time, combination of the MFDM solution algorithm with the MLPG5 formulation. Moreover, some extensions of the original Atluri's concept will be also proposed and tested.

5 Basic Meshless Finite Difference Method solution approach

The basic MFDM solution approach consists of the following steps:

- generation of the cloud of nodes,
- cloud of nodes topology determination,
- MFD star determination,
- function discretization (selection of degrees of freedom) and Moving Weighted Least Squares approximation,

- generation of the MFD operators,
- numerical integration (for the global formulations),
- generation of the MFD equations,
- discretization of the boundary conditions,
- solution of the Simultaneous Algebraic System of Equations (SAE),
- appropriate postprocessing.

The bases and the recent state of the art in the research on the MFDM, as well as several possible directions of its development are briefly presented in [14,25]. Although the MFDM is the oldest, and possibly the most developed meshless method, its solution approach is still being currently developed. The latest MFDM extensions include the higher order approximation based on correction terms, multipoint approach, a-posteriori error estimation as well as an adaptation approach, and are presented in [15–17, 19–26].

Independently of the bvp formulation type applied, one always starts from the generation of a cloud of nodes, that may be irregularly scattered, but usually are without any imposed structure on them, like finite element or regular mesh, and have no mapping restrictions to the regularized stencil. In such a cloud of nodes, some additional ones may be easily added, removed or shifted, if necessary, causing only small changes in nodes structure.

Basically any nodes generator might be applied. However, it is very convenient to use a nodes generator specially designed for the MFDM, e.g. the Liszka type [12, 13, 18], that is based on the nodes density control. Nodes are 'sieved out' from the regular very dense background mesh, according to a prescribed density. Although such a generator provides arbitrarily irregular cloud of nodes, it is useful to determine its topology afterwards. This includes generation of the subdomains prescribed to nodes – i.e. Voronoi polygons (in 2D or polyhedrons in 3D), and (in 2D) the Delaunay triangles – placed between nodes (Fig.4). The topology information may be applied for star generation and/or for integration purposes. In the MLPG analysis, such subdomains may be also used as the supports for the test function.

Once the nodes for MFD stars are selected (e.g. using topology oriented criteria like Voronoi neighbors), the local approximation of the unknown function is performed at every point of interest (node, Gauss point). It is done using the Taylor series expansion, and the MWLS approximation [11, 14]. It is crucial to the method that the MFD star may consist of more nodes (m) than the minimum required to provide the approximation order (p). Evaluation of the MWLS approximation requires the minimisation of the weighted error functional

$$J = (\boldsymbol{q} - \boldsymbol{P}\boldsymbol{D}u)^t \boldsymbol{W}^2 (\boldsymbol{q} - \boldsymbol{P}\boldsymbol{D}u) \tag{6}$$

where $\boldsymbol{P}$ - interpolants matrix, $\boldsymbol{D}u$ – derivatives vector (up to the p-th order), $\boldsymbol{q}$ –MFD star nodal values vector, and $\boldsymbol{W}$ – diagonal weights matrix. Weighting functions are singular for the central node of the MFD equations. In this way, interpolation is enforced there. As a consequence, the essential boundary conditions are satisfied without any additional techniques.

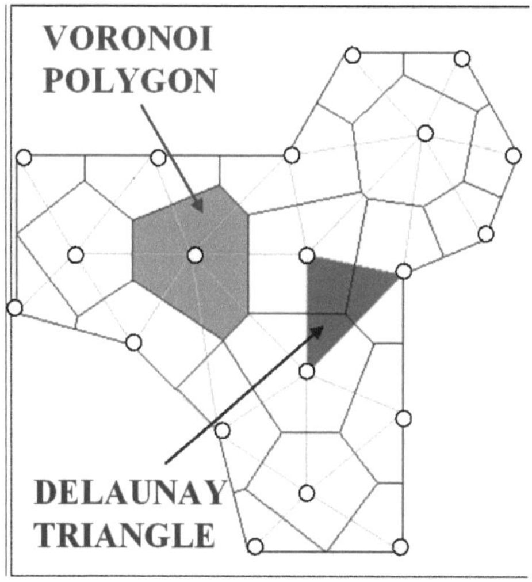

Fig. 4. Topology of an irregular cloud of nodes

After minimisation of (6), one obtains the complete set of derivatives $\boldsymbol{K}$, up to the p-th order

$$\boldsymbol{D}u = \boldsymbol{K}\boldsymbol{q} \quad , \quad \boldsymbol{K} = (\boldsymbol{P}^t\boldsymbol{W}^2\boldsymbol{P})^{-1}\boldsymbol{P}^t\boldsymbol{W}^2 \tag{7}$$

which are, in fact, the coefficients for the local approximation. Moreover, the approximation error may be easily estimated, by considering several additional terms of the Taylor series expansion. It is worth stressing that other meshless methods, use equivalent polynomials for the function approximation [3], instead of the truncated Taylor series. However, though the results of the approximation are the same, the polynomial approach does not provide at once so much valuable information (e.g. about local errors and derivatives).

More deails concerning the MWLS approximation and study of various parameters which have influence on its quality are given in [11, 14, 25]. Some extensions of the MWLS approximation, like use of the generalised degrees of freedom or local constrains are presented in [14].

In the case of global formulations integration is required. The following techniques may be used:

- integration around nodes over the Voronoi polygons, which is the best solution for the even order differential operators,
- integration between the nodes over the Delaunay triangles (2D), which produces the most accurate results for the odd order differential operators,
- integration on a background mesh, independent of the nodes distribution,
- integration over the zones of influence of the weighting functions of the MWLS approximation.
- integration over the local subdomains (MLPG).

Generation of the MFD operators, which appears in formulations (1)÷(3) is done using the MWLS approximation, and appropriate formulae composition. Generation of the MFD equations depends on the formulation type. In the case of local formulation (1) one may use the collocation technique, whereas in the case of global formulations (2) and (3), one has to minimise the energy functional or use the relevant variational principle.

The essential boundary conditions are automatically satisfied by using singular weights in the MWLS approximation. However, discretization of the natural or mixed boundary conditions usually requires additional MFD approximation on the boundary. Such approximation may use only internal nodes from the domain, but it is of poor quality then. Introducing additional, external, fictitious nodes or generalised degrees of freedom may raise the approximation quality on the boundary.

The MFDM solution approach ends with analysis of the S(L)AE. It is most convenient to use a solver which takes advantages of the method's nature, like the multigrid approach. The final postprocessing of the results is performed using once again the MWLS approximation.

There are many extensions of the basic MFDM solution approach. Among them one may mention:

- cloud of nodes adaptation [18,23–25],
- various MWLS extensions [14],
- higher order approximation [15–17,19–25]
- a-posteriori error analysis [1,9,23–25],
- multigrid solution approach [5,18],
- MFDM in various bvp formulations, including the MFDM/MLPG combinations [26],
- MFDM on the differential manifold,
- MFDM / FEM combinations [10],
- Experimental and numerical data smoothing [8].

Some of them are discussed here in a more detailed way.

6 Combination of the MFDM and MLPG5

In the standard bvp global formulations (2)÷(3), both the test and trial functions are prescribed by sets of their nodal values. The trial function is approximated on the MFD star, whereas the test function needs interpolation over the integration cell.

Here proposed is the combination of the MFDM solution approach with the original Atluri's concept of the MLPG5 [2]. The variational principle is satisfied locally then on the Voronoi polygons prescribed to each node. In such a way, the local subdomains do not overlap and they cover the whole domain. The test function is constant over the integration cell (Voronoi polygon), and does not need approximation. Only the trial function is prescribed by the set of nodal values. It is approximated on the MFD star using the MWLS approach.

7 Higher order approximation based on correction terms

The solution quality may be improved by increasing the number of nodes or by raising the order of local approximation. This may be done using Higher Order MFD operators [4, 7], generalised degrees of freedom [14], multipoint approach [6, 15–17] or Higher Order Approximation (HOA, [19–25], based on correction terms. The last approach will be discussed here.

Instead of providing new nodes or degrees of freedom into the simple MFD operator, considered are some additional terms. They result from the Taylor series expansions of the simple MFD operator coefficients. Besides the Higher Order derivatives $\boldsymbol{D}u^{(H)}$, they may contain also the $\boldsymbol{S}$ingularity $\boldsymbol{c}^t\boldsymbol{S}$ or discontinuity ($\boldsymbol{J}$ump) terms $\boldsymbol{e}^t\boldsymbol{J}$ ($\boldsymbol{c}$,$\boldsymbol{e}$ - appropriate MWLS coefficients). Higher Order derivatives may be calculated using appropriate formulae composition and use of the basic MFD solution, corresponding to the simple, not improved MFD operator.

The Higher Order Approximation (HOA) concept is based on splitting the MWLS approximation terms into two parts, namely the low (L) and higher order (H) ones

$$\boldsymbol{P}^t\boldsymbol{D}u^{(L)} \quad + \quad \left(\boldsymbol{P}^{(H)}\right)^t \boldsymbol{D}u^{(H)} - \boldsymbol{c}^t\boldsymbol{S} - \boldsymbol{e}^t\boldsymbol{J} \quad = \quad \boldsymbol{q} \tag{8}$$

These additional Higher Order terms are treated as the known values. In such a way, the final results (derivatives up to the p-th order) depend on the nodal values, and on the correction terms $\boldsymbol{\Delta}$ mentioned above

$$\boldsymbol{D}u^{(L)} = \boldsymbol{K}\boldsymbol{q} - \boldsymbol{\Delta}, \quad \boldsymbol{\Delta} = \boldsymbol{K}\left[\left(\boldsymbol{P}^{(H)}\right)^t \boldsymbol{D}u^{(H)} - \boldsymbol{c}^t\boldsymbol{S} - \boldsymbol{e}^t\boldsymbol{J}\right] \tag{9}$$

It is assumed here, that the approximation order is raised to $2p$–th. The Higher Order derivatives are calculated in the most accurate manner then.

The Higher Order MFD solution is obtained in two steps. In the first step, only the low order part (7) is taken into account and the basic, low order solution $\boldsymbol{u}^{(L)}$ is obtained. Then, after the implicit postprocessing, values of the correction terms (9) are calculated, using formulae composition of the low order solution. They modify the right hand sides of the MFD equations, leaving the coefficient matrix unchanged. The new improved Higher Order solution $\boldsymbol{u}^{(H)}$ is exact within the approximation order assumed ($2p$-th) and, in general, does not depend on the quality of the MFD operator.

Higher Order solution may be applied in many aspects of the MFDM approach, especially for improving

- solution quality inside the domain [19],

- solution quality on the boundary [20],
- the a-posteriori error analysis [21–25],
- the adaptation criteria [23–25],
- and for modification of the multigrid approach [13, 18, 23–25].

Special attention is paid here to the a-posteriori error analysis.

8 A-posteriori error analysis

Once the exact (true) solution $u^{(T)}$ is known e.g. for the benchmark problems, one may calculate the exact solution errors for both the MFD solutions at every point of the domain and/or its boundary

$$\begin{cases} e_i^{(LT)} = u_i^{(L)} - u_i^{(T)} \\ e_i^{(HT)} = u_i^{(H)} - u_i^{(T)} \end{cases} \tag{10}$$

Replacing the true solution $u^{(T)}$ by the approximate Higher Order one $u^{(H)}$, one obtains a local estimation of the Low Order solution

$$e_i^{(LH)} = u_i^{(L)} - u_i^{(H)} \tag{11}$$

Similarly, the exact residual error, defined for the continues solution $\bar{u}$, based on approximation of the exact nodal values $u_i^{(T)}$

$$\bar{r} = L\bar{u} - f \tag{12}$$

may be locally estimated using either the standard low order estimation

$$r_i^{(L)} = Lu_i - f_i \tag{13}$$

or the improved Higher Order residual formula

$$r_i^{(H)} = Lu_i - \Delta_i - f \tag{14}$$

The second one will be used for the adaptation criteria of new nodes.

Solution errors (10) may be also measured over a chosen subdomain or over the whole domain. The numerical integration is involved then. Once a reference solution is applied instead of the analytical one, one gets the so called error estimator denoted here as η. There are several types of such global estimators, commonly defined for and applied in the Finite Element Method analysis [1, 9]. They all differ in the form of the reference solution.

For the group of hierarchical estimators, the solution obtained either from a denser mesh ($h \to h/2$, h-type) or from the raised approximation order ($p \to p+1$, p-type) is needed. Here, we propose the use of the Higher Order MFD solution as the reference. It provides estimation of the $2p$-th order, and does not require solution of discretised boundary value problem either for a new discretization or a new Higher Order approximation.

For the group of smoothing estimators, the difference between rough and smoothed solutions derivatives is investigated (like in the Z-Z estimator, [9]). We propose here to use the Low and Higher Order solution derivatives instead of separate smoothing procedures.

The last group of residual estimator, considered here, is based on the residual error distribution (12). These estimators may be of explicit or implicit type. The improved Higher Order estimation of the residual error (14) may be applied here.

It is worth stressing here that the Higher Order approach in the error analysis, discussed above, is of general character. Though it was developed for the MFDM approach, it may also be used e.g. in the FEM analysis, yielding then the error estimation of superior quality, when compared to the results obtained by other techniques.

9 Adaptive solution approach

The following h-adaptation strategy is adopted: the residual error is examined at points which belong to the cloud of nodes of one rank denser than the recent one but do not belong to the last cloud [12–14, 18]. Nodes generation criteria are based on the improved estimation of the residual error

$$r_x = \frac{|b(Lu - \Delta, v) - l(v)|}{|b_{\max}|} > \beta \cdot r_{\max} \quad , \quad \beta \in [0,1] \tag{15}$$

as well as limitation of abrupt mesh density ρ_{ij} changes (e.g. limitation of its gradient).

$$\eta_{ij} = \frac{\sqrt{\Omega_i} - \sqrt{\Omega_j}}{\rho_{ij}} \geq \eta_{adm} \quad , \quad \rho_{ij}^2 = (x_i - x_j)^2 + (y_i - y_j)^2 \tag{16}$$

Here, $\beta \cdot r_{\max}$ is a error threshold level, η_{adm} denotes the admissible density change, whereas the Ω_i, Ω_j are the areas of the Voronoi polygons prescribed to nodes "i" and "j".

10 Error indicators

On each irregular cloud of nodes, calculated is the a so called error indicator $(\bar{h}, \bar{e})$. It was introduced [20] in several ways for a simple, and effective way of evaluation of the global error. It is based on the local errors evaluated at many points in the whole domain, and is representation of the domain. As it was shown in our previous works [20, 21], the best results were obtained for error indicator defined as the simple centre of gravity

$$\bar{h} = \frac{1}{N}\sum_i h_i \quad , \bar{e} = \frac{1}{N}\sum_i |e_i| \tag{17}$$

of the group of points (h_i, e_i) given in the (h, e) coordinate system (h – local nodes modulus, e – local error level). In adaptation process, each irregular cloud of nodes has its own representative pair (17). In this way, the convergence rate of solution and residual errors for the set of adaptive meshes (clouds) may be simply estimated, using the linear regression of indicated $(\bar{h}, \bar{e})$ error data, calculated for each mesh.

The general MFDM/MLPG5 solution approach discussed above will be tested now on 1D and 2D benchmarks.

11 HO MFDM / MLPG5 approach in 1D

Let us consider the following local formulation of the 1D b.v.p.

$$\begin{cases} w''(x) = f(x) & in\ \Omega = [0,4] \quad , \quad w \in C^2 \\ w = \bar{w} & on\ \partial\Omega = \{0,4\} \end{cases} \tag{18}$$

We may derive from there the variational non-symmetric global formulation

$$\int_0^4 w''v \cdot dx = \int_0^4 f \cdot v \cdot dx \quad , \quad v \in H^0, w \in H_0^2 + \tilde{w} \tag{19}$$

as well as variational symmetric global formulation

$$-\int_0^4 w'v' \cdot dx + w'v|_0^4 = \int_0^4 f \cdot v \cdot dx \quad , \quad v \in H^1, w \in H_0^1 + \tilde{w} \tag{20}$$

In the standard variational approach both global formulations require approximation of both test (v), and trial functions (w) on the integration cell $[a, b]$ (Fig.5). However, in the MLPG5 case, one has

$$v(x) = \begin{cases} 1 & , \quad a \le x \le b \\ 0 & , \quad x < a\ lub \quad x > b \end{cases} \quad \rightarrow \quad v'(x) = 0, \quad v''(x) = 0$$

which produces the discretized form of the non-symmetric formulation (19)

$$\sum_{l=1}^{N_g} \omega_l \left[w''_{(l)} - f(x_l) \right] = 0 \quad , \quad w''_{(l)} = \sum_{j=1}^{m_w} m_{3,j}^{(w)} w_{j(l)} - \Delta_{3,l} \quad , \quad k = 2, .., n-1 \tag{21}$$

and discretization of the symmetric formulation (20)

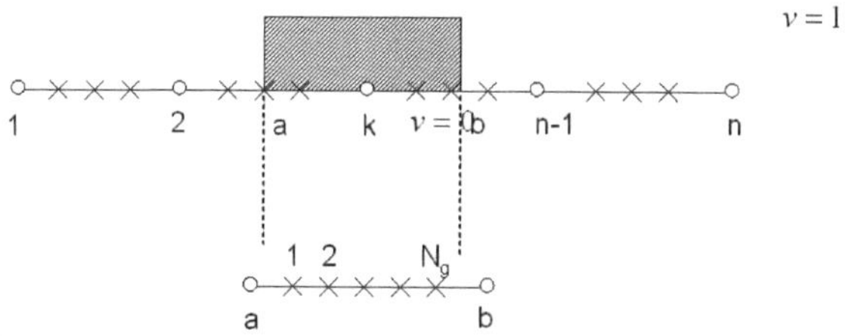

Fig. 5. MLPG5 in 1D case

$$\sum_{j=1}^{m_w} m_{2,j}^{(w)} w_{j(b)} - \Delta_{2,b}^{(w)} - \sum_{j=1}^{m_w} m_{2,j}^{(w)} w_{j(a)} + \Delta_{2,a}^{(w)} = J_k \sum_{l=1}^{N_g} \omega_l \cdot f(x_l), \ \ k = 2, ..., n-1 \tag{22}$$

Here, and in the following algorithms $p = 2$ is the standard approximation order, and $2p = 4$ is the approximation order with correction terms Δ, N_g is the number of the Gauss points inside the node interval (equivalent to the Voronoi polygon in 2D), J_k, ω_l, x_l are the Jacobian, integration weight, and the Gauss point respectively, while $m_{i,j}^{(w)}$ are the coefficients of the MFD formulas. Notice, that the test function does not appear in the above discrete forms since it has a constant value. Discretization of the second non-symmetric formulation

$$\int_0^4 wv'' \cdot dx + (w'v - wv')|_0^4 = \int_0^4 f \cdot v \cdot dx \quad , \quad v \in H^2, w \in H_0^1 + \tilde{w} \tag{23}$$

also leads to the discrete form (22), since all the derivatives of the test function vanish.

12 HO MFDM / MLPG5 approach in 2D

Let us consider the following local formulation ($w \in C^2$)

$$\begin{array}{l} \nabla^2 w(x,y) = f(x,y) \quad in \ \Omega \\ w = \bar{w} \quad on \ \partial\Omega \end{array} \tag{24}$$

and two equivalent variational formulations, namely the non-symmetric

$$\int_\Omega (w_x'' + w_y'')\, v d\Omega = \int_\Omega f \cdot v \cdot d\Omega \quad , \quad v \in H^0, w \in H_0^2 + \tilde{w} \tag{25}$$

and the symmetric one

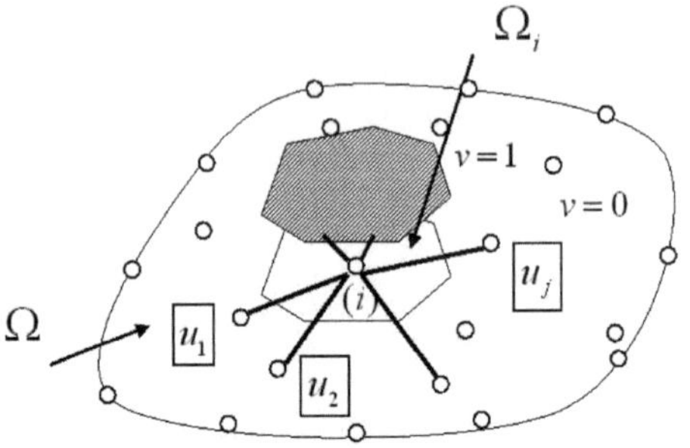

Fig. 6. MLPG5 in 2D case

$$-\int_{\Omega}\left(w'_x v'_x + w'_y v'_y\right) d\Omega + \int_{\partial\Omega} v \cdot \left(n_x \cdot w'_x + n_y \cdot w'_y\right) d\partial\Omega = \int_{\Omega} f \cdot v \cdot d\Omega \quad , \quad v \in H_0^1, w \in H_0^1 + \tilde{w} \tag{26}$$

Following the MLPG5 assumptions, the test function has a constant value over the Voronoi polygons, prescribed to each node x_i , $i = 1, ..., n$ (Fig.6), and zero elsewhere

$$x_i \quad \rightarrow \quad \Omega_i : \quad v(x,y) = \begin{cases} 1, & (x,y) \in \Omega_i \\ 0, & (x,y) \notin \Omega_i \end{cases} , \quad i = 1, 2, ..., n \tag{27}$$

This form of the test function leads to the following discrete variational forms

$$J_k \sum_{l=1}^{N_g} \omega_l \left[u''_{xx(l)} + u''_{yy(l)} - f(x_l, y_l)\right] = 0 \quad , \quad u_{(l)}^{(k)} = \sum_{j=1}^{m_u} m_{k,j}^{(u)} u_{j(l)} - \Delta_{k(l)} \quad , \quad k = 1, ..., n \tag{28}$$

and

$$J_k \sum_{l=1}^{N_g} \omega_l v_{(l)} \left(f(x_l, y_l) + n_x u'_{x(l)} + n_y u'_{y(l)}\right) = 0 \quad , \quad k = 1, ..., n \tag{29}$$

13 Extensions of the MFDM / MLPG5 solution approach

Applied was the original Atluri's concept [2] of MLPG5 in which

- the variational principle (4) is satisfied over a local subdomain prescribed to each node,
- the test function is constant (Heaviside) over each local subdomain.

There are possible various extensions of this approach, using the features of the meshless approximation and discretization. Below presented is a proposed classification, collecting already discussed concepts, and some new ones. This classification concerns:

- integration subdomain
  - global (one test function is given in the whole domain),
  - local (over each subdomain prescribed to node) – MLPG5,
  - on the patch of a local subdomains (e.g. triangles),
- integration scheme
  - around nodes (over the Voronoi polygons in 2D – MLPG,
  - between nodes (over the Delaunay triangles in 2D),
  - over the independent mesh,
  - over the zones of influence of the weighting functions,
  - over the local subdomains,
- the order of the local interpolation of the test function
  - constant around nodes (MLPG5),
  - linear between the nodes,
  - higher order,
- types of degrees of freedom of test function
  - test function values only,
  - generalised degrees of freedom (single derivatives or the whole differential operators values).

The approach in which the test function is linear between the nodes (over the Delaunay triangle in 2D) will be discussed in more detailed way. Following the classification proposed by Atluri, consequently it will be named MLGP7.

14 The MFDM / MLPG7 approach

Considered will be here the MLPG7 and MFDM solution approach. The integration is performed over the triangular subdomains, between the nodes. As in the MLPG5, such type of domain partition guarantees coverage of the whole domain using not overlapping subdomains.

In 1D case, such assumptions lead to the linear interpolation of the test function between the nodes (Fig.7).

It may be written in the following form

$$v(x) = \begin{cases} (a_k v_k + a_{k+1} v_{k+1})\, x + b_k v_k + b_{k+1} v_{k+1} & , \quad x_k \le x \le x_{k+1} \\ 0 \quad , \quad x < x_k \ or \quad x > x_{k+1} \end{cases} \quad \rightarrow$$

$$\begin{cases} v'(x) = \begin{cases} a_k v_k + a_{k+1} v_{k+1} & , \quad x_k \le x \le x_{k+1} \\ 0 \quad , \quad x < x_k \ or \quad x > x_{k+1} \end{cases} \\ v''(x) = 0 \end{cases}$$

$$a_k = \tfrac{1}{x_k - x_{k+1}}, \quad b_k = -a_k x_{k+1}, \quad a_{k+1} = \tfrac{1}{x_{k+1} - x_k}, \quad b_{k+1} = -a_{k+1} x_k \tag{30}$$

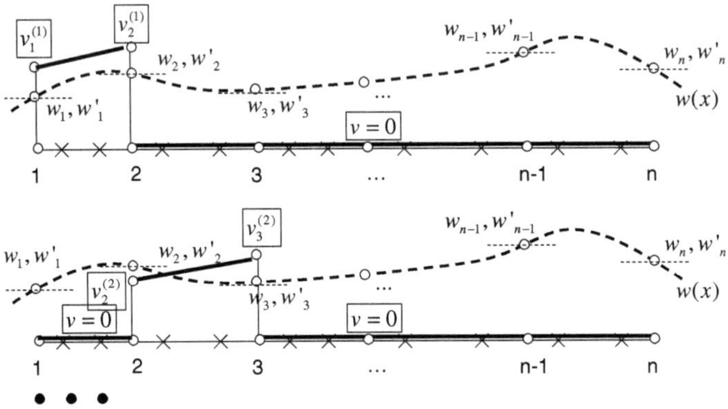

Fig. 7. Concept of the MLPG7 – 1D case

Notice that the first derivative of the test function is non-zero. This leads to the following discrete forms of the variational principles (19÷20)

$$J_k \sum_{l=1}^{N_G} \omega_l \left(w''_{j(l)} - f(x_l) \right) \left[\left(a_k v_k^{(k)} + a_{k+1} v_{k+1}^{(k)} \right) x_l + b_k v_k^{(k)} + b_{k+1} v_{k+1}^{(k)} \right] = 0 \tag{31}$$

$$J_k \sum_{l=1}^{N_G} \omega_l \left[-w'_{j(l)} \left(a_k v_k^{(k)} + a_{k+1} v_{k+1}^{(k)} \right) - f(x_l) \left(x_l + b_k v_k^{(k)} + b_{k+1} v_{k+1}^{(k)} \right) \right] + \left[w'v \right]_{x_k}^{x_{k+1}} = 0 \tag{32}$$

where $k = 1, 2, ..., n-1$. Moreover, the discretization of the second non-symmetric formulation (23) is possible now

$$- J_k \sum_{l=1}^{N_G} \omega_l f(x_l) \left(x_l + b_k v_k^{(k)} + b_{k+1} v_{k+1}^{(k)} \right) + \left[w'v - v'w \right]_{x_k}^{x_{k+1}} = 0 \tag{33}$$

and it differs from the formula (32).

In 2D the test function

$$\begin{cases} w_1, w_2, ..., w_i, ..., w_M \\ v = \begin{cases} 0 \quad in \quad \Omega \backslash \Omega_i \\ \sum_{i=0}^{2} v_{k+i} N_{k+i}(x, y) \quad in \quad \Omega_i \end{cases} \end{cases} , \quad i = 1, ..., M \tag{34}$$

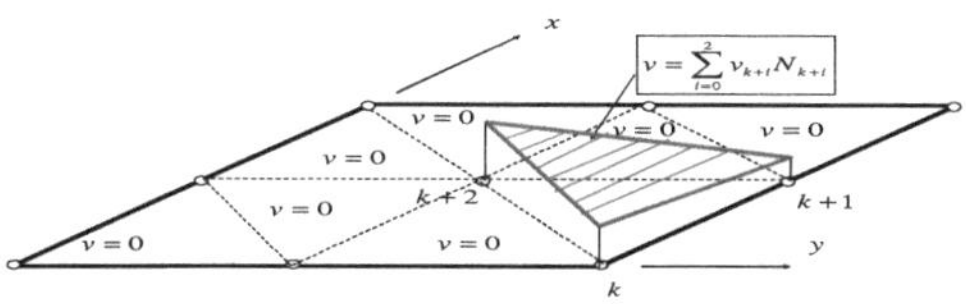

Fig. 8. Concept of the MLPG7 in 2D case

is prescribed by the linear shape functions using the three nodal values in each triangle (Fig.8)

Those v_{k+i} values are independent for each triangle. Note that the number of triangles (M) may be greater than the number of nodes (n), which leads to the overdetermined set of equations, for the 2D and 3D cases. The trial function is approximated in the standard manner, on the MFD star using the MWLS approximation.

The discrete forms of the variational principles used here are as follows $(t = 1, ..., M)$

$$J_t \sum_{l=1}^{N_g} \omega_l \left(w''_{xx(l)} + w''_{yy(l)} - f(x_l, y_l) \right) \left(x_l \sum_{i=0}^{2} a_i v_{i+t} + y_l \sum_{i=0}^{2} b_i v_{i+t} + \sum_{i=0}^{2} c_i v_{i+t} \right) = 0 \tag{35}$$

$$\begin{aligned} & -J_t \sum_{l=1}^{N_g} \omega_l \left(w'_{x(l)} \sum_{i=0}^{2} a_i v_{i+t} + w'_{y(l)} = \sum_{i=0}^{2} b_i v_{i+t} \right) + \\ & + \sum_{k=1}^{3} J_{t,k} \sum_{l=1}^{N_g} \omega_l \left(w'_{x(l)} n_x + w'_{y(l)} n_y \right) \cdot \left(x_l \sum_{i=0}^{2} a_i v_{i+t} + y_l \sum_{i=0}^{2} b_i v_{i+t} + \sum_{i=0}^{2} c_i v_{i+t} \right) = \\ & = J_i \sum_{l=1}^{N_g} \omega_l \cdot f(x_l, y_l) \left(x_l \sum_{i=0}^{2} a_i v_{i+t} + y_l \sum_{i=0}^{2} b_i v_{i+t} + \sum_{i=0}^{2} c_i v_{i+t} \right) \end{aligned} \tag{36}$$

The discretization of the appropriate second non-symmetric variational form is also possible.

15 Numerical examples

Several aspects of the combined MFDM / MLPG solution approach were examined. The most interesting are

- comparison of the results obtained from the standard variational principle discretization versus those obtained by means of the MLPG5,
- examination of the Low and Higher Order solution qualities,
- examination of the local and global solution, and residual error estimates,

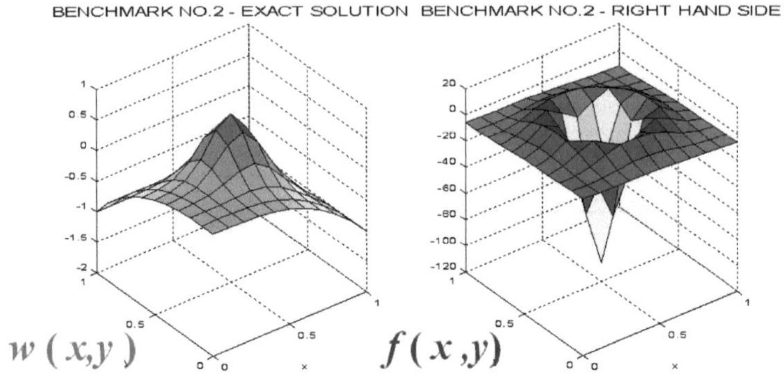

Fig. 9. The 2D benchmark test

- h-adaptation approach for the Higher Order MFDM/MLPG5 solution,
- examination and comparison of the convergence rates of the Low and Higher Order solutions,
- comparison of the results obtained from the MFDM/MLPG5, and the MFDM/MLPG7 approaches.

Let us consider the 2D Poisson problem (24) with an appropriate right hand side resulting from the analytical solution

$$w(x,y) = -x^3 - y^3 + \exp\left(-\left(\frac{x-0.5}{0.2}\right)^2 - \left(\frac{y-0.5}{0.2}\right)^2\right) \tag{37}$$

defined on the $\Omega = [(x,y), \quad 0 \leq x \leq 1, \quad 0 \leq y \leq 1]$.

Both the analytical solution (37), and the right hand side are presented in Fig.9.

In the first phase of analysis, the regular mesh with 64 nodes was applied.

Below presented are the exact (true) solution errors, namely Low and Higher Order ones, evaluated by means of the Higher Order MFDM / MLPG5 approach for the non-symmetric (Fig.10), and symmetric (Fig.12) formulations. Respectively, Fig.11 and Fig.13 present the local estimation of the solution error for those formulations. The mean and maximum error norms are shown on each graph.

Results for the MLPG5 are slightly better than the ones obtained for the standard formulations and they require less computational effort.

In the next step, the global error estimation of the symmetric MLPG5 formulation was analysed. Estimators η were evaluated on the nodal subdomains, and for the whole domain as well. The results were compared to the exact global error $\bar{e}$, using the effectivity index

$$i = 1 + \frac{|\eta - \bar{e}|}{\bar{e}} \tag{38}$$

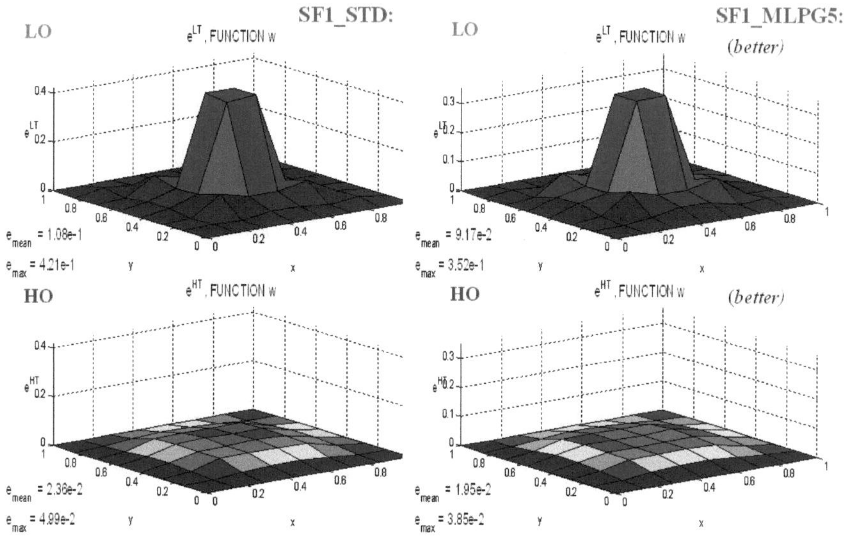

Fig. 10. The exact Low and Higher Order solution errors for the MLPG5 non-symmetric for

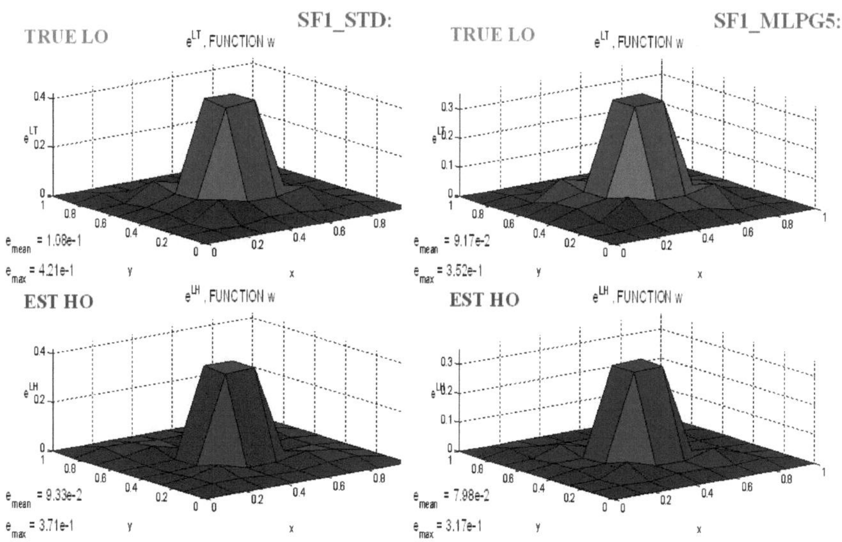

Fig. 11. The Low Order solution error and its Higher Order estimation for the MLPG5 non-symmetric form

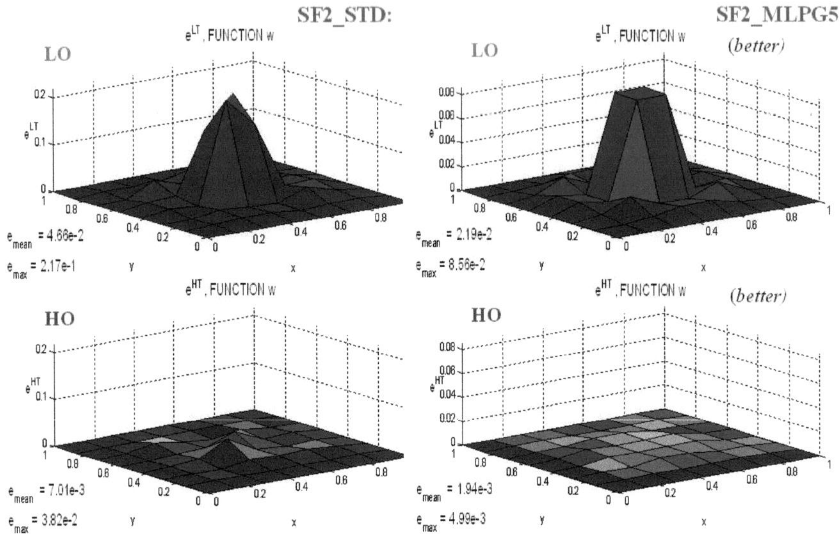

Fig. 12. The exact Low and Higher Order solution errors for the MLPG5 symmetric form

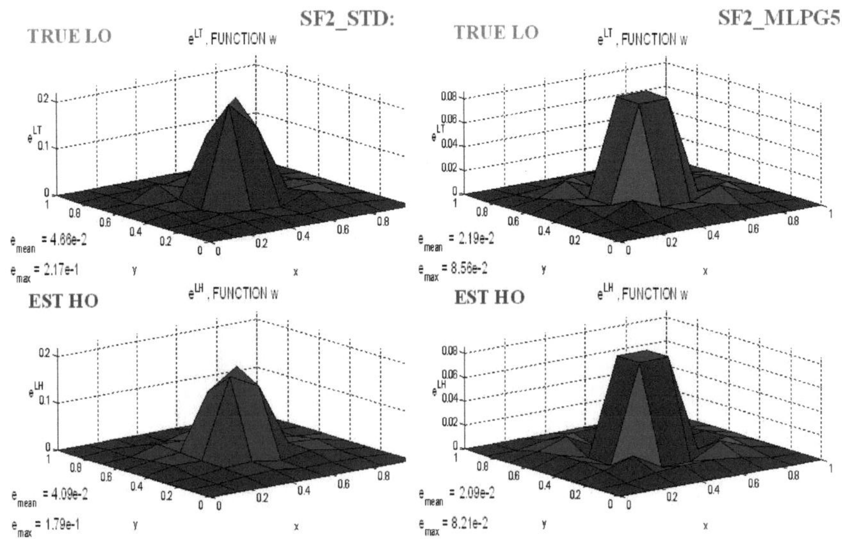

Fig. 13. The Low solution error and its Higher Order estimation for the MLPG5 symmetric form

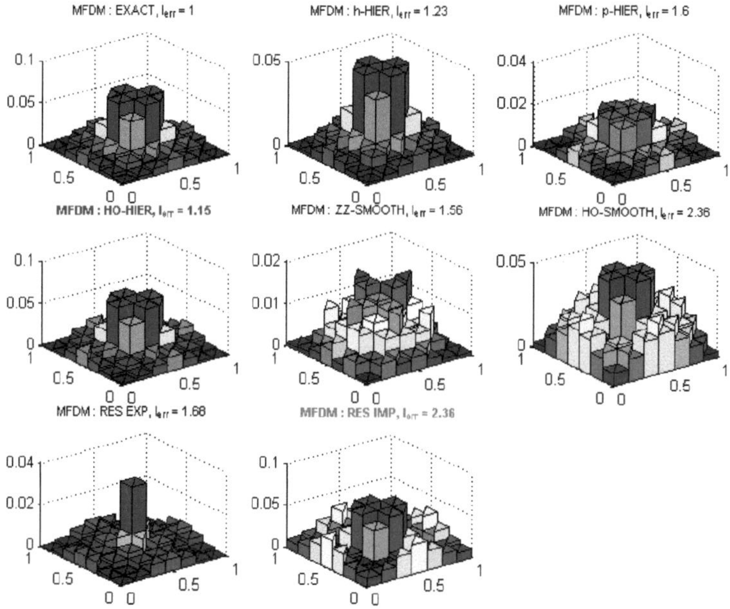

Fig. 14. Global error estimation for the 2D benchmark test

Results for seven different types of estimators are collected in Fig.14. The first graph presents the exact error, followed by (going by the rows), three hierarchical estimators (h-type, p-type, HO-type), two smoothing ones (ZZ-type, HO-type) and two residual ones (namely explicit and implicit type). The best results (the smallest values of effectivity index (38)) were obtained for the HO-type hierarchical estimator. Moreover, it required less computational effort than most of the other estimators.

Finally, using the above error estimation, an adaptation process was performed starting from the coarse regular mesh (16 nodes). The final strongly irregular cloud consists of 179 nodes. These two and several intermediate clouds of nodes are shown in Fig.15. Notice that the nodes concentrate in zones where the exact solution, and the right hand side exhibit the largest magnitude of gradients.

Fig.16 shows the convergence of both Low, and Higher Order solution errors (the first graph), as well as the residual errors (the second graph). Appropriate convergence rates proved that Higher Order errors decrease over 100 times faster than the Low Order ones.

The last comparison was carried out on the chosen irregular cloud of 64 nodes, taken from the previously generated set. The exact Low Order solution errors obtained for several formulations (local, variational non-symmetric,

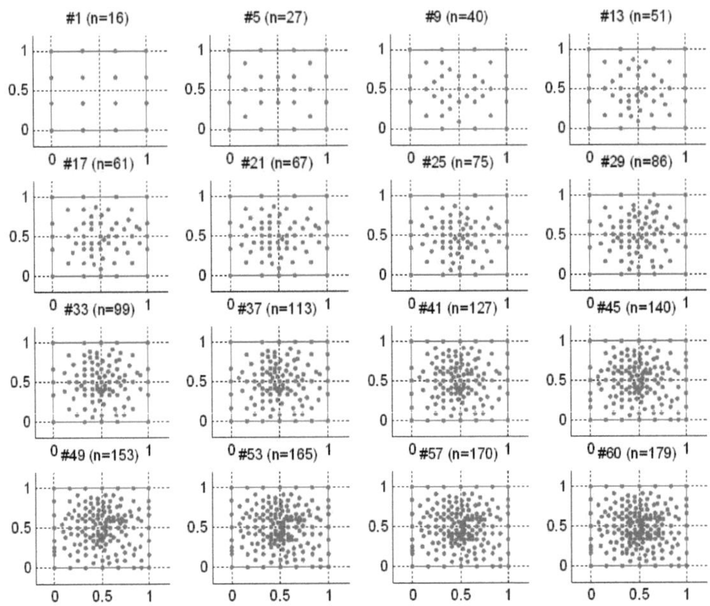

Fig. 15. Irregular clouds of nodes generated adaptively

variational symmetric, MLPG5, MLPG7) of the b.v.p. are presented in Fig.17, whereas the corresponding exact Higher Order solution errors are shown in Fig.18. Notice the presence of the MLPG7 formulation, giving the best results out of all variational formulations.

16 Final remarks

Presented was combination of the Meshless Finite Difference Method (MFDM) with the original Atluri's concept of the Meshless Local Petrov-Galerkin formulation (MLPG) in which approximation of an unknown function is prescribed by its nodal values only. The variational principle is then satisfied locally in subdomains, whereas the test function may be different from the trial function.

Applied was the MLPG5 formulation [2, 26] in which the test function is assumed as the Heaviside step function. The MLPG5 was combined for the first time ever with the Meshless Finite Difference Method, based on arbitrarily irregular clouds of nodes, and the MWLS approximation.

Presented and applied were also recent developments in the MFDM, including

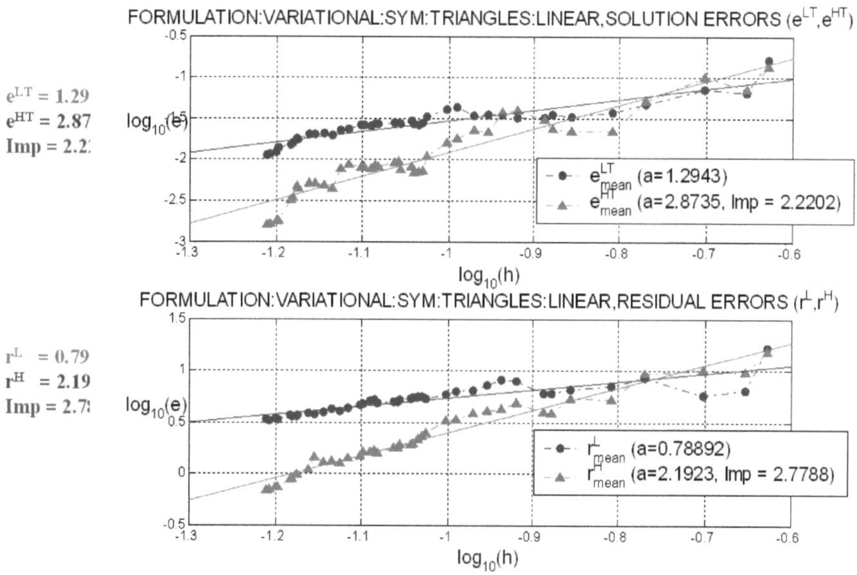

Fig. 16. Convergence on the set of irregular clouds of nodes, obtained for the error indicator (17).

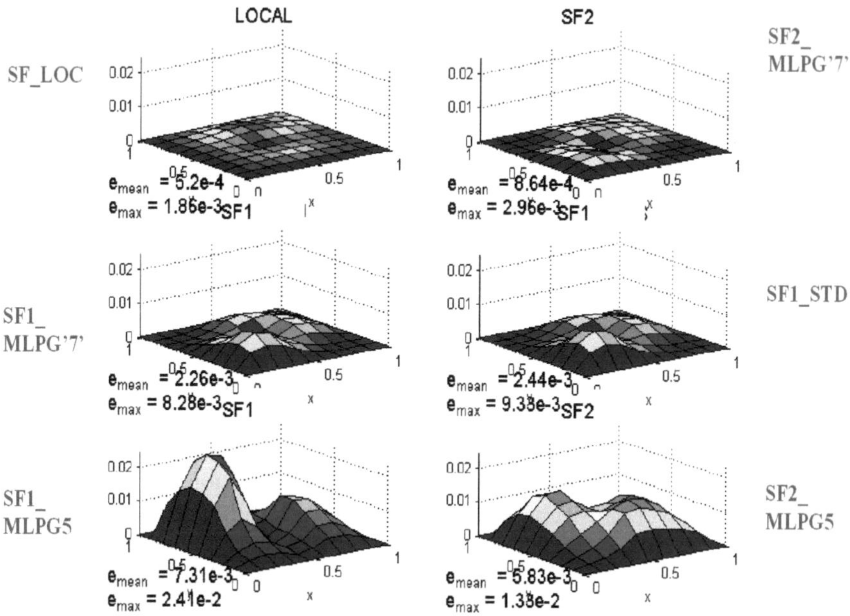

Fig. 17. Comparison of the exact Low Order solution errors

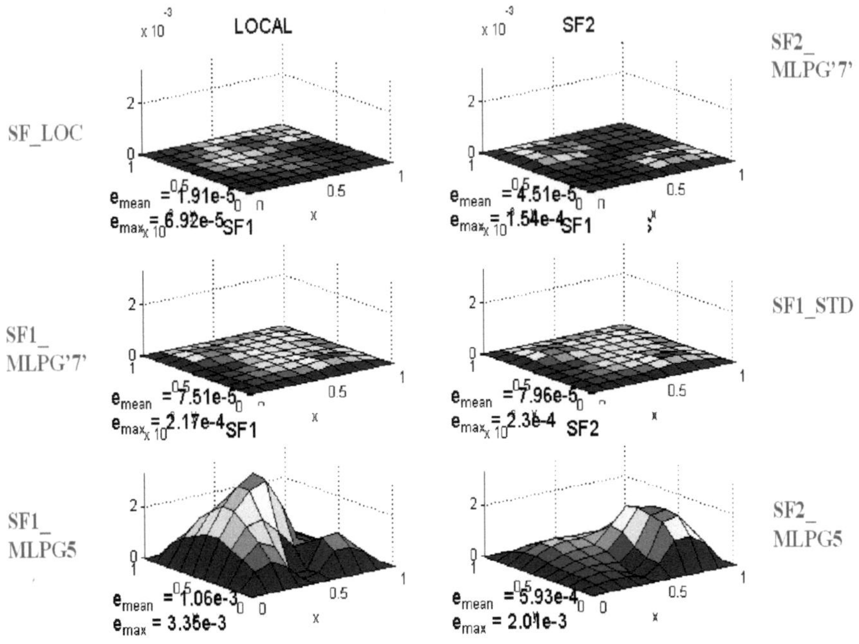

Fig. 18. Comparison of the exact Higher Order solution errors

- higher order approximation,
- improved a-posteriori error analysis,
- improved adaptive solution approach.

Especially interesting were the results of preliminary investigation of the MLPG5 and MFDM combination. The numerical results obtained so far are very encouraging regarding their precision and efficiency.

Numerical 1D and 2D tests reflected comparable quality of the MLPG5/MFDM results with the results obtained by using standard b.v.p. formulations (e.g. local or Galerkin ones). However, the MLPG5/MFDM needs less computational effort, due to the simplicity of the test function and its supporting subdomain. High solution quality was obtained when using Higher Order terms.

Besides they provide superior quality reference solution that may be applied in the global error estimators. High rate of both the solution, and residual error improvement was observed, when examining the Low and Higher Order convergence rates evaluated on the set of irregular adaptively generated clouds of nodes. It is worth noticing here that the error estimation approach discussed here may be also applied to other numerical solutions, e.g. to the FE ones.

Future plans include

- further testing of the MLPG5 and MLPG7 formulations towards various extensions of the original MLPG concept,
- combination of these and the other MLPG formulations with the MFDM solution approach,
- combination of the Higher Order approximation based on correction terms with the Finite Element Method, especially for the improved a-posteriori solution error estimation,
- engineering applications.

References

1. Ainsworth M, Oden JT, *A-posteriori error estimation in finite element analysis*, Comp. Meth Appl. Mech Engng (1997), 142:1-88.
2. Atluri S.N. *The Meshless Method (MLPG) for Domain & Bie Discretizations*, Tech Science Press, 2004.
3. Belytchko T., *Meshless methods: An overview and recent developments.* Comp Meth Appl Mech Engng (1996), 139:3-47.
4. Benito J.J., Ureña F., Gavete L., Alonso B. *Application of the GFDM to improve the approximated solution of 2nd order partial differential equations*, ICCES MM 2007.
5. Brandt A., *Multi-level Adaptive Solutions To Boundary Value Problems*, Math. Comp., 31, 1977, pp 333 - 390.
6. Collatz L., *The Numerical Treatment of Differential Equations*, Springer, Berlin, 1966.
7. Hackbush W., *Multi-Grid Methods and Applications*, Springer – Verlag, Berlin, 1985.
8. Karmowski W., Orkisz J., *A physically based method of enhancement of experimental data – concepts, formulation and application to identification of residual stresses*, Proc IUTAM Symp on Inverse Problems in Engng Mech, Tokyo 1992; On Inverse Problems in Engineering Mechanics, Springer Verlag, 1993, pp 61-70.
9. Krok J., *A New Formulation of Zienkiewicz-Zhu a Posteriori Error Estimators Without Superconvergence Theory*, Proceedings *Third MIT Conference on Computational Fluid and Solid Mechanics*, Cambridge, Ma, USA, June 14-17, 2005.
10. Krok J., Orkisz J., *A Discrete Analysis of Boundary-Value Problems with Special Emphasis on Symbolic Derivation of Meshless FDM/FEM Models, Computer Methods in Mechanics CMM*, June 19-22, 2007, Spala, Lodz, Poland.
11. Lancaster P, Salkauskas K, *Surfaces generated by moving least-squares method*, Math Comp (1981), 55:141-158.
12. Liszka T, *An interpolation method for an irregular net of nodes*, Int J Num Meth Eng (1984), 20:1599-1612.
13. Liszka T, Orkisz J. *The Finite Difference Method at arbitrary irregular grids and its applications in applied mechanics*, (1980), Comp Struct 11:83-95
14. Orkisz J., Finite Difference Method (Part III), in Handbook of Computational Solid Mechanics, M.Kleiber (Ed.) Springer-Verlag, Berlin, (1998), 336-431.
15. Orkisz J., *Higher Order Meshless Finite Difference Approach*, 13th *Inter-Institute Seminar for Young Researchers*, Vienna, Austria, October 26-28, 2001.

16. Orkisz J., Jaworska I., *Some Concepts of 2D Multipoint HO Operators for the Meshless FDM Analysis, ICCES Special Symposium On Meshless Methods*, 15-17 June 2007, Patras, Greece.
17. Orkisz J., Jaworska I., Milewski S., *Meshless finite difference method for higher order approximation, Third International Workshop Meshfree Methods for Partial Differential Equations*, September 12-15, 2005, Bonn, Germany.
18. Orkisz J., Lezanski P, Przybylski P., *Multigrid approach to adaptive analysis of bv problems by the meshless GFDM.* IUTAM/IACM Symposium On Discrete Methods In Structural Mechanics II, Vienna, 1997.
19. Orkisz J., Milewski S., *On higher order approximation in the MFDM metod*, Elsevier (editor: K.J. Bathe), Proceedings *Third MIT Conference on Computational Fluid and Solid Mechanics*, Cambridge, Ma, USA, June 14-17, 2005.
20. Orkisz J., Milewski S., *Higher order approximation approach in meshless finite difference analysis of boundary value problems, the 16th International Conference on Computer Methods in Mechanics* CMM-2005, June 21 – 24, 2005, Czestochowa, Poland.
21. Orkisz J., Milewski S., *On a-posteriori error analysis in Higher Order approximation in the Meshless Finite Difference Method, ICCES Special Symposium On Meshless Methods*, 14-16 June 2006, Dubrovnik, Croatia.
22. Orkisz J., Milewski S., *Recent advances in the Higher Order Approximation in the Meshless Finite Difference Method, 7th World Congress on Computational Mechanics*, Los Angeles, California, July 16 - 22, 2006.
23. Orkisz J., Milewski S, *Recent advances in a-posteriori error estimation based on the Higher Order correction terms in the Meshless Finite Difference Method, ICCES Special Symposium On Meshless Methods*, 15-17 June 2007, Patras, Greece.
24. Orkisz J., Milewski S, *Higher Order approximation multigrid approach in the Meshless Finite Difference Method, Computer Methods in Mechanics CMM*, June 19-22, 2007, Spala, Lodz, Poland.
25. Orkisz J., Milewski S. *Higher order a-posteriori error estimation in the Meshless Finite Difference Method, Meshfree Methods for Partial Differential Equations IV*, Lecture Notes in Computational Science and Engineering, Springer, M. Griebel and M. A. Schweitzer (eds.), (2008), 189 – 213
26. Orkisz J., Milewski S. *Higher Order discretization of the Meshless Local Petrov Galerkin formulations*, Proceedings *CMM-2009 – Computer Methods in Mechanics*, 18–21 May 2009, Zielona Gora, Poland.
27. Schweitzer M.A., *Meshfree methods for partial differential equations, Computer Methods in Mechanics CMM*, June 19-22, 2007, Spala, Lodz, Poland.

Treatment of general domains in two space dimensions in a Partition of Unity Method

Marc Alexander Schweitzer* and Maharavo Randrianarivony†

Institut für Numerische Simulation, Universität Bonn, Wegelerstr. 6, D-53115 Bonn, Germany
schweitzer@ins.uni-bonn.de, randrian@ins.uni-bonn.de

Summary. This paper is concerned with the approximate solution of partial differential equations using meshfree Galerkin methods on arbitrary domains in two space dimensions which we assume to be given as CAD/IGES data. In particular we focus on the particle-partition of unity method (PPUM) yet the presented technique is applicable to most meshfree Galerkin methods. The basic geometric operation employed in our cut-cell type approach is the intersection of a planar NURBS patch and an axis-aligned rectangle. Note that our emphasis is on the speed of the clipping operations since these are invoked frequently while trying to attain a small number of patches for the representation of the intersection. We present some first numerical results of the presented approach.

Key words: meshfree method, partition of unity method, complex domain

1 Introduction

One major advantage of meshfree methods (MM) over traditional mesh-based approximation approaches is that the challenging task of mesh generation can (in principle) be avoided. Thus the numerical treatment of partial differential equations (PDE) on complex time-dependent domains can be simplified substantially by MM.

Collocation techniques for instance are essentially independent of the computational domain. They employ a collection of points for the discretization process only and require no explicit access to the domain or the boundary. In meshfree Galerkin methods however we have to integrate the respective bilinearform and linearform over the domain Ω (and parts of the boundary $\partial\Omega$).

*This work was supported in part by the Sonderforschungsbereich 611 *Singular phenomena and scaling in mathematical models* funded by the *Deutsche Forschungsgemeinschaft*.

†Hausdorff Center for Mathematics

M. Griebel, M.A. Schweitzer (eds.), *Meshfree Methods for Partial Differential Equations V*, Lecture Notes in Computational Science and Engineering 79, DOI 10.1007/978-3-642-16229-9_2,

Hence, we must be concerned with the issues of a meshfree domain representation and the efficient numerical integration of the meshfree shape functions over Ω and $\partial\Omega$. These issues often lead to the perception that a background mesh is needed in meshfree Galerkin methods. This is in fact not the case, we must rather compute an appropriate decomposition of the domain into pairwise disjoint cells which respect the algebraic structure of the employed shape functions as well as the geometry of the domain. Thus, we must be able to compute this decomposition efficiently *on the fly* since the construction of the meshfree shape functions is independent of the domain and the domain may change during the simulation.

In the following we present a two step procedure for the efficient computation of a decomposition in two space dimensions for the particle–partition of unity method (PPUM) [9, 10, 20] where we assume the domain to be given as CAD/IGES data. Note that our approach is easily extendable to other meshfree Galerkin methods.

The remainder of this paper is organized as follows: In section 2 we give a short review of the construction of meshfree shape functions in the PPUM which is essentially independent of the domain Ω. Moreover we introduce the employed weak formulation and an initial decomposition which respects the algebraic structure of the constructed shape functions and covers the domain and its boundary with pairwise disjoint cells. The main contribution of this paper, the treatment of general domains in two space dimensions, is discussed in section 3. Here, we introduce the fundamental IGES entities used in our implementation for the domain representation and present an efficient cell-clipping approach for the computation of the intersection of an axis-aligned rectangle with a NURBS patch. First numerical results are reported in section 4 before we conclude with some remarks in section 5.

2 Particle–Partition of Unity Method

In this section let us shortly review the core ingredients of the PPUM, see [9, 10, 20] for details. In a first step, we need to construct a PPUM space V^{PU}, i.e., we need to specify the PPUM shape functions $\varphi_i \vartheta_i^n$. With these shape functions, we then set up a sparse linear system of equations $A\tilde{u} = \hat{f}$ via the classical Galerkin method where we realize essential boundary conditions via Nitsche's method [16]. The arising linear system is then solved by our multilevel iterative solver [10].

An arbitrary function $u^{\mathrm{PU}} \in V^{\mathrm{PU}}$ is defined as the linear combination

$$u^{\mathrm{PU}}(x) = \sum_{i=1}^{N} \varphi_i(x) u_i(x) \quad \text{with} \quad u_i(x) = \sum_{m=1}^{d_i} u_i^m \vartheta_i^m(x) \tag{1}$$

and the respective PPUM space V^{PU} is defined as

$$V^{\mathrm{PU}} := \sum_{i=1}^{N} \varphi_i V_i \quad \text{with} \quad V_i := \operatorname{span}\langle \vartheta_i^m \rangle. \tag{2}$$

Here, we assume that the functions φ_i form an admissible partition of unity (PU) on the domain Ω; i.e., the supports $\omega_i := \operatorname{supp}(\varphi_i)$ cover the complete domain Ω and its boundary $\partial\Omega$, and refer to the spaces V_i with $\dim(V_i) = d_i$ as local approximation spaces. Hence, the shape functions employed in the PPUM are the products $\varphi_i \vartheta_i^m$ of a PU function φ_i and a local basis function ϑ_i^m.

This abstract approach is the basis of any partition of unity method [2,3] such as e.g. the generalized/extended finite element method (GFEM/XFEM) [15,25–27]. The key difference between our PPUM and the GFEM/XFEM is that the PU in GFEM/XFEM is usually constructed by classical FE shape functions ϕ_i based on some kind of mesh which may also encode the (discrete) computational domain Ω or may simply cover Ω. In the PPUM, however, the PU is constructed from independent point data only; i.e. it is always independent of the representation of the computational domain Ω.

The fundamental construction principle employed in [9] for the construction of the PU $\{\varphi_i\}$ is a d-binary tree. Based on the given point data $P = \{x_i \,|\, i = 1, \ldots, \hat{N}\}$, we sub-divide a bounding-box $\mathcal{C}_\Omega \supset \Omega$ of the domain Ω until each cell

$$\mathcal{C}_i = \prod_{l=1}^{d} (c_i^l - h_i^l, c_i^l + h_i^l)$$

associated with a leaf of the tree contains at most a single point $x_i \in P$, see Figure 1. We obtain an overlapping cover $C_\Omega := \{\omega_i\}$ from this tree by defining the cover patches ω_i by

$$\omega_i := \prod_{l=1}^{d} (c_i^l - \alpha h_i^l, c_i^l + \alpha h_i^l), \quad \text{with } \alpha > 1. \tag{3}$$

Note that we define a cover patch ω_i for leaf-cells $\mathcal{C}_i$ that contain a point $x_i \in P$ as well as for *empty* cells that do not contain any point from P.

To obtain a PU on a cover C_Ω with $N := \operatorname{card}(C_\Omega)$ we define a weight function $W_i : \Omega \to \mathbb{R}$ with $\operatorname{supp}(W_i) = \omega_i$ for each cover patch ω_i by

$$W_i(x) = \begin{cases} \mathcal{W} \circ T_i(x) & x \in \omega_i \\ 0 & \text{else} \end{cases} \tag{4}$$

with the affine transforms $T_i : \overline{\omega}_i \to [-1,1]^d$ and $\mathcal{W} : [-1,1]^d \to \mathbb{R}$ the reference d-linear B-spline. By simple averaging of these weight functions we obtain the Shepard functions

$$\varphi_i(x) := \frac{W_i(x)}{S(x)}, \quad \text{with} \quad S(x) := \sum_{l=1}^{N} W_l(x). \tag{5}$$

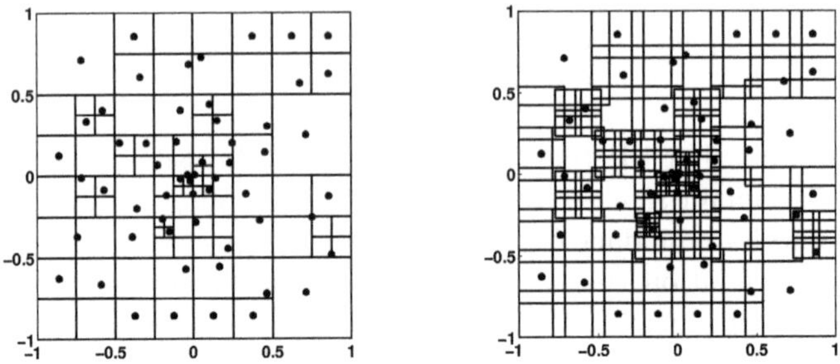

Fig. 1. Point cloud and induced tree decomposition (left) and the resulting overlapping patches (3) with $\alpha = 1.3$ (right).

We refer to the collection $\{\varphi_i\}$ with $i = 1, \ldots, N$ as an admissible partition of unity since there hold the relations

$$\begin{aligned} 0 \leq \varphi_i(x) \leq 1, \qquad & \sum_{i=1}^{N} \varphi_i \equiv 1 \text{ on } \overline{\Omega}, \\ \|\varphi_i\|_{L^\infty(\mathbb{R}^d)} \leq C_\infty, \qquad & \|\nabla\varphi_i\|_{L^\infty(\mathbb{R}^d)} \leq \frac{C_\nabla}{\operatorname{diam}(\omega_i)} \end{aligned} \tag{6}$$

with constants $0 < C_\infty < 1$ and $C_\nabla > 0$ so that the assumptions on the PU for the error analysis given in [3] are satisfied by our PPUM construction.

Furthermore, the PU (5) based on the cover C_Ω obtained from the scaling of a tree decomposition with (a particular choice of) $\alpha \in (1, 2)$ satisfies

$$\mu(\{x \in \omega_i \,|\, \varphi_i(x) = 1\}) \approx \mu(\omega_i),$$

i.e., the PU has the flat-top property, see [12, 23]. This ensures that the product functions $\varphi_i\vartheta_i^n$ are linearly independent and stable, provided that the employed local approximation functions ϑ_i^n are stable with respect to $\{x \in \omega_i \,|\, \varphi_i(x) = 1\}$ (and ω_i) [22].

With the help of the shape functions $\varphi_i\vartheta_i^n$ we then discretize a PDE in weak form

$$a(u, v) = \langle f, v \rangle$$

via the classical Galerkin method to obtain a discrete linear system of equations $A\tilde{u} = \hat{f}$. Since our general PPUM shape functions do not satisfy essential boundary conditions by construction we employ Nitsche's method for their implementation.[3] This approach, for instance, yields the bilinear form

$$a_\beta(u, v) := \int_\Omega \nabla u \cdot \nabla v \; dx - \int_{\partial\Omega} \big((\partial_n u)v + u(\partial_n v)\big) \; ds + \beta \int_{\partial\Omega} uv \; ds \tag{7}$$

and the associated linear form

[3] Note, however, that there is also a conforming treatment of boundary conditions for the PPUM [24].

$$\langle l_\beta, v\rangle := \int_\Omega fv\,dx - \int_{\partial\Omega} g(\partial_n v)\,ds + \beta\int_{\partial\Omega} gv\,ds \tag{8}$$

for the Poisson model problem

$$\begin{aligned} -\Delta u &= f \text{ in } \Omega \subset \mathbb{R}^D, \\ u &= g \text{ on } \partial\Omega, \end{aligned} \tag{9}$$

where the regularization parameter β is used to enforce the definiteness of the bilinear form (7) with respect to the employed finite dimensional discretization space and can be computed automatically, see [11, 20, 21] for details.

2.1 Numerical Integration

Note that the PU functions (5) in the PPUM are in general piecewise rational functions only. Therefore, the use of an appropriate numerical integration scheme is indispensable in the PPUM as in most meshfree Galerkin approaches [1, 4, 5, 8, 10].

In the FEM the (numerical) integration of the weak form of the considered PDE is simpler than in meshfree methods due to the availability of a mesh. This fact often leads to the perception that a mesh is required for numerical integration also in meshfree methods and that therefore meshfree Galerkin methods are not truly meshfree. However, a mesh is not required for the reliable and stable numerical integration of the weak form. We only need an appropriate decomposition of the integration domain into cells with pairwise disjoint interiors (a far more general partitioning of the domain than a mesh). Observe for instance that we can in principle allow for hanging nodes of arbitrary order in the union of these cells—a property that is usually not acceptable for FE meshes to ensure inter-element continuity of the shape functions. Thus, our construction is a much simpler task than full-blown mesh-generation.

Recall that the global regularity of our PPUM shape functions $\varphi_i\vartheta_i^m \in V^{\mathrm{PU}}$ is dominated by the regularity of the PU functions φ_i of (5), i.e.

$$\varphi_i(x) := \frac{W_i(x)}{S(x)}, \quad \text{with} \quad S(x) := \sum_{l=1}^{N} W_l(x).$$

Thus, let us first focus on the PU functions φ_i and how we can construct a decomposition $\{T_\alpha\}$ of its support ω_i such that $\varphi_j|_{T_\alpha}$ is smooth (i.e. of arbitrary regularity) for all $\omega_j \in C_i := \{\omega_l \in C_\Omega \,|\, \omega_i \cap \omega_l \neq \emptyset\}$. To this end, we consider a patch $\omega_i \cap \partial\Omega = \emptyset$ and carry out the differentiation in (7) and (8) and introduce the shorthand notation $G_i := \nabla W_i S - W_i \nabla S$, $\Omega_i := \Omega \cap \omega_i$, $\Omega_{i,j} := \Omega_i \cap \omega_j$, and $\Gamma_{i,j} := \partial\Omega \cap \overline{\omega}_i \cap \overline{\omega}_j$ to obtain the integrals

$$a(\varphi_j\vartheta_j^m, \varphi_i\vartheta_i^n) = \int_{\Omega_{i,j}} S^{-4}G_iG_j\ \vartheta_i^n\vartheta_j^m\ dx + \int_{\Omega_{i,j}} S^{-2}W_iW_j\ \nabla\vartheta_i^n\nabla\vartheta_j^m\ dx + \int_{\Omega_{i,j}} S^{-3}\Big(G_iW_j\ \vartheta_i^n\nabla\vartheta_j^m + W_iG_j\ \nabla\vartheta_i^n\vartheta_j^m\Big)\ dx \tag{10}$$

for the stiffness matrix and the integrals

$$\langle f, \varphi_i\vartheta_i^n\rangle_{L^2} = \int_{\Omega_i} S^{-1}W_i\vartheta_i^n f\ dx \tag{11}$$

for the right-hand side. Thus, we need to assess the regularity of the functions S, W_j and G_j for all $\omega_j \in C_i$ to construct a decomposition $\{T_\alpha\}$ of the domain, i.e. of each patch ω_i, such that their restrictions to the interiors of the cells T_α are smooth functions. In essence this means that all weight functions W_j must be smooth on the cells T_α. Our weight functions W_j however are (tensor products of) splines and therefore only piecewise smooth functions. Hence, our decomposition $\{T_\alpha\}$ must resolve the overlapping supports of weight functions as well as the piecewise character of the weight functions.

Recall that we obtained the patches ω_i of our cover C_Ω from a tree-decomposition of a bounding box of the domain Ω. Thus, there is an initial pairwise disjoint decomposition $\{\mathcal{C}_i\}_{i=1}^N$ which covers the domain, i.e. $\overline{\Omega} \subset \bigcup_{i=1}^N \overline{\mathcal{C}_i}$, available. Hence, we must only be concerned with the (minimal) refinement of the cells $\mathcal{C}_i$ such that the restrictions of all weight functions W_j (and local approximation functions ϑ_j^n) are arbitrarily smooth on the refined cells $T_{i,\alpha}$. Observe that this refined decomposition is easily computable on the fly since we must consider the intersections of axis-aligned rectangles only due to our construction. Thus, we obtain a decomposition into axis-aligned rectangular cells.

First we split a cell $\mathcal{C}_i$ into disjoint rectangular sub-cells according to its intersections with $\omega_j \in C_i$, then we consider the piecewise character of the respective spline weight functions W_j to define the decomposition $\{T_{i,\alpha}\}$, see Figures 2 and 3.[4] Thus, all PU functions φ_j satisfy $\varphi_j|_{T_{i,\alpha}} \in C^\infty(T_{i,\alpha})$.

Remark 1. This decomposition is sufficient if we employ polynomial local approximation space $V_j = \mathcal{P}^{p_j}$ in our PPUM only. It only remains to select an integration rule on the cells $T_{i,\alpha}$ considering the bilinear form and the maximal polynomial degree $p_i + p_j$ of the integrands.

In the case of a non-polynomial enrichment, i.e. $V_j = \mathcal{P}^{p_j} + \mathcal{E}_j$, we must also consider the discontinuities and singularities of the enrichment functions either by the choice of appropriate integration rules or additional refinement of the decomposition $\{T_{i,\alpha}\}$.

Remark 2. Note that the construction above is suitable for all domain integrals that involve the assembled shape functions $\varphi_i\vartheta_i^m$ for a specific choice of

[4] Observe however that for the flat-top region of our PU functions it is sufficient to construct a single cell $T_{i,\alpha}$.

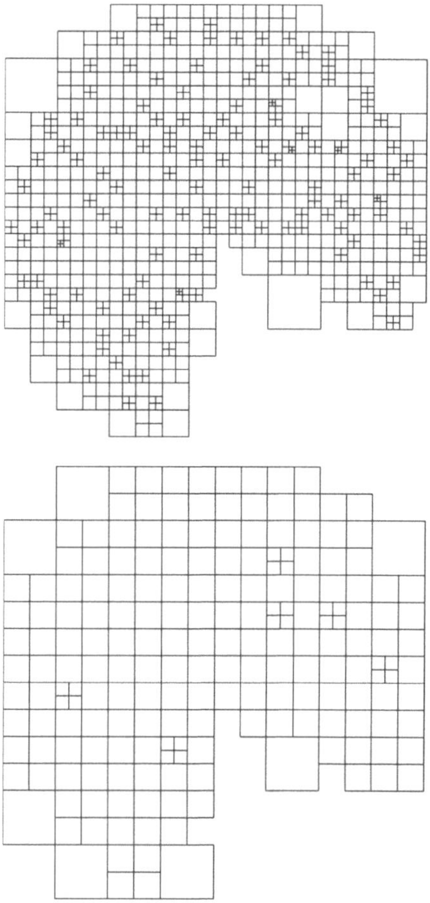

Fig. 2. Initial decomposition based on the tree-cells $\mathcal{C}_j$ only (top: level 7, bottom: level 5). The respective tree decomposition was generated by sampling with Halton points.

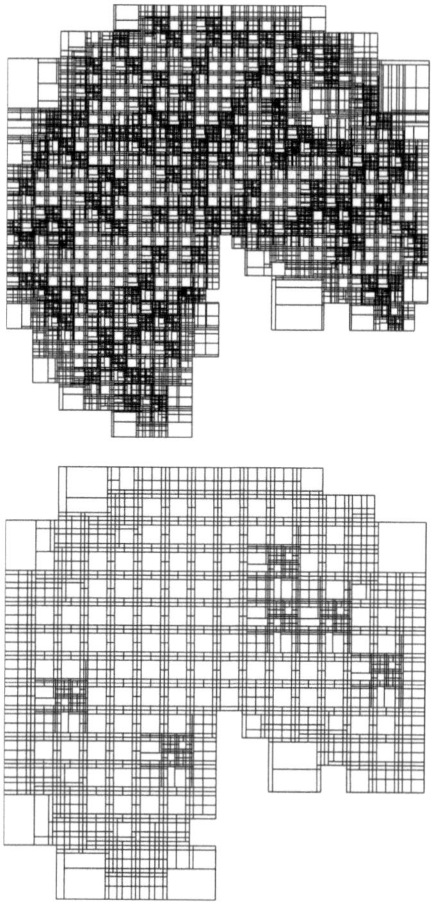

Fig. 3. Refined decomposition (top: level 7, bottom: level 5).

α in (3) and weight function $\mathcal{W}$. In the multilevel PPUM however we may also need to compute certain operators locally; i.e. just using the local approximation functions ϑ_i^m on ω_i or on the subset $\omega_{\mathrm{FT},i} := \{x \in \omega_i \,|\, \varphi_i(x) = 1\}$, see [10, 22]. These local operators can be integrated with much less integration cells since they are independent of the PU functions φ_i. An appropriate decomposition for these local integrals can be obtained easily with a variant of the above construction where we consider only the overlapping patches but not the weight functions.

The rectangular cells $T_{i,\alpha}$ obviously cover the complete domain Ω and its boundary $\partial\Omega$, however, they are not aligned with $\partial\Omega$ since our PPUM construction is completely independent of the geometry. For the definition of a refined decomposition $\{T_{i,\alpha,\partial\Omega}\}$ with boundary aligned integration cells $T_{i,\alpha,\partial\Omega}$ from the above decomposition $\{T_{i,\alpha}\}$ we must now be concerned with the geometry of the domain and its representation in our meshfree PPUM implementation.

3 Realization on General Domains

In this section we are concerned with the efficient application of the PPUM approach on general multiply connected domains Ω in two space dimensions. To this end, we will assume that the domain Ω is given as a CAD object described by a collection of IGES entities and our PPUM discretization process will essentially operate directly on this description of Ω. The fundamental geometric operation needed for this approach is the efficient clipping of the domain against an axis-aligned rectangle. Recall however that we will compute the integration cells on the fly thus this operation must be rather fast. Therefore, we employ a two-step procedure: First we perform a setup step where we decompose the general multiply connected domain Ω into a collection of convex quadrilateral NURBS patches $\{\mathcal{P}\}$. Note that this setup step is independent of the discretization process and can be pre-computed. Based on this collection of NURBS patches $\{\mathcal{P}\}$ we can now compute the intersection of an arbitrary axis-aligned rectangle $\mathcal{R}$ with the domain Ω simply via the intersections of $\mathcal{R}$ with the NURBS patches $\{\mathcal{P}\}$; an operation which is substantially faster than directly computing the intersection $\mathcal{R} \cap \Omega$.

3.1 Domain Representation

Let us first summarize the employed geometry representation in two dimensions and its CAD digitization using the IGES format. Here, the fundamental objects are B-spline and NURBS curves. In order to introduce the B-spline basis, we consider any constant integer $k \geq 2$ which controls the smoothness of the spline and a knot sequence $\zeta_0, ..., \zeta_{n+k}$ such that $\zeta_{i+k} \geq \zeta_i$. The usual definition of B-spline basis functions [14,17] with respect to the knot sequence $(\zeta_i)_i$ is

$$N_i^k(t) = (\zeta_{i+k} - \zeta_i)[\zeta_i, ..., \zeta_{i+k}](\cdot - t)_+^{k-1} \tag{12}$$

where$[\zeta_i, ..., \zeta_{i+k}]$ denotes the forward divided difference and $(\cdot - t)_+^k$ is the truncated power function

$$(x - t)_+^k := \begin{cases} (x-t)^k & \text{if} \quad x \geq t, \\ 0 & \text{if} \quad x < t. \end{cases} \tag{13}$$

By induction, one can show that the above definition is equivalent to

$$N_i^1(t) := \begin{cases} 1 & \text{if} \quad t \in [\zeta_i, \zeta_{i+1}), \\ 0 & \text{otherwise}, \end{cases} \tag{14}$$

$$N_i^k(t) := \left(\frac{t - \zeta_i}{\zeta_{i+k-1} - \zeta_i} \right) N_i^{k-1}(t) + \left(\frac{\zeta_{i+k} - t}{\zeta_{i+k} - \zeta_{i+1}} \right) N_{i+1}^{k-1}(t). \tag{15}$$

A B-spline curve f with control points $\mathbf{d}_i \in \mathbb{R}^2$ with respect to the above knot sequence is defined as

$$f(t) = \sum_{i=0}^{n} \mathbf{d}_i N_i^k(t) \text{ for all } t \in [\zeta_0, \zeta_{n+k}]. \tag{16}$$

To ensure that the B-spline curve f is *open*, we assume that the knot sequence is *clamped*; i.e. there holds

$$\zeta_0 = \cdots = \zeta_{k-1} < \zeta_k \leq \zeta_{k+1} \leq \cdots \leq \zeta_n < \zeta_{n+1} = \cdots = \zeta_{n+k}. \tag{17}$$

The above assumption (17) ensures that the initial and final control points are interpolated such that the curve begins and ends at the control points. The case of rational splines or NURBS is given as

$$f(t) = \frac{\sum_{i=0}^{n} w_i \mathbf{d}_i N_i^k(t)}{\sum_{i=0}^{n} w_i N_i^k(t)} \text{ for all } t \in [\zeta_0, \zeta_{n+k}] \tag{18}$$

and the weights $\{w_i\}$ are assumed to be in $]0, 1]$.

As a CAD input for the PPUM implementation, we accept a multiply connected domain Ω. To this end, let us assume that there are univariate smooth functions $\boldsymbol{\kappa}_i^j$ defined on $[e_i^j, f_i^j] \subset \mathbb{R}$ with $\mathcal{B}_j^i = \boldsymbol{\kappa}_i^j([e_i^j, f_i^j])$ which encode the boundary $\partial\Omega$; i.e. there holds

$$\partial\Omega = \bigcup_{i=0}^{N} \bigcup_{j=0}^{n_i} \mathcal{B}_j^i. \tag{19}$$

Moreover, we make the convention that the external boundary

$$\Gamma_0 := \bigcup_{j=0}^{n_0} \mathcal{B}_j^0$$

is traversed in counter clockwise direction and the internal boundaries $\Gamma_p := \bigcup_{j=0}^{n_p} \mathcal{B}_j^p$ for $p = 1, ..., N$ are traversed in clockwise direction, compare Figure 4(a). A realistic example of the above description is shown in Figure 4(b) where the control polygons are identified by the dashed lines. The above representation are practically realized with the help of the IGES format. It is a CAD standard written in structured records specified as IGES entities which are stored in five sections. Note that we have restricted ourselves to IGES 144 where the most important geometric items are summarized in Table 1 since a complete IGES implementation is rather cumbersome and usually not

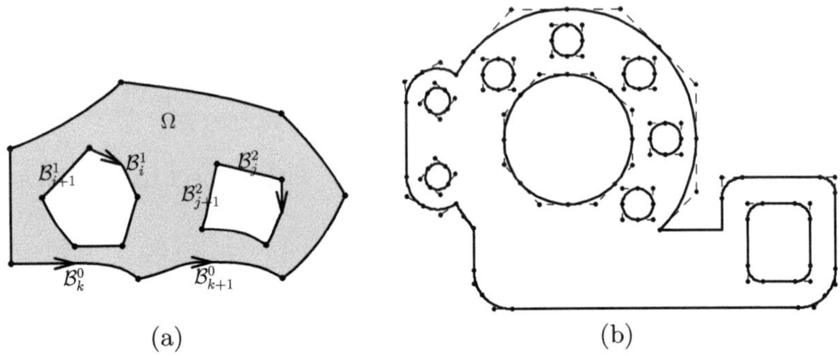

Fig. 4. (a) The boundary of $\Omega \subset \mathbb{R}^2$ is composed of the images $\mathcal{B}^i_j$ of several curves $\boldsymbol{\kappa}^j_i$. (b) A realistic CAD model where the control points are identified by large dots.

IGES Entities	ID numbers	IGES-codes
Line	110	`LINE`
Circular arc	100	`ARC`
Polynomial/rational B-spline curve	126	`B_SPLINE`
Composite curve	102	`CCURVE`
Polynomial/rational B-spline surface	128	`SPLSURF`
Transformation matrix	124	`XFORM`

Table 1. Appropriate IGES entities for 2D curved multiply connected domains.

necessary. Moreover, we will describe our approach assuming that all $\boldsymbol{\kappa}^j_i$ are B-spline or NURBS curves, since all practical curves including circular arcs and lines can be represented as such.

Our geometric objective consists of the following two tasks when we are given a rectangle $\mathcal{R}$. First, we need to find the intersection $\mathcal{I}$ of $\mathcal{R}$ with the multiply connected domain Ω. Second, if that intersection $\mathcal{I}$ is not empty, then we decompose it into several four-sided patches π_i and we find a mapping from the unit square to each π_i. In addition to those two points, we need that those operations are very efficient and robust because they need to be applied very often for the PPUM application. Since clipping a rectangle against a NURBS patch is easier than clipping against the whole domain Ω, we adopt the following two-stage approach. First, we determine the intersection of the domain Ω with a a coarse subdivision $G = \cup_i R_i$ of a bounding box of Ω as illustrated in Figure 6 (compare section 2). That is, we intersect each rectangle R_i of G against the domain Ω. We represent that intersection as a union of NURBS patches $\mathcal{P}^i_j$ such that $\mathcal{D}_i := R_i \cap \Omega = \cup_j \mathcal{P}^i_j$ and $\Omega = \cup_i \mathcal{D}_i = \cup_{i,j} \mathcal{P}^i_j$. Second, upon availability of the results of this setup phase using G, the clipping of any rectangle $\mathcal{R}$ against Ω amounts to the clipping of $\mathcal{R}$ against $R_i \cap \Omega$, thus against the relevant $\mathcal{P}^i_j$. Performing the setup phase has several advantages. First, it serves as a coarsest level in the PPUM method. Second,

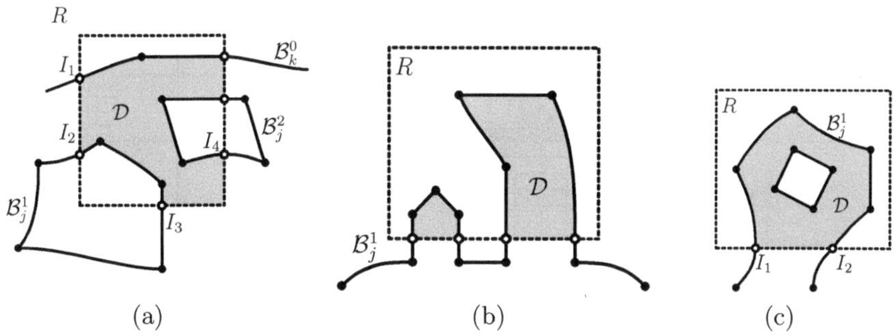

Fig. 5. Clipped region: (a) $\mathcal{D}$ is simply connected, (b) $\mathcal{D}$ is disconnected and has several connected components, (c) $\mathcal{D}$ is multiply connected.

it gives a fast location to determine which rectangles R_i are relevant to make each clipping fast.

3.2 Clipping a curved multiply connected domain

Let us suppose that we have a rectangle $R \in G$ which intersects the domain Ω. We want to briefly discuss how to identify the boundary of the intersection $\mathcal{D} = R \cap \Omega$. Several situations regarding the connectivity of $\mathcal{D}$ may be encountered. First, $\mathcal{D}$ can be simply connected as illustrated in Figure 5(a) where the shaded region defines the clipped domain $\mathcal{D}$. On the other hand, it may be disconnected. Note that this case can occur even if the original domain Ω is simply connected as displayed in Figure 5(b). Moreover, the clipped region $\mathcal{D}$ may contain some holes and is therefore multiply connected. Combinations of these situations can also be encountered. That is, $\mathcal{D}$ has several connected components and some of them are multiply connected.

The determination of the clipped region $\mathcal{D}$ is as follows. The first step consists of finding the boundary curves $\Gamma_{s_1},\ldots,\Gamma_{s_N}$ which intersect R. Then, we must identify the corresponding intersection points I_p as illustrated in Figure 5 to which we assign an additional marker indicating whether the respective curve is *entering* or *leaving* the clipping rectangle $R \in G$ through the point I_p. With this information at hand, we start from any intersection point e.g. I_1 and we distinguish two cases. First, if that intersection point is of type *leaving*, we traverse the boundary of the rectangle $R \in G$ counter clockwise until another intersection point e.g. I_2 is met. In the case that I_1 is of *entering* type in which we suppose that I_1 is the intersection of Γ_{s_1} and R, we traverse Γ_{s_1} according to its original orientation. That is to say, the traversal is counter clockwise if Γ_{s_1} encodes an external boundary whereas it is in clockwise direction if Γ_{s_1} is an internal boundary of Ω. Again, this traversal is done until we meet another intersection point. We repeat this process until we return to the initial intersection point I_1. At this stage, we have generated one connected component of $\mathcal{D}$.

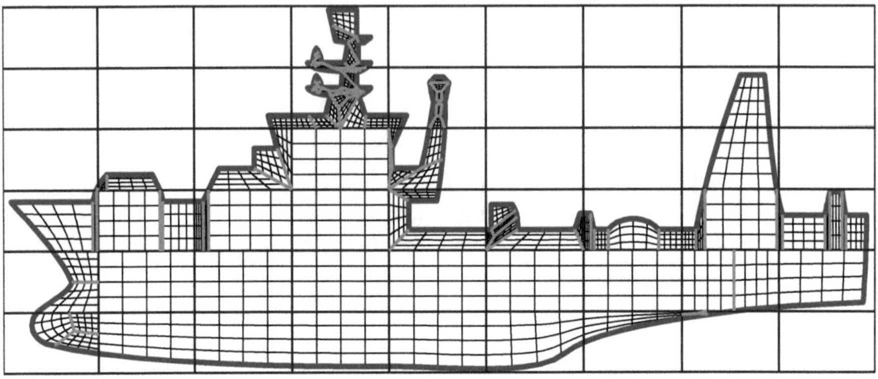

Fig. 6. Setup phase on the coarsest level: intersection against a coarse decomposition G in form of several NURBS patches.

If all intersection points have been traversed already, the intersection is completely determined and we terminate. Otherwise, we remove those intersection points which have been traversed and we repeat the same procedure based on the remaining intersection points in order to find the other connected components of $\mathcal{D}$. After all intersection points have been dealt with, we have constructed a collection of simply connected components of $\mathcal{D}$. If the original domain Ω contains some holes, we need to perform a few additional steps. For each internal curve Γ_p of Ω, we test if it is completely located inside the rectangle R. If so, we test further whether Γ_p is inside one connected component of $\mathcal{D}$ and we insert it there in the positive case. After those steps, we obtain the correct intersection by the union of several possibly multiply connected components of $\mathcal{D}$.

The above description requires the process of intersection between a NURBS curve $\mathcal{C}$ and a rectangle (see Figure 7(a)) which we briefly summarize now. Obviously, this task can be reduced to intersecting an infinite line $\mathcal{L}$ and a curve $\mathcal{C}$. Without loss of generality, we suppose that the line $\mathcal{L}$ is horizontal. We denote by $\mathbb{H}^+$ (resp. $\mathbb{H}^-$) the half-plane having positive (resp. negative) ordinates. The search for the intersections consists in examining the position of the control points with respect to $\mathbb{H}^+$ and $\mathbb{H}^-$. If the first and the last control points of $\mathcal{C}$ are located on different half-planes, then there is surely an intersection. If all control points are in one half-plane, no intersection point exists. Note that it is possible that there are some control points on both half-planes while the curve is completely inside one half-plane. To treat that ambiguous case, we apply a subdivision which is the process of splitting a NURBS curve $\mathcal{C}$ at a parameter value t_0 so as to obtain two curves which are again described in NURBS representation. One way of doing this is by means of discrete B-splines [6]. If the knot sequence of the original curve is defined in $[a, b]$, then those of the resulting curves are respectively in $[a, t_0]$ and $[t_0, b]$.

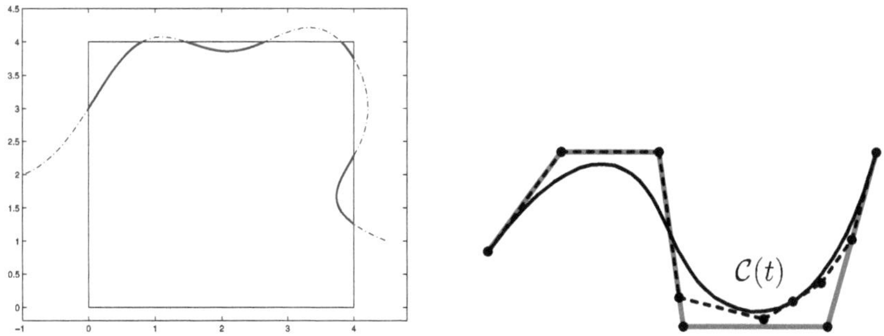

(a) (b)

Fig. 7. (a) Fast clipping operation which works even in the case where only a little part is inside or outside the rectangle. (b) A B-spline curve split into two B-spline curves.

An illustration is shown in Figure 7(b). A repeated application of subdivisions then yields the coordinates of the intersections.

3.3 Decomposition and parametrization

Now, we assume that we have a multiply connected domain $\mathcal{D} = \Omega \cap R$ as in Figure 5 and we would like to briefly describe the way of obtaining its decomposition into four-sided patches $\mathcal{P}_j$. It is beyond the scope of this paper to completely describe that decomposition. We will summarize only the most important steps and refer the reader to [18] for details. First, we take a coarse polygonal approximation $\mathbb{P}$ of the domain $\mathcal{D}$. For the case of a simply connected polygon $\mathbb{P}$, we have shown that it is always possible to chop off one quadrilateral (which is not necessarily convex) by inserting at most one internal node. By recursively applying this approach, one can generate a quadrangulation of $\mathbb{P}$. In the case of a multiply connected polygon $\mathbb{P}$, we need to insert cuts. That is, we join two vertices which are located on an interior curve and on an exterior one respectively. Note that in most cases, several possible cuts can be inserted. We have devised an algorithm [18] for choosing the optimal direction and the position of cuts to be inserted automatically. A drawback of this approach is that we may obtain some quadrilaterals which are non-convex so that we must employ some additional steps to convert the nonconvex quadrilaterals into convex ones. To obtain the decomposition of $\mathcal{D}$ from $\mathbb{P}$, we simply replace every straight boundary edge of the quadrilaterals by the corresponding curvilinear part from $\mathcal{D}$. Note however that we must be concerned with issues like corner smoothing or boundary interference [18]. The number of the our-sided patches $\mathcal{P}_j$ such that $\mathcal{D} = \cup_j \mathcal{P}_j$ constructed by this approach is not minimal but small.

Now, we want to generate a mapping onto the four-sided subdomains $\mathcal{P}_j$ which result from the above process. Let $\boldsymbol{\alpha}$, $\boldsymbol{\beta}$, $\boldsymbol{\gamma}$, $\boldsymbol{\delta}$: $[0,1] \longrightarrow \mathbf{R}^2$ be four

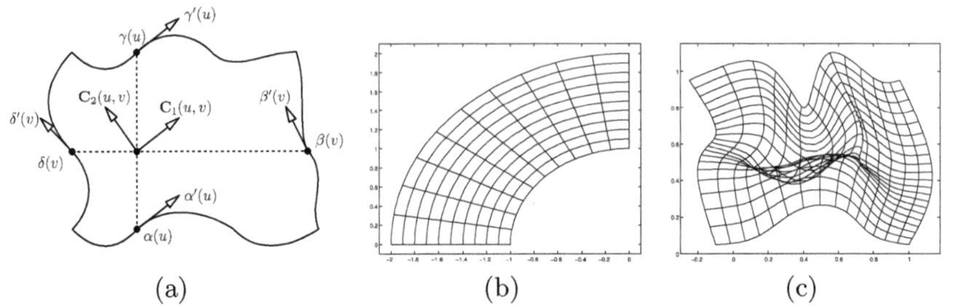

Fig. 8. (a) Tangents on a four sided domain for Coons patch. (b) Diffeomorphic Coons patches. (c) Undesired overspill phenomena.

$\mathcal{C}^1[0,1]$ curves that satisfy the compatibility conditions at the corners such as $\boldsymbol{\alpha}(0) = \boldsymbol{\delta}(0)$, $\boldsymbol{\alpha}(1) = \boldsymbol{\beta}(0)$, $\boldsymbol{\gamma}(0) = \boldsymbol{\delta}(1)$, $\boldsymbol{\gamma}(1) = \boldsymbol{\beta}(1)$. We assume that besides those common points, there are no further intersection points. Since our method of generating a map from the unit square to the four-sided domain S bounded by $\boldsymbol{\alpha}$, $\boldsymbol{\beta}$, $\boldsymbol{\gamma}$, $\boldsymbol{\delta}$ is based on transfinite interpolation, we briefly recall some basic facts about this technique.

We are interested in generating a parametric surface $\mathbf{x}(u,v)$ defined on the unit square $[0,1]^2$ such that the boundary of the image of $\mathbf{x}$ coincides with the given four curves:

$$\begin{array}{lll} \mathbf{x}(u,0) = \boldsymbol{\alpha}(u) & \mathbf{x}(u,1) = \boldsymbol{\gamma}(u) & \forall\, u \in [0,1] \\ \mathbf{x}(0,v) = \boldsymbol{\delta}(v) & \mathbf{x}(1,v) = \boldsymbol{\beta}(v) & \forall\, v \in [0,1]\,. \end{array} \tag{20}$$

This transfinite interpolation problem can be solved by a first order Coons patch whose construction involves the operators

$$(\mathcal{Q}_1\mathbf{x})(u,v) := F_0(v)\mathbf{x}(u,0) + F_1(v)\mathbf{x}(u,1) \tag{21}$$
$$(\mathcal{Q}_2\mathbf{x})(u,v) := F_0(u)\mathbf{x}(0,v) + F_1(u)\mathbf{x}(1,v) \tag{22}$$

where the so-called blending functions F_0 and F_1 denote two arbitrary smooth functions satisfying

$$F_i(j) = \delta_{ij} \quad i,j = 0,1 \quad \text{and} \qquad F_0(t) + F_1(t) = 1 \quad \forall t \in [0,1], \tag{23}$$

i.e. form a univariate PU. Obviously, there is much freedom in the choice of F_0 and F_1, throughout this paper we employ a linear blending. Now, a Coons patch $\mathbf{x}$ can be defined [7] by the relation

$$\mathcal{Q}_1 \oplus \mathcal{Q}_2(\mathbf{x}) = \mathbf{x}, \qquad \text{where} \qquad \mathcal{Q}_1 \oplus \mathcal{Q}_2 := \mathcal{Q}_1 + \mathcal{Q}_2 - \mathcal{Q}_1\mathcal{Q}_2. \tag{24}$$

It follows that $\mathbf{x}$ is of the form

$$\mathbf{x}(u,v) = -\begin{bmatrix} -1 \\ F_0(u) \\ F_1(u) \end{bmatrix}^T \begin{bmatrix} \mathbf{0} & \mathbf{x}(u,0) & \mathbf{x}(u,1) \\ \mathbf{x}(0,v) & \mathbf{x}(0,0) & \mathbf{x}(0,1) \\ \mathbf{x}(1,v) & \mathbf{x}(1,0) & \mathbf{x}(1,1) \end{bmatrix} \begin{bmatrix} -1 \\ F_0(v) \\ F_1(v) \end{bmatrix}. \tag{25}$$

The above Coons representation can be converted into B-spline or NURBS form provided that the four boundary curves are B-spline or NURBS curves. In Figure 8, we illustrate that for simple cases a Coons patch is already diffeomorphic. However, when the boundary curves become too wavy, like in Figure 8(c), we observe overlapping isolines indicating that the mapping is not invertible. We will need the notion of discrete B-splines to formulate some of our subsequent results. If $\mathbf{t} = (t_i)$ is a subknot of $\boldsymbol{\tau} = (\tau_i)$, then $N_j^{k,\mathbf{t}} = \sum_i b_{j,k}^{\boldsymbol{\tau},\mathbf{t}}(i) N_i^{k,\boldsymbol{\tau}}$ where $b_{j,k}^{\boldsymbol{\tau},\mathbf{t}}$ are the discrete B-splines given by the recurrence relations:

$$\begin{aligned} b_{j,1}^{\boldsymbol{\tau},\mathbf{t}}(i) &:= N_j^{1,\mathbf{t}}(t_i) \\ b_{j,k}^{\boldsymbol{\tau},\mathbf{t}}(i) &:= \omega_{i,k,\mathbf{t}}(\tau_{i+k-1}) b_{j,k-1}^{\boldsymbol{\tau},\mathbf{t}}(i) + (1 - \omega_{j+1,k,\mathbf{t}}(\tau_{i+k-1})) b_{j+1,k-1}^{\boldsymbol{\tau},\mathbf{t}}(i) \end{aligned} \tag{26}$$

where $\omega_{i,k,\mathbf{t}}(u) := (u - t_i)/(t_{i+k-1} - t_i)$. Below, we present some conditions on the boundary curves that guarantee the regularity of the Coons map. The linear independence of tangents on opposite curves (see Figure 8(a)) in conjunction with a second condition that controls the curvatures of the boundary curves, are sufficient for the regularity of $\mathbf{x}$. We suppose first that the boundary curves $\boldsymbol{\alpha}$, $\boldsymbol{\beta}$, $\boldsymbol{\gamma}$, $\boldsymbol{\delta}$ are B-spline curves with control points $\boldsymbol{\alpha}_i$, $\boldsymbol{\beta}_i$, $\boldsymbol{\gamma}_i$, $\boldsymbol{\delta}_i$. The opposite curves $\boldsymbol{\alpha}$ and $\boldsymbol{\gamma}$ are supposed to be defined on the knot sequence $\mathbf{t}^u = (t_i^u)$ while $\boldsymbol{\beta}$ and $\boldsymbol{\delta}$ on $\mathbf{t}^v = (t_i^v)$:

$$\boldsymbol{\alpha}(t) = \sum_{i=0}^{n_u} \boldsymbol{\alpha}_i N_i^{k_u}(t), \qquad \boldsymbol{\beta}(t) = \sum_{i=0}^{n_v} \boldsymbol{\beta}_i N_i^{k_v}(t), \tag{27}$$

$$\boldsymbol{\gamma}(t) = \sum_{i=0}^{n_u} \boldsymbol{\gamma}_i N_i^{k_u}(t), \qquad \boldsymbol{\delta}(t) = \sum_{i=0}^{n_v} \boldsymbol{\delta}_i N_i^{k_v}(t). \tag{28}$$

Since the orders of opposite curves are different in general, we use the discrete B-spline techniques in (26) to obtain equal order representations. To ensure that the first and the last control points are interpolated, we assume that the knot sequences $\mathbf{t}^u = (t_i^u)$ and $\mathbf{t}^v = (t_i^v)$ are clamped as in (17). Moreover, let us assume that the blending function F_1 is expressed in Bézier form such that $F_1(t) = \sum_{i=0}^n \phi_i B_i^n(t) = 1 - F_0(t)$ and introduce $F := \max\{S^1, S^2\}$ where

$$S^1 := \max_{i=0,\cdots,n_v} \{\rho \|\boldsymbol{\beta}_i - \boldsymbol{\delta}_i\|\} \qquad \text{and} \qquad S^2 := \max_{i=0,\cdots,n_u} \{\rho \|\boldsymbol{\gamma}_i - \boldsymbol{\alpha}_i\|\}. \tag{29}$$

Furthermore, we define

$$\lambda_i := (k_u - 1)/(t_{i+k_u}^u - t_{i+1}^u) \quad \text{and} \quad \mu_j := (k_v - 1)/(t_{j+k_v}^v - t_{j+1}^v) \tag{30}$$

for all $i = 1, \cdots, n_u$, $j = 1, \cdots, n_v$ and introduce the expressions

$$\begin{aligned} A_{ij} &:= \lambda_i \mu_j \det[\boldsymbol{\alpha}_{i+1} - \boldsymbol{\alpha}_i, \boldsymbol{\delta}_{j+1} - \boldsymbol{\delta}_j], \; B_{ij} := \lambda_i \mu_j \det[\boldsymbol{\alpha}_{i+1} - \boldsymbol{\alpha}_i, \boldsymbol{\beta}_{j+1} - \boldsymbol{\beta}_j], \\ C_{ij} &:= \lambda_i \mu_j \det[\boldsymbol{\gamma}_{i+1} - \boldsymbol{\gamma}_i, \boldsymbol{\delta}_{j+1} - \boldsymbol{\delta}_j], \; D_{ij} := \lambda_i \mu_j \det[\boldsymbol{\gamma}_{i+1} - \boldsymbol{\gamma}_i, \boldsymbol{\beta}_{j+1} - \boldsymbol{\beta}_j], \end{aligned}$$

and

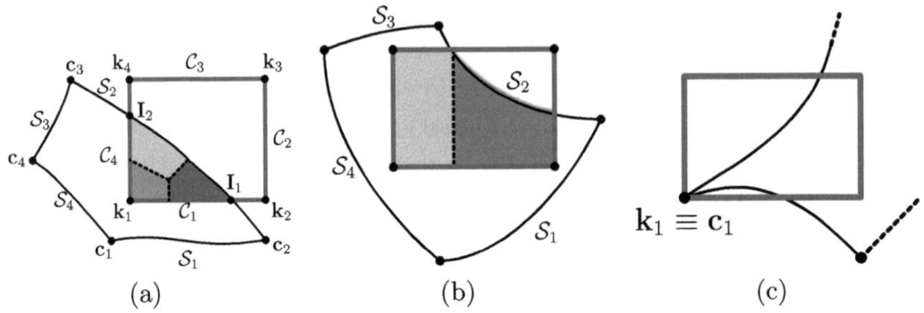

Fig. 9. (a) Special case where $\mathcal{S}_2$ intersects $\mathcal{C}_1$ and $\mathcal{C}_4$ and $\mathbf{k}_1$ in $\mathcal{P}$. (b) Special case where $\mathbf{k}_1$, $\mathbf{k}_2$, $\mathbf{k}_4$ are inside $\mathcal{P}$ while $\mathcal{C}_2 \cap \mathcal{S}_2 \neq \emptyset$ and $\mathcal{C}_3 \cap \mathcal{S}_2 \neq \emptyset$. (c) Nodal coincidence.

$$\tau := \min_{i,j}\{A_{ij}, B_{ij}, C_{ij}, D_{ij}\}. \tag{31}$$

Let M be a constant such that

$$\begin{aligned} \lambda_i\|(1-\phi_j)(\boldsymbol{\alpha}_i - \boldsymbol{\alpha}_{i-1}) + \phi_j(\boldsymbol{\gamma}_i - \boldsymbol{\gamma}_{i-1})\| \leq M \\ \mu_l\|(1-\phi_j)(\boldsymbol{\delta}_l - \boldsymbol{\delta}_{l-1}) + \phi_j(\boldsymbol{\beta}_l - \boldsymbol{\beta}_{l-1})\| \leq M, \end{aligned} \tag{32}$$

for all $i = 1, \cdots, n_u$; $l = 1, \cdots, n_v$ and $j = 0, \cdots, n$. Suppose that A_{ij}, B_{ij}, C_{ij}, D_{ij} are all positive for all $i = 0, \cdots, n_u - 1$ and $j = 0, \cdots, n_v - 1$. Then the condition $2MF + F^2 < \tau$ is sufficient [13] for $\mathbf{x}$ to be a diffeomorphism. More efficient results for checking regularity are detailed in [13, 18] by using adaptive subdivisions. We used a method [13] for treating curves which are not necessarily in the form (27) and (28).

3.4 Rectangle-NURBS clipping

This section will discuss the fast process of NURBS-decomposition of the intersection between a rectangle $\mathcal{R}$ and a NURBS patch $\mathcal{P}$ which does not present an overspill phenomenon as in Figure 8(c). Note that the process here is different from the one in Section 3.2 and Section 3.3. Of course, one can apply the method there but our main objective here, apart from finding a result, is to make that intersection process fast because it has to be applied very often in PPUM simulation. Let the four curve sides of $\mathcal{P}$ be $\mathcal{S}_1$, $\mathcal{S}_2$, $\mathcal{S}_3$, $\mathcal{S}_4$ and its corners be $\mathbf{c}_1$, $\mathbf{c}_2$, $\mathbf{c}_3$, $\mathbf{c}_4$. Similarly, the sides and the corners of $\mathcal{R}$ are respectively $\mathcal{C}_i$ and $\mathbf{k}_i$. The process consists in distinguishing many special cases depending on several factors: (1) intersection of $\mathcal{S}_i$ with $\mathcal{C}_j$, (2) position of the corner $\mathbf{c}_i$ with respect to $\mathcal{R}$, (3) position of the corner $\mathbf{k}_i$ with respect to $\mathcal{P}$. We need to implement a program where each special case is individually treated. It is beyond the scope of this paper to describe all possible special cases. In Figure 9(a) and Figure 9(b), we display two special cases. For the first situation, the patch side $\mathcal{S}_2$ intersects rectangle sides $\mathcal{C}_1$ and $\mathcal{C}_4$ while no corners $\mathbf{c}_i$ are included in $\mathcal{R}$ and the corner $\mathbf{k}_1 \in \mathcal{P}$. For the second situation,

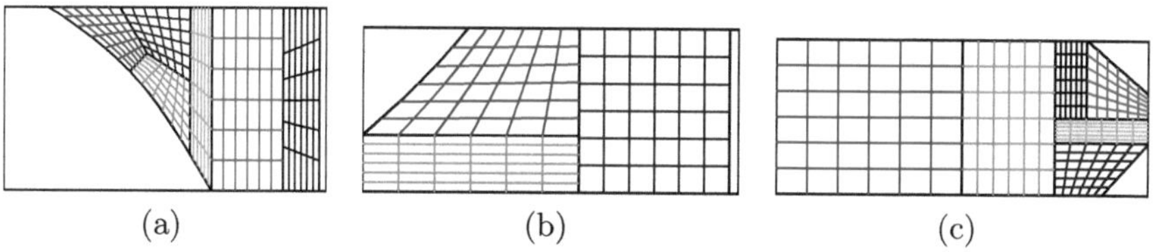

Fig. 10. Recursively applying some special cases.

three corners $\mathbf{k}_i$ are inside the patch while $\mathcal{S}_2$ intersects $\mathcal{C}_2$ and $\mathcal{C}_3$. In practice, about 15 cases are sufficient if none of the corners $\mathbf{c}_i$, $\mathbf{k}_j$ coincide as in Figure 9(c) and if we have $\mathbf{c}_i \notin \mathcal{C}_j$ and $\mathbf{k}_i \notin \mathcal{S}_j$ for all $i, j = 1, ..., 4$. More cases must be implemented to treat those latter cases which are not a rare situation for a simulation on practical CAD models. The practical difficulty is to come up to a fast and efficient point location method inside a NURBS patch and curve-curve intersections. If the rectangle $\mathcal{R}$ is too large then we split it into two rectangles $\mathcal{R}_1$ and $\mathcal{R}_2$ and apply the same method to each subrectangle $\mathcal{R}_i$. One chooses between vertical or horizontal splitting whichever gives the better shape (closer to a square) for the sub-rectangles. Some results of such recursive splitting are displayed in Figure 10. Note that the resulting NURBS patches are not globally continuous [19] but that does not create any problem for the PPUM approach. Problems related to curvature may occur in those special cases if the curves are too wavy. In such a situation, one has to apply NURBS subdivisions.

4 Numerical Experiments

The former geometric processing has been implemented in C/C++ and was integrated in our PPUM implementation. As a reference example of a CAD model, we use the exterior domain of an aircraft, see Figure 11. Let us now show some numerical results about the clipping of a NURBS patch $\mathcal{P}$ by a rectangle $\mathcal{R}$ as described in Section 3.4. To quantify the distortion of the bounding curves from being straight, we used the following distortion gauge $\mathcal{G}(\mathcal{P})$. For a NURBS curve $\mathcal{S}$ which has control points $\mathbf{d}_i$ for $i = 0, ..., n$ and which starts at $\mathbf{A}$ and terminates at $\mathbf{B}$, we define

$$\mathcal{G}(\mathcal{S}) := \Big|\ell(\mathcal{S}) - \|\mathbf{A} - \mathbf{B}\|\Big| \sum_{i=0}^{n} \mathrm{proj}[\mathbf{d}_i, \mathcal{L}(\mathbf{A}, \mathbf{B})] \tag{33}$$

where $\ell(\mathcal{S})$ designates the chord length of the control polygon while proj $[\mathbf{x}, \mathcal{L}(\mathbf{A}, \mathbf{B})]$ denotes the projection of a point $\mathbf{x} \in \mathbf{R}^2$ onto the line $\mathcal{L}(\mathbf{A}, \mathbf{B})$ passing through $\mathbf{A}$ and $\mathbf{B}$. For a NURBS patch $\mathcal{P}$ having four boundary curves $\mathcal{S}_1$,...,$\mathcal{S}_4$, we define the distortion gauge to be $\mathcal{G}(\mathcal{P}) := \sum_{i=1}^{4} \mathcal{G}(\mathcal{S}_i)$. That is, for a NURBS patch $\mathcal{P}$ which is a perfect convex quadrilateral, the distortion gauge $\mathcal{G}(\mathcal{P})$ is zero.

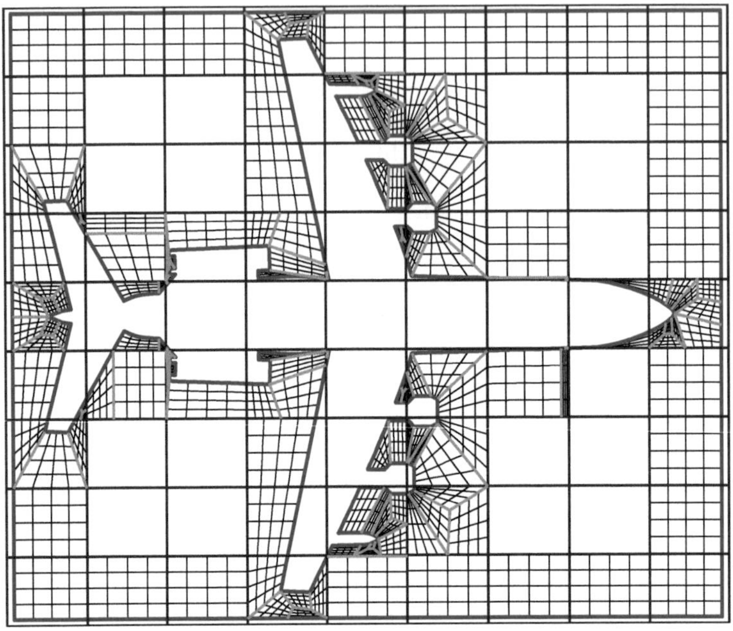

Fig. 11. Setup phase for an exterior domain.

Table 2. Performance of clipping operations for 2000 intersections with respect to average number of patches *np* (Fixed distortion gauge=7.693015).

Size of rectangles	*np*
0.00-0.10	1.009
0.10-0.20	1.041
0.20-0.30	1.103
0.30-0.40	1.146
0.40-0.50	1.179
0.50-0.60	1.246
0.60-0.70	1.295
0.70-0.80	1.338
0.80-0.90	1.390
0.90-1.00	1.441

Table 3. Performance of clipping operations for 2000 intersections with respect to average number of patches *np*.

Distortion gauge	*np*
0.000000	1.000000
0.095194	1.329500
0.734081	1.571786
2.347471	1.945000
5.220395	2.069104
9.528208	2.331331
15.386027	2.413327

First, we would like to examine the number of resulting NURBS patches. Table 2 gathers some numerical results from 2000 clipping operations. The first column presents the ratio of the area of the rectangle $\mathcal{R}$ with respect to the area of the original NURBS patch $\mathcal{P}$. The rectangles are chosen randomly using the condition that the intersections are not empty. We observe that the average number of resulting patches are significantly small.

As a second test, we generate a NURBS patch $\mathcal{P}$ whose distortion coefficient $\mathcal{G}(\mathcal{P})$ can be changed. We investigate the average number of patches in clipping operations in terms of the distortion coefficient. The NURBS patch is

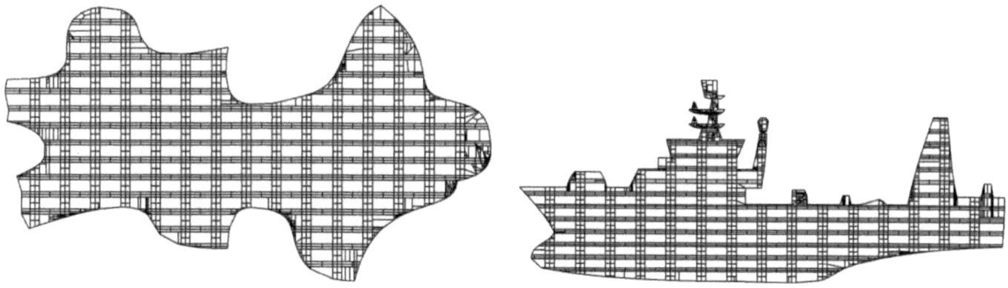

Fig. 12. Wireframe representation of integration cells for two different reference domains on level 4. All interior cells are affine.

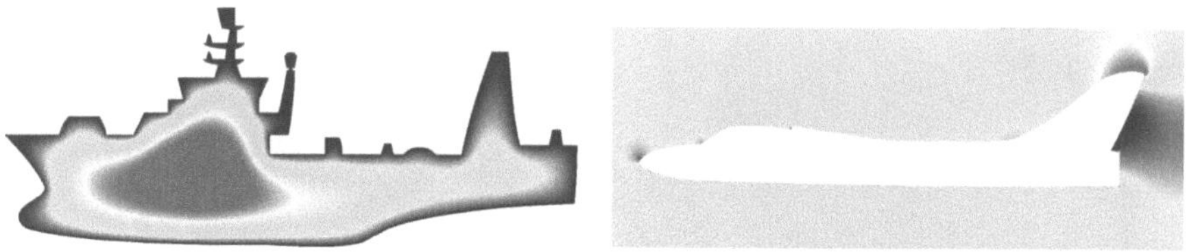

Fig. 13. Contour plot of approximate solution to problem (34) with homogeneous Dirichlet boundary conditions and $f = 1$ on level 8 (left). Contour plot of computed pressure for a potential flow problem (34) with inflow boundary conditions at the left boundary on level 8.

chosen so that when $\mathcal{G}(\mathcal{P})$ vanishes, $\mathcal{P}$ coincides to a rectangle. In Tab. 3, we display the results of such tests. We observe that the number of patches for the intersections is reasonably small even when the distortion gauge is already practically large.

Finally, we present some approximation results with the PPUM on general domains. To this end, we consider a simple diffusion problem

$$\begin{aligned} -\Delta u &= f \text{ in } \Omega, \\ u &= g_D \text{ on } \Gamma_D \subset \partial\Omega, \\ \partial_n u &= g_N \text{ on } \Gamma_N := \partial\Omega \setminus \Gamma_D \end{aligned} \tag{34}$$

on three different realistic domains in two space dimensions, see Figures 13 and 14, and a linear elasticity model problem

$$\begin{aligned} -\mathbf{div}\,\boldsymbol{\sigma}(u) &= f \quad \text{in } \Omega, \\ u &= g_D \quad \text{on } \Gamma_D \subset \partial\Omega, \\ \boldsymbol{\sigma}(u)\cdot n &= g_N \quad \text{on } \Gamma_N := \partial\Omega \setminus \Gamma_D, \end{aligned} \tag{35}$$

see Figure 15.

We consider a sequence of uniformly refined covers C_Ω^k with $\alpha = 1.3$ in (3) and local polynomial spaces $\mathcal{P}^{p_{i,k}}$ on all levels $k = 1, \dots, J$ in this paper. From the plots depicted in Figure 12 we see that only a small number of integration cells must be intersected with the boundary and that the total number of integration cells is increased only slightly. Thus, the compute time spent in the

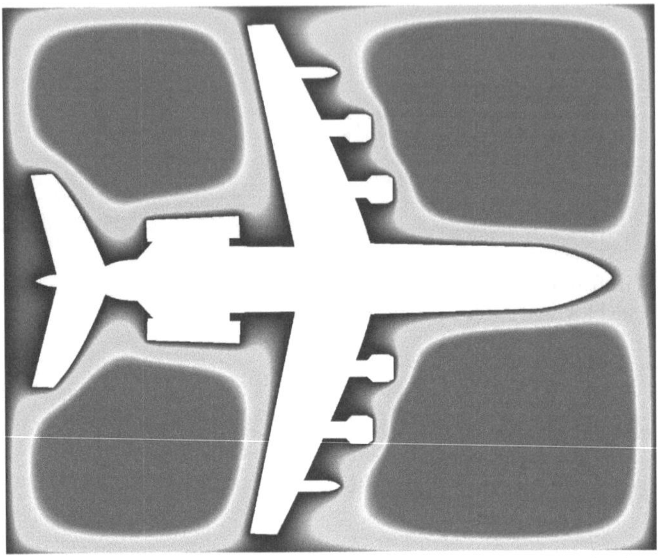

Fig. 14. Contour plot of approximation to problem (34) with homogeneous Dirichlet boundary conditions and $f = 1$ on level 9.

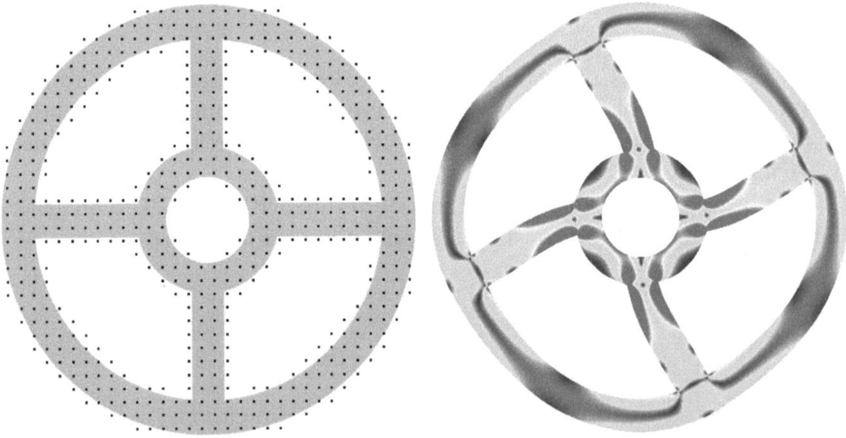

Fig. 15. Computational domain and particle distribution on level 5 considered in (35) (left). Here, we apply tangential tractions on the outer ring and homogeneous Dirichlet boundary conditions along the inner ring. Contour plot of the computed von Mises stress on the deformed configuration on level 9 (right).

assembly of the linear system is almost unaffected by the geometric complexity of the domain. However, the compute time spent in the processing of the domain, i.e. in the computation of the intersections, is currently the most time consuming step; it takes about 70% of the total compute time — which is comparable with the situation in FEM.

Recall that we construct a cover patch ω_i, i.e. a PU function φ_i, for each tree-cell $\mathcal{C}_i$ which satisfies $\mathcal{C}_i \cap \Omega \neq \emptyset$. Thus, close to the boundary we may have to deal with PU functions φ_i whose support ω_i intersects the domain $\omega_i \cap \Omega$ barely. Due to this issue we cannot ensure that all the PU functions have the flat-top property and we may experience a deterioration of the condition number of the stiffness matrix. How to overcome this issue is the subject of current work and will be discussed in a forthcoming paper. Here, we simply employ our multilevel solver [9] as a preconditioner for a conjugate gradient solver applied to the possibly ill-conditioned arising linear system.

The measured asymptotic convergence rate of a $V(1,1)$-preconditioned CG solver in our experiments varies between 0.25 and 0.80 depending e.g. on the number of patches ω_i with very small intersections $\mathcal{C}_i \cap \Omega$. The respective rate using a $V(5,5)$-cycle as a preconditioner however was already very stable at roughly 0.1 up to level 9 with about 500.000 degrees of freedom.

5 Concluding Remarks

We presented a general approach to the treatment of arbitrary domains in two space dimensions with meshfree Galerkin methods in this paper. We have implemented the proposed scheme in the PPUM and presented some first numerical results which clearly demonstrate the viability of our approach.

There are two main challenges which are currently being investigated. The compute time for the processing of the geometry must be further reduced to allow for an on the fly use of the presented approach which is essential for a direct coupling of the simulation engine to a CAD system. Moreover, the impact of very small intersections on the conditioning of the basis and stiffness matrix must be analyzed in detail.

References

1. I. Babuška, U. Banerjee, and J. E. Osborn, *Survey of Meshless and Generalized Finite Element Methods: A Unified Approach*, Acta Numerica, (2003), pp. 1–125.
2. I. Babuška and J. M. Melenk, *The Partition of Unity Finite Element Method: Basic Theory and Applications*, Comput. Meth. Appl. Mech. Engrg., 139 (1996), pp. 289–314. Special Issue on Meshless Methods.
3. ——, *The Partition of Unity Method*, Int. J. Numer. Meth. Engrg., 40 (1997), pp. 727–758.
4. S. Beissel and T. Belytschko, *Nodal Integration of the Element-Free Galerkin Method*, Comput. Meth. Appl. Mech. Engrg., 139 (1996), pp. 49–74.
5. J. S. Chen, C. T. Wu, S. Yoon, and Y. You, *A Stabilized Conforming Nodal Integration for Galerkin Mesh-free Methods*, Int. J. Numer. Meth. Engrg., 50 (2001), pp. 435–466.

6. E. Cohen, T. Lyche, and R. Riesenfeld, *Discrete B-Splines and Subdivision Techniques in Computer Aided Geometric Design and Computer Graphics*, Computer Graphics and Image Processing, 14 (1980), pp. 87–111.
7. S. Coons, *Surfaces for Computer Aided Design of Space Forms*, tech. report, Department of Mechanical Engineering in MIT, 1967.
8. J. Dolbow and T. Belytschko, *Numerical Integration of the Galerkin Weak Form in Meshfree Methods*, Comput. Mech., 23 (1999), pp. 219–230.
9. M. Griebel and M. A. Schweitzer, *A Particle-Partition of Unity Method—Part II: Efficient Cover Construction and Reliable Integration*, SIAM J. Sci. Comput., 23 (2002), pp. 1655–1682.
10. ———, *A Particle-Partition of Unity Method—Part III: A Multilevel Solver*, SIAM J. Sci. Comput., 24 (2002), pp. 377–409.
11. ———, *A Particle-Partition of Unity Method—Part V: Boundary Conditions*, in Geometric Analysis and Nonlinear Partial Differential Equations, S. Hildebrandt and H. Karcher, eds., Springer, 2002, pp. 517–540.
12. ———, *A Particle-Partition of Unity Method—Part VII: Adaptivity*, in Meshfree Methods for Partial Differential Equations III, M. Griebel and M. A. Schweitzer, eds., vol. 57 of Lecture Notes in Computational Science and Engineering, Springer, 2006, pp. 121–148.
13. H. Harbrecht and M. Randrianarivony, *From Computer Aided Design to Wavelet BEM*, Journal of Computing and Visualization in Science, 13 (2010), pp. 69–82.
14. J. Hoschek and D. Lasser, *Grundlagen der geometrischen Datenverarbeitung*, Teubner, 1989.
15. N. Moës, J. Dolbow, and T. Belytschko, *A Finite Element Method for Crack Growth without Remeshing*, Int. J. Numer. Meth. Engrg., 46 (1999), pp. 131–150.
16. J. Nitsche, *Über ein Variationsprinzip zur Lösung von Dirichlet-Problemen bei Verwendung von Teilräumen, die keinen Randbedingungen unterworfen sind*, Abh. Math. Sem. Univ. Hamburg, 36 (1970–1971), pp. 9–15.
17. H. Prautzsch, W. Boehm, and M. Paluszny, *Bézier and B-spline Techniques*, Springer, 2002.
18. M. Randrianarivony, *Geometric Processing of CAD Data and Meshes as Input of Integral Equation Solvers*, ph.d. thesis, Technische Universität Chemnitz, 2006.
19. ———, *On Global Continuity of Coons Mappings in Patching CAD Surfaces*, Computer Aided Design, 41 (2009), pp. 782–791.
20. M. A. Schweitzer, *A Parallel Multilevel Partition of Unity Method for Elliptic Partial Differential Equations*, vol. 29 of Lecture Notes in Computational Science and Engineering, Springer, 2003.
21. ———, *Meshfree and Generalized Finite Element Methods*, Habilitationsschrift, Institut für Numerische Simulation, Universität Bonn, 2008.
22. ———, *Stable Enrichment and Local Preconditioning in the Particle–Partition of Unity Method*, tech. report, Sonderforschungsbereich 611, Rheinische Friedrich-Wilhelms-Univeristät Bonn, 2008.
23. ———, *An Adaptive hp-Version of the Multilevel Particle–Partition of Unity Method*, Comput. Meth. Appl. Mech. Engrg., 198 (2009), pp. 1260–1272.
24. ———, *An Algebraic Treatment of Essential Boundary Conditions in the Particle–Partition of Unity Method*, SIAM J. Sci. Comput., 31 (2009), pp. 1581–1602.

25. T. Strouboulis, I. Babuška, and K. Copps, *The Design and Analysis of the Generalized Finite Element Method*, Comput. Meth. Appl. Mech. Engrg., 181 (2000), pp. 43–69.
26. T. Strouboulis, K. Copps, and I. Babuška, *The Generalized Finite Element Method*, Comput. Meth. Appl. Mech. Engrg., 190 (2001), pp. 4081–4193.
27. T. Strouboulis, L. Zhang, and I. Babuška, *Generalized Finite Element Method using mesh-based Handbooks: Application to Problems in Domains with many Voids*, Comput. Meth. Appl. Mech. Engrg., 192 (2003), pp. 3109–3161.

Sampling Inequalities and Support Vector Machines for Galerkin Type Data

Christian Rieger[1]

Institute for Numerical Simulation & Hausdorff Center for Mathematics, University of Bonn, Wegelerstr. 6, 53115 Bonn, Germany
rieger@ins.uni-bonn.de

Summary. We combine the idea of *sampling inequalities* and Galerkin approximations of weak formulations of partial differential equations. The latter is a well-established tool for finite element analysis. We show that sampling inequalities can be interpreted as Pythagoras law in the energy norm of the weak form. This opens the way to consider regularization techniques known from machine learning in the context of finite elements. We show how sampling inequalities can be used to provide a deterministic worst case error estimate for reconstruction problems based on Galerkin type data. Such estimates suggest an *a priori* choice for regularization parameter(s).

Key words: Galerkin Methods, Reproducing Kernel Hilbert Spaces, Sampling Inequalities, Regularization, Support Vector Regression

1 Introduction

A differentiable function cannot attain arbitrarily large values anywhere in a bounded domain if both its values on a sufficiently dense discrete set and its derivatives are bounded. This qualitative observation has been made quantitative by *sampling inequalities* [11, 14, 22]. Instead of point evaluations one can also consider various kinds of discrete data [15]. We show that *sampling inequalities* arise naturally in the context of variational formulations of partial differential equations if one considers Galerkin-type data [15]. In particular, we will show that sampling inequalities in the energy norm of the weak formulation are nothing but the Pythagoras law with respect to the inner product induced by the weak form. Here, instead of point evaluations (see [20] for Euklidean domains and [8] for results on spheres) the discrete data is assumed to be generated by the energy inner product. A typical example for Galerkin type data is

$$S_{a_P}(f) := \left(\int_\Omega \nabla f(x) \nabla \phi_1(x) dx, \ldots, \int_\Omega \nabla f(x) \nabla \phi_N(x) dx \right)^T \in \mathbb{R}^N$$

M. Griebel, M.A. Schweitzer (eds.), *Meshfree Methods for Partial Differential Equations V*, Lecture Notes in Computational Science and Engineering 79,
DOI 10.1007/978-3-642-16229-9_3,

for a sufficiently smooth function f with some orthonormal test functions $\{\phi_i\}$. There are two motivations for our approach: The first one is simply that such weak data occurs in finite element methods as well. Hence, we do not have to generate new types of data. The second reason to consider this kind of data is that we do not need to assume continuous point evaluations.

It is worth mentioning that the Sobolev's embedding theorem yields a continuous embedding of the Sobolev space $W_2^k(\Omega)$ on a sufficiently smooth domain $\Omega \subset \mathbb{R}^d$ into the space of continuous functions only if $k > d/2$, where k denotes the smoothness and d the space dimension. This condition becomes more and more restrictive as the space dimension grows although this problem is often neglected. In particular, when radial basis functions are applied it is often stated that the methods would work in arbitrary space dimensions assuming implicitly high regularity [17]. This last remark seems to make the results presented here interesting for other applications in a classical machine learning context.
The sampling inequalities can be applied to provide a deterministic error analysis for various regularized reconstruction problems involving Galerkin data, one important example being least squares regression which is related to spline smoothing [19]. The regularization of finite element methods in a much more general and more theoretical context can be found in [9]. There, however, the main focus is on continuous norms. Our results may be interpreted as a dicretization of results in [9].

The remainder of the manuscript is organized as follows: In Section 2 we briefly review some known sampling inequalities and give a short idea of their proofs. In Section 3 we shall present a sampling inequality which is nothing but a version of Pythagoras' law. In Section 4, we give some applications of sampling inequalities to the deterministic error analysis of certain regularized problems. This is to be understood as a general tool which is capable to treat a large class of regularization networks [6, 16–18].

2 Review on sampling inequalities

In the univariate setting, sampling inequalities arise naturally from the fundamental theorem of calculus. We shall use the notation of Sobolev spaces from [1]. In the following, we will focus on the Hilbert space case $p = 2$. We recall the definition of Sobolev spaces in that particular case.

Definition 1 ([1]). *Let $\Omega \subset \mathbb{R}^d$ be a domain. We define the Sobolev spaces of integer orders $k \in \mathbb{N}$ as*

$$W_2^k(\Omega) = \{f \in L_2(\Omega) : f \text{ has weak derivatives } D^\alpha f \in L_2(\Omega) \text{ of order } |\alpha| \le k\}$$

with the norm

$$\|u\|_{W_2^k(\Omega)} := \left(\sum_{|\alpha|\leq k} \|D^\alpha u\|^2_{L_2(\Omega)} \right)^{1/2} .$$

For fractional smoothness $s = k + \sigma$ with $0 < \sigma < 1$ and $k \in \mathbb{N}$ we define the semi-norm

$$|u|_{W_2^s(\Omega)} := \left(\sum_{|\alpha|=k} \int_\Omega \int_\Omega \frac{|D^\alpha u(x) - D^\alpha u(y)|^2}{\|x-y\|_2^{d+2\sigma}} dxdy \right)^{1/2} ,$$

and set

$$W_2^s(\Omega) := \left\{ u \in L_2(\Omega) \ : \ \left(\|u\|^2_{W_2^k(\Omega)} + |u|^2_{W_2^s(\Omega)} \right)^{1/2} < \infty \right\} .$$

In this section, we essentially repeat the calculations from [15]. We assume a sufficiently smooth function f on an interval $[a,b]$ and a discrete ordered set of centers $X = \{x_1, \ldots, x_N\} \subset [a,b]$ with

$$a = x_1 < x_2 < \cdots < x_{N-1} < x_N = b.$$

Then for any point $x \in [a,b]$ and the closest point $x_j \in X$

$$f(x) = \int_{x_j}^x f'(t)\, dt + f(x_j) \quad , \text{ i.e.,}$$

$$|f(x)| \leq \sqrt{|x - x_j|} \sqrt{\int_{x_j}^x |f'(t)|^2 \, dt} + |f(x_j)| \ .$$

If we define the fill distance

$$h := h_{X,[a,b]} := \sup_{x\in[a,b]} \min_{x_j \in X} |x - x_j|$$

we find the sampling inequality

$$\|f\|_{L_\infty([a,b])} \leq \sqrt{h} |f|_{W_2^1[a,b]} + \|f|_X\|_{\ell_\infty(X)} \ .$$

Here, the weak continuous $L_\infty([a,b])$-norm of the function f is bounded by a combination of two terms. The first is the stronger continuous $W_2^1([a,b])$-norm weighted in terms of the fill-distance, and the second term penalizes the discrete values sampled on the finite set X.

We now focus on sampling inequalities for Sobolev spaces on sufficiently smooth bounded domains $\Omega \subset \mathbb{R}^d$ with $d \geq 1$. The strong sampling of a function $f \in W_p^s(\Omega)$ for $p > 1$ is formalized by means of a *sampling operator*. For a given discrete set $X = \{x_1, \ldots, x_N\} \subset \Omega$, we define

$$S_X : W_p^s(\Omega) \to \mathbb{R}^N, \tag{1}$$
$$f \mapsto \left(\delta_{x_1}(f), \ldots, \delta_{x_N}(f)\right)^T = \left(f(x_1), \ldots, f(x_N)\right)^T,$$

which is well-defined for $s > d/p$. The multivariate discretization measure is given as the fill distance which is the Hausdorff distance from X to Ω

$$h_{X,\Omega} := \sup_{x\in\Omega} \min_{x_j\in X} \|x - x_j\|_2 \,. \tag{2}$$

Various *strong sampling inequalities*, i.e., sampling inequalities involving point evaluations, for functions $u \in W_p^k(\Omega)$ from Sobolev spaces $W_p^k(\Omega)$ with $1 < p < \infty$ and $k > d/p$, or with $p = 1$ and $k \geq d$ on domains $\Omega \subset \mathbb{R}^d$ have been obtained in the last years. Narcowich, Ward and Wendland considered the case of functions with scattered zeros [12] and showed how the following results can be used in the deterministic error analysis of interpolatory numerical reconstruction processes.

Theorem 1 (see [12]). *There exist positive constants C and h_0 such that the inequality*

$$|u|_{W_q^m(\Omega)} \leq C h^{k-m-d\left(\frac{1}{p}-\frac{1}{q}\right)_+} |u|_{W_p^k(\Omega)}$$

holds for all functions $u \in W_p^k(\Omega)$ with $k - m > d/p$ and $S_X(u) = 0$ on arbitrary discrete sets X whose fill distance $h := h_{X,\Omega}$ in the sense of (2) satisfies $h \leq h_0$. The constants C, h_0 may depend on q, m, p, k, Ω, and d, but not on X, h or u.

In [22] this result was generalized to functions with arbitrary values on scattered locations:

Theorem 2 (see [22]). *We assume $1 \leq q \leq \infty$, $\alpha \in \mathbb{N}_0^d$, $k \in \mathbb{N}$, and $1 \leq p < \infty$ with $k > |\alpha| + d/p$ if $p > 1$, or with $k \geq |\alpha| + d$ if $p = 1$. Then there are constants C, $h_0 > 0$ such that*

$$\|D^\alpha u\|_{L_q(\Omega)} \leq C\left(h^{k-|\alpha|-d\left(\frac{1}{p}-\frac{1}{q}\right)_+} |u|_{W_p^k(\Omega)} + h^{-|\alpha|} \|S_X u\|_{\ell_\infty(\mathbb{R}^N)}\right)$$

holds for all $u \in W_p^k(\Omega)$ and all discrete sets $X \subset \Omega$ with associated sampling operators S_X from (1) and fill distance $h := h_{X,\Omega} \leq h_0$.

A similar theorem, where the sampled data is measured in a weaker discrete ℓ_p-norm was provided by Madych [11]. A result valid also for unbounded domains with applications to spline interpolation and smoothing was given by Arcangéli, López di Silanes and Torrens in [2]. In all cases the sampling order, i.e., the power of h in front of the continuous Sobolev-norm depends only on the smoothness difference of the two continuous (semi–)norms involved.

2.1 Proof Sketch

A standard way to prove sampling inequalities follows the lines of [12] and [22]. The basic ingredient is a (local) stable polynomial reproduction. We shall denote by $\pi_k(\mathbb{R}^d)$ the space of all d-variate polynomials of degree at most k.

Definition 2. *[21, Definition 3.1] A process that defines for every set $X = \{x_1, \ldots, x_N\} \subset \Omega$ a family of functions $a_j = a_j^X : \Omega \to \mathbb{R}$, $1 \leq j \leq N$, provides a local polynomial reproduction of degree k if there exist constants h_0, C_1, C_2 such that*

- $\sum_{j=1}^N a_j(x)p(x_j) = p(x)$ *for all* $p \in \pi_k(\mathbb{R}^d)|_\Omega$
- $\sum_{j=1}^N |a_j(x)| \leq C_1$ *for all* $x \in \Omega$
- $a_j(x) = 0$ *if* $\|x - x_j\|_2 > C_2 h_{X,\Omega}$, *all* $x \in \Omega$ *and all* $1 \leq j \leq N$

is satisfied for all X with $h_{X,\Omega} \leq h_0$.

The existence of stable polynomial reproductions is not obvious. It is clear from simple dimension arguments that given the set X is unisolvent for $\pi_k(\Omega) := \pi_k(\mathbb{R}^d)|_\Omega$, i.e., the only polynomial vanishing at all points in X is the zero-polynomial, we can form a Lagrange basis with cardinality $|X|$. In this case, however, we cannot expect the constant C_1 to be bounded, since it is nothing but the usual Lebesgue-constant for polynomial interpolation. One way out is to spend more than $|X|$ functions and to use the freedom gained by this oversampling to bound the Lebesgue-constant C_1. A short and elegant way to do so is based on a so-called norming set argument, see [10,12]. We will not go into details here but simply state that one basically needs to bound the norm of the inverse of the sampling operator restricted to $\pi_k(\Omega)$. To be precise, we consider

$$\begin{aligned} S_X|_{\pi_k(\Omega)} \,:\, \pi_k(\Omega) &\to \mathbb{R}^N, \\ p &\mapsto \left(\delta_{x_1}(p), \ldots, \delta_{x_N}(p)\right)^T = \left(p(x_1), \ldots, p(x_N)\right)^T, \end{aligned}$$

and we have to uniformly bound [21, p. 27]

$$\left\|S_X^{-1}\right\|_{(\mathbb{R}^N, \|\cdot\|_{\ell_\infty}) \to (\pi_k(\Omega), \|\cdot\|_{L_\infty})} .$$

Such a bound clearly implies that X is unisolvent for $\pi_k(\Omega)$. Once this bound is established one can invoke a general theorem [21, Theorem 3.4] to deduce the existence of a stable polynomial reproduction for X. It is a well-known basic principle (see [7] for instance), that with a local stable polynomial reproduction we can define a quasi-interpolant

$$\mathcal{I}_x(S_X(f)) = \sum_{j=1}^N a_j(x) f(x_j) \quad \text{for } f \in C(\Omega) .$$

The operator $\mathcal{I}_{(\cdot)} \circ S_X$ is exact on the space of d-variate polynomials $\pi_k(\Omega)$, i.e., for all $x \in \Omega$

$$\mathcal{I}_x(S_X(p)) = \sum_{j=1}^{N} a_j(x)p(x_j) = p(x) \quad \text{for all } p \in \pi_k(\Omega) \; .$$

This implies that the quasi-interpolant inherits the local approximation quality of the space of d-variate polynomials $\pi_k(\Omega)$. In particular, we see that for a fixed $x \in \Omega$ the norm of $\mathcal{I}_x : \mathbb{R}^N \to \mathbb{R}$ is given by

$$L_{\mathcal{I}}(x) := \sup_{\substack{y \in \mathbb{R}^N \\ \|y\|_{\ell_\infty}=1}} \left| \sum_{j=1}^{N} a_j(x) y_j \right| = \sum_{j=1}^{N} |a_j(x)| \; ,$$

which is uniformly bounded by C_1 for a stable polynomial reproduction. Finally, we can follow [12] and [22] or several other places, to get

$$\begin{aligned}
|f(x)| &\leq |f(x) - p(x)| + |p(x)| = |f(x) - p(x)| + |\mathcal{I}_x(S_X(p))| \\
&\leq |f(x) - p(x)| + |\mathcal{I}_x(S_X(p-f))| + |\mathcal{I}_x(S_X(f))| \\
&\leq (1 + L_{\mathcal{I}}(x)) \, \|f - p\|_{L_\infty(\Omega)} + L_{\mathcal{I}}(x) \|f\|_{\ell_\infty(X)} \; .
\end{aligned}$$

Using local polynomial approximation results [4], this leads to a local sampling inequality which is then carried over to the global domain Ω by a careful covering argument due to Duchon [5]. We recalled the proof sketch to highlight how much work is needed to bound the Lebesgue-constant in this setting. In the next section we shall present a different approach to sampling inequalities which leads to bounded Lebesgue-constants automatically.

3 Sampling Inequalities based on Weak Formulations

We are seeking sampling inequalities involving weak discrete data instead of the usual point evaluations. We shall give some motivation from which kinds of problems such data arises. Let V and H be real separable Hilbert spaces with V compactly embedded into H. We shall denote the inner product on the respective spaces by $(\cdot,\cdot)_V$ and $(\cdot,\cdot)_H$. Let $a(\cdot,\cdot) : V \times V \to \mathbb{R}$ be a symmetric bi-linear form such that there are positive constants $c_C, c_E > 0$ satisfying

$$\begin{aligned}
a(u,v) &\leq C_C \|u\|_V \|v\|_V \quad \text{for all } u, v \in V \\
a(u,u) &\geq C_E \|u\|_V^2 \quad \text{for all } u \in V \; .
\end{aligned} \tag{3}$$

Galerkin type data arises naturally from discretizations of variational problems of the form

$$\text{for } f \in H \text{ find } u \in V \text{such that} \quad a(u,v) = (f,v)_H \quad \text{for all } v \in V \; . \tag{4}$$

To be more concrete, we have the following two model problems from [3] in mind. Let $\Omega \subset \mathbb{R}^d$ be a bounded domain with sufficiently smooth boundary

$\partial\Omega$. For given $f \in L_2(\Omega)$, the Poisson-problem with homogeneous Dirichlet boundary conditions reads [3, Section 8.4.1]

$$-\Delta u = f \quad \text{in } \Omega \quad \text{and} \quad u|_{\partial\Omega} \equiv 0\,. \tag{5}$$

The weak formulation is given by

$$a_P(u,v) := \int_\Omega \nabla u(x)\nabla v(x)dx \quad \text{and} \quad F(v) := (f,v)_{L_2(\Omega)} = \int_\Omega f(x)v(x)dx$$

where $a_P(\cdot,\cdot)$ satisfies the assumptions (3) for $V = W^1_{0,2}(\Omega)$. The additional subscript 0 stands for the restriction to functions with vanishing trace. Here, $H = L_2(\Omega)$. The boundary conditions have to be incorporated into the function spaces.

The second model problem is the Helmholtz equation with natural boundary conditions [3, Section 8.4.3]

$$-\Delta u + u = f \quad \text{in } \Omega \quad \text{and} \quad \frac{\partial u}{\partial \nu}|_{\partial\Omega} \equiv 0 \tag{6}$$

where ν is the outer unit-normal to $\partial\Omega$. The weak formulation is given by

$$a_H(u,v) := \int_\Omega \nabla u(x)\nabla v(x) + u(x)v(x)dx \quad \text{and}$$
$$F(v) := (f,v)_{L_2(\Omega)} = \int_\Omega f(x)v(x)dx$$

where $a_H(\cdot,\cdot)$ satisfies the assumptions (3) for $V = W^1_2(\Omega)$, and $H = L_2(\Omega)$. In this case, the boundary conditions are directly incorporated into the variational formulation. The remainder of the manuscript, however, is not limited to these two special model problems.

3.1 Sampling inequalities based on Pythagoras law

We consider a Galerkin-approximation [15] of the elliptic problem (4). For a given N-dimensional trial space $V_N \subset V$, we build a sampling inequality based on the sampling operator

$$S_a : V \to \mathbb{R}^N, \quad f \mapsto \left(a\left(f,\phi_1\right),\ldots,a\left(f,\phi_N\right)\right)^T \in \mathbb{R}^N\,, \tag{7}$$

where $\{\phi_1,\ldots,\phi_N\}$ is an $a(\cdot,\cdot)$-orthonormal system spanning $V_N \subset V$. For the Poisson model problem this means

$$S_{a_P}(f) := \left(\int_\Omega \nabla f(x)\nabla\phi_1(x)dx,\ldots,\int_\Omega \nabla f(x)\nabla\phi_N(x)dx\right)^T \in \mathbb{R}^N$$

for all $f \in W^1_{0,2}(\Omega)$. In Section 4 we study the reconstruction of an unknown function from this kind of discrete data. To derive error bounds for the reconstruction we shall employ sampling inequalities which we now discuss. Similar sampling inequalities can be found in [15, Section 3.4.2] and [14] (see also the references therein). The main difference is that we consider now estimates in the energy norm. The following theorem can be seen as a variant of the famous Céa-Lemma [20] which is a standard tool in finite element analysis.

Theorem 3. *Let $(V, a(\cdot,\cdot))$ be a Hilbert space with inner product $a(\cdot,\cdot)$ and denote by $V_N = \text{span}\,\{\phi_1, \ldots, \phi_N\} \subset V$ an N-dimensional trial space. Suppose that $\{\phi_1, \ldots, \phi_N\}$ is an orthonormal system with respect to the bi-linear form $a(\cdot,\cdot)$, i.e.,*

$$a(\phi_j, \phi_k) = \delta_{j,k} \quad \text{for all } j,k \in \{1, \ldots, N\} \ .$$

Then for every $f \in V$

$$\|f\|_a^2 := a(f,f) = \min_{s \in V_N} \|f-s\|_a^2 + \sum_{j=1}^{N} |a(f,\phi_j)|^2 \ . \tag{8}$$

Proof. The proof works by standard arguments from linear algebra. Similar arguments can be found in almost every textbook on finite elements showing the Céa-Lemma, see e.g., [4]. For $f \in V$ denote by $s_f^\star$ the $a(\cdot,\cdot)$-orthogonal projection of f onto V_N. Due to the orthonormality of $\{\phi_1, \ldots, \phi_N\}$ we get

$$s_f^\star = \sum_{j=1}^{N} a(s_f^\star, \phi_j)\phi_j = \sum_{j=1}^{N} a(f, \phi_j)\phi_j \ ,$$

which implies

$$\left\|s_f^\star\right\|_a^2 = a\left(s_f^\star, s_f^\star\right) = \sum_{j=1}^{N} \left|a\left(s_f^\star, \phi_j\right)\right|^2 = \sum_{j=1}^{N} \left|a\left(f, \phi_j\right)\right|^2 = a\left(s_f^\star, f\right) \ .$$

Then Eq. (8) is the usual Pythagoras law [23]

$$a\left(f,f\right) = a\left(f - s_f^\star, f - s_f^\star\right) + a\left(s_f^\star, s_f^\star\right) \ .$$

We point out that Eq. (8) contains a best approximation error which can be formalized in terms of *Jackson inequalities* (cf. [13, Example 7]). We say that the pair $\left(V, W_2^k\right)$ satisfies a *Jackson inequality* with respect to the family $\{V_N\}$ if there is a constant $C > 0$ and a sequence of positive numbers $h\left(V_N, k\right)$ with $h\left(V_N, k\right) \to 0$ if $N \to \infty$ such that for all $f \in W_2^k$

$$\min_{s \in V_N} \|f-s\|_a \leq C h(V_N, k) \|f\|_{W_2^k(\Omega)} \ . \tag{9}$$

The discretization parameter $h(V_N, k)$ typically behaves like $h(V_N, k) \sim N^{-(k-1)/d}$ for quasi-uniform data [21, Proposition 14.1]. Here we assume $\|\cdot\|_a \sim \|\cdot\|_{W_2^1}$ which is reasonable in the setting of second order elliptic partial differential equations as outlined in (5) and (6).

Corollary 1. *Suppose that the pair (V, W_2^k) satisfies a* Jackson inequality *with respect to $\{V_N\}$. Then*

$$\|f\|_a^2 \leq C^2 h^2(V_N, k)\|f\|_{W_2^k(\Omega)}^2 + \sum_{j=1}^{N} |a(f, \phi_j)|^2 \qquad \textit{for all } f \in V .$$

We note that in Corollary 1 we need not assume $V \subset C(\Omega)$ since point evaluations are avoided. Further, the constant in front of the discrete term equals unity. This is remarkable since we had to spend a lot of work to bound this constant in the previous settings.

4 Regularization and Machine Learning

We choose the probably most direct approach toward least-squares regularization problems. We denote the stiffness matrix by

$$A_\Phi := \begin{pmatrix} a(\phi_1, \phi_1) & \dots & a(\phi_1, \phi_N) \\ \vdots & \vdots & \vdots \\ a(\phi_N, \phi_1) & \dots & a(\phi_N, \phi_N) \end{pmatrix} \in \mathbb{R}^{N \times N} ,$$

where we skip for the moment the orthonormality assumption. The following theorem provides an example of spline-smoothing [19, Theorem 1.3.1], see also [9].

Theorem 4. *Let $V_N = span\ \{\phi_1, \dots, \phi_N\} \subset V$ be an N-dimensional subspace and let $\lambda > 0$. For a given $F = (f_1, \dots, f_N)^T \in \mathbb{R}^N$, the solution $s_f^\star$ to the infinite dimensional optimization problem*

$$\min_{v \in V} \sum_{j=1}^{N} \left(a\left(v, \phi_j\right) - f_j\right)^2 + \lambda \|v\|_a^2 \tag{10}$$

is contained in V_N, i.e., $s_f^\star \in V_N$. The coefficients c_i from the representation $s_f^\star = \sum_{j=1}^N c_j \phi_j$ can be computed as solution of the linear system

$$\left(A_\Phi + \lambda Id_{N \times N}\right) c = F .$$

The proof of Theorem 4 works along the lines of the usual proof of representer theorems, see e.g., [18, Theorem 4.2] or [19, Theorem 1.3.1].

Following [24, Theorem 2], the solvability of infinite dimensional problems of the form (10) is addressed in two steps. A representer theorem reduces the problem to a finite-dimensional one, which then can be treated by standard tools from quadratic optimization theory.

Unfortunately, the minimization in Theorem 4 does not yield some higher regularity, since no higher derivatives are penalized. Usually, one expects the function f to be reconstructed from its data $S_a(f) \in \mathbb{R}^N$ to reside in a regularity space $R \subset V \subset H$, where each embedding is supposed to be compact. The additional regularity is gained from the embedding $R \subset V$. A typical choice in our context would be $R \subset W_2^k(\Omega)$.

We employ the notion of a reproducing kernel Hilbert space, see e.g. [17,18,21] and references therein.

Definition 3. *Let $\mathcal{H}(\Omega)$ be a Hilbert space of functions $f : \Omega \to \mathbb{R}$. A function $K : \Omega \times \Omega \to \mathbb{R}$ is called reproducing kernel of $\mathcal{H}(\Omega)$, if*

- *$K(y, \cdot) \in \mathcal{H}(\Omega)$ for all $y \in \Omega$ and*
- *$f(y) = (f, K(y, \cdot))_{\mathcal{H}(\Omega)}$ for all $f \in \mathcal{H}(\Omega)$ and all $y \in \Omega$.*

There is a one-to-one correspondence between positive semi-definite kernels and reproducing kernel Hilbert spaces [21, Chapter 10]. In particular, there are different kernels for the Hilbert spaces W_2^k and $W_{0,2}^k$, respectively. We shall in the following focus on spaces $\mathcal{H}_k \sim W_2^k(\mathbb{R}^d)$ with $k > d/2$. Let $K(\cdot, \cdot)$ be the reproducing kernel of $\mathcal{H}_k \sim W_2^k(\mathbb{R}^d)$ with $k > d/2$. Then K is a radial function, i.e., there is a univariate function $\tilde{K}$ depending on $r = \|x - y\|_2$ such that $K(x, y) = \tilde{K}(r)$ and K is positive definite. Furthermore, there is an explicit formula

$$K(x,y) = \tilde{K}(r) = \frac{2^{1-k}}{\Gamma(k)} r^{k-\frac{d}{2}} \kappa_{\frac{d}{2}-k}(r) , \tag{11}$$

where κ is the modified Bessel function of the third kind [21, Theorem 6.13]. Though Eq. (11) defines the kernel on the whole Euclidean space $\mathbb{R}^d$, the restriction to a sufficiently smooth domain $\Omega \subset \mathbb{R}^d$ gives rise to the reproducing kernel of $\mathcal{H}_k(\Omega) \sim W_2^k(\Omega)$ [21, Section 10.7].

We use the notation $\mu_j = a(\cdot, \phi_j) \in V'$, where V' denotes the dual space of V. Then the Riesz-representer with respect to the inner product of $\mathcal{H}_k(\Omega) \subset V$ is given by

$$K_j^a = \mu_j^{(z)} K(z, \cdot) := x \mapsto a\left(K(x, \cdot), \phi_j(\cdot)\right) .$$

The notation $\mu^{(z)}$ indicates that μ acts with respect to the variable z (cf. [21]). The Riesz-representer gives rise to a generalized Gramian matrix [21, Theorem 16.7]

$$A = \left(\mu_j^{(x)} \mu_\ell^{(y)} K(x, y)\right)_{j,\ell=1,\dots,N} = \left((K_j^a, K_\ell^a)_{\mathcal{H}_k(\Omega)}\right)_{j,\ell=1,\dots,N} .$$

This matrix is symmetric and positive definite provided the functionals $a(\cdot, \phi_i)$ are linearly independent [21, Theorem 16.7], which holds true if the functions $\{\phi_i\}$ are linearly independent. The following theorem is a special case of [21, Theorem 16.1] adapted to our notation.

Theorem 5. *Suppose* $F = (f_1, \ldots, f_N)^T \in \mathbb{R}^N$, *and* $\lambda > 0$. *A solution* $s_f^\star$ *of*

$$\min_{v \in W_2^k(\Omega)} \sum_{j=1}^{N} \left(a\left(v, \phi_j\right) - f_j\right)^2 + \lambda \|v\|^2_{W_2^k(\Omega)}$$

is contained in span $\left\{K_j^a(\cdot)\right\}$, *i.e.,* $s_f^\star = \sum_{j=1}^N \hat{s}_j K_j^a$ *where* K *is the reproducing kernel of* $W_2^k(\Omega)$. *The coefficients* $\hat{s}$ *of* $s_f^\star = \sum_{j=1}^N \hat{s}_j K_j^a$ *are given as solution of the linear system*

$$\left(A_{\Phi,K} + \lambda Id_{N\times N}\right)\hat{s} = F\ .$$

Proof. This is again just a usual representer theorem. In the general setting of optimal recovery, it can be seen as a special case of [21, Theorem 16.1].

Now we can use the above-mentioned embedding $W_2^k(\Omega) \subset W_2^1(\Omega) = V$ to yield some approximation rates. We consider the reconstruction of an unknown function $f \in W_2^k(\Omega)$ from its Galerkin-data

$$F := \begin{pmatrix} f_1 \\ \vdots \\ f_N \end{pmatrix} = \begin{pmatrix} a(f, \phi_1) \\ \vdots \\ a(f, \phi_N) \end{pmatrix} = S_a(f) \tag{12}$$

where $\{\phi_i\}$ form an $a(\cdot,\cdot)$-orthonormal system (cf. (7)).

Theorem 6. *Suppose that the pair* $\left(V, W_2^k\right)$ *satisfies a* Jackson inequality *with respect to* $\{V_N\}$ *(cf. (9)). For* $f \in W_2^k(\Omega)$ *and* $\lambda > 0$ *denote by* $s_{f,\lambda}^\star$ *a solution of the infinite dimensional optimization problem*

$$\min_{v \in W_2^k(\Omega)} \sum_{j=1}^{N} \left(a\left(v, \phi_j\right) - f_j\right)^2 + \lambda \|v\|^2_{W_2^k(\Omega)}\ .$$

Then there is a constant C *such that for all* $f \in W_2^k$ *and all* $\lambda > 0$

$$\left\|f - s_{f,\lambda}^\star\right\|_a \leq C\left(h(V_N, k) + \sqrt{\lambda}\right)\|f\|_{W_2^k(\Omega)}\ .$$

Proof. The proof is completely analogous to [22, Proposition 3.6] and [14, Section 7.3]. We recall it for the readers' convenience. We control the regularity norm of the reconstruction using the minimality property of $s_{f,\lambda}^\star$ via

$$\begin{aligned}\lambda\left\|s_{f,\lambda}^\star\right\|^2_{W_2^k(\Omega)} &\leq \sum_{j=1}^{N}\left(a\left(s_{f,\lambda}^\star, \phi_j\right) - f_j\right)^2 + \lambda\|s_{f,\lambda}^\star\|^2_{W_2^k(\Omega)} \\ &= J(s_{f,\lambda}^\star) \leq J(f) = \lambda\|f\|^2_{W_2^k(\Omega)}\ .\end{aligned}$$

Similarly the discrete term $\|S_a(s^\star_{f,\lambda}) - F\|_{\ell_2(N)}$ is bounded via

$$\begin{aligned}\left\|S_a(s^\star_{f,\lambda} - f)\right\|^2_{\ell_2(N)} &= \sum_{j=1}^{N} \left(a\left(s^\star_{f,\lambda}, \phi_j\right) - f_j\right)^2 \leq J(s^\star_{f,\lambda}) \\ &\leq J(f) = \lambda \|f\|^2_{W_2^k(\Omega)} \,.\end{aligned}$$

Now we can invoke Corollary 1 to get with universal constants $C > 0$

$$\begin{aligned}\|f - s^\star_{f,\lambda}\|^2_a &\leq Ch^2(V_N, k)\|f - s^\star_{f,\lambda}\|^2_{W_2^k(\Omega)} + \sum_{j=1}^{N} \left|a(f - s^\star_{f,\lambda}, \phi_j)\right|^2 \\ &\leq C\left(h(V_N, k)^2 + \lambda\right) \|f\|^2_{W_2^k(\Omega)} \,.\end{aligned}$$

Corollary 2. *With the notation of Theorem 6, for $\lambda = h(V_N, k)^2$ we obtain the usual error estimate.*

Acknowledgement. The author would like to thank Barbara Zwicknagl and Robert Schaback many helpful and stimulating discussions.

References

1. R.A. Adams, *Sobolev Spaces*, Pure and Applied Mathematics A (65), Academic Press, London, 1975.
2. R. Arcangéli, M.C. López di Silanes, and J.J. Torrens, *An extension of a bound for functions in Sobolev spaces, with applications to (m, s)-spline interpolation and smoothing*, Numer. Math. **107(2)** (2007), 181–211.
3. K. Atkinson and W. Han, *Theoretical numerical analysis. a functional analysis framework*, Texts in applied mathematics, Springer, 2005.
4. S. Brenner and L. Scott, *The Mathematical Theory of Finite Element Methods*, Springer, New York, 1994.
5. J. Duchon, *Sur l'erreur d' interpolation des fonctions de plusieurs variables par les D^m-splines.*, Rev. Française Automat. Informat. Rech. Opèr. Anal. Numer. **12** (1978), 325–334.
6. F. Girosi, *An equivalence between sparse approximation and support vector machines*, Neural Computation **10 (8)** (1998), 1455–1480.
7. T. Hangelbroek, F. J. Narcowich, and J. D. Ward, *Kernel approximation on manifolds I: Bounding the lebesgue constant*, SIAM Journal on Mathematical Analysis **42** (2010), no. 4, 1732–1760.
8. K. Jetter, J. Stöckler, and J. D. Ward, *Norming stes and scattered data approximation on spheres*, Approximation Theory IX, Vol. II: Computational Aspects, Vanderbilt University Press, 1998, 137–144.
9. Andrew Knyazev and Olof Widlund, *Lavrentiev regularization + Ritz approximation = uniform finite element error estimates for differential equations with rough coefficients*, Math. Comp. **72** (2003), no. 241, 17–40 (electronic). MR MR1933812 (2003i:65107)

10. Q. T. Le Gia, *Galerkin approximation for elliptic pdes on spheres*, Journal of Approximation Theory **130** (2004), 123–147.
11. W. R. Madych, *An estimate for multivariate interpolation II*, J. Approx. Theory **142** (2006), 116–128.
12. F.J. Narcowich, J.D. Ward, and H. Wendland, *Sobolev bounds on functions with scattered zeros, with applications to radial basis function surface fitting*, Mathematics of Computation **74** (2005), 743–763.
13. P. Oswald, *Frames and space splittings in hilbert spaces*, Survey Lecture given at the Universities Bonn and Lancaster in 1997 available via http://www.faculty.jacobs-university.de/poswald, 1997.
14. C. Rieger, *Sampling inequalities and applications*, Ph.D. thesis, Universität Göttingen, 2008.
15. C. Rieger, R. Schaback, and B. Zwicknagl, *Sampling and stability*, Mathematical Methods for Curves and Surfaces, Lecture Notes in Computer Science, vol. 5862, Springer, New York, 2010, pp. 347–369.
16. C. Rieger and B. Zwicknagl, *Deterministic error analysis of support vector machines and related regularized kernel methods*, Journal of Machine Learning Research **10** (2009), 2115–2132.
17. R. Schaback and H. Wendland, *Kernel Techniques: From Machine Learning to Meshless Methods*, Acta Numerica **15** (2006), 543–639.
18. B Schölkopf and A.J. Smola, *Learning with kernels - Support Vector Machines, Regularisation, and Beyond*, MIT Press, Cambridge, Massachusetts, 2002.
19. G. Wahba, *Spline Models for Observational Data, CBMS-NSF*, Regional Conference Series in Applied Mathematics, Siam, Philadelphia, 1990.
20. H. Wendland, *Meshless Galerkin methods using radial basis functions*, Math. Comput. **68** (1999), 1521–1531.
21. H. Wendland, *Scattered data approximation*, Cambridge Monographs on Applied and Computational Mathematics, Cambridge University Press, Cambridge, 2005.
22. H. Wendland and C. Rieger, *Approximate interpolation*, Numerische Mathematik **101** (2005), 643–662.
23. D. Werner, *Funktionalanalysis*, Springer-Lehrbuch, Springer, Berlin, 2002.
24. D. X. Zhou, *Derivative reproducing properties for kernel methods in learning theory*, Journal of computational and Applied Mathematics **220** (2008), 456–463.

Meshfree Vectorial Interpolation Based on the Generalized Stokes Problem

Csaba Gáspár[1]

Széchenyi István University, P.O.Box 701, H-9007 Győr, Hungary
gasparcs@sze.hu

Summary. A vectorial interpolation problem is considered. In addition to the interpolation conditions taken at discrete points, a global, divergence-free condition is also prescribed. Utilizing the idea of the multi-elliptic interpolation, the divergence-free interpolation problem is converted to a generalized Stokes problem. To numerically solve this new problem, an Uzawa-type method and the method of fundamental solutions are proposed. In the second method, a linear system with large and dense matrix is to be solved, while in the first method, this problem is avoided.

Key words: divergence-free interpolation, generalized Stokes problem, fundamental solution, multi-elliptic interpolation

1 Introduction

Vectorial interpolation problems can be considered a generalization of the scalar scattered data interpolation problems. Such problems frequently appear in flow problems, meteorological models etc. The vector field (typically velocity field in the applications) is assumed to satisfy a global condition e.g. it should be divergence-free. If the velocity components are interpolated independently, such a global condition cannot be fulfilled in general.

Narcowich and Ward [13] used a matrix-valued conditionally positive definite function to generate a divergence-free interpolant. The method was applied and generalized by Lowitzsch [12] and Fuselier [6] not only for divergence-free but also for curl-free interpolation. Dudu and Rabut [4] treated the interpolation problem via the minimization of a special seminorm containing both the divergence and the rotation of the vector field. All of these techniques convert the original interpolation problem to a large linear system of equations which is often severely ill-conditioned. In contrast to these approaches, Gáspár [8] proposed a multi-elliptic interpolation for the potential or the stream function of the field, which circumvents the problem of the large and ill-conditioned matrices and converts the interpolation problem to

M. Griebel, M.A. Schweitzer (eds.), *Meshfree Methods for Partial Differential Equations V*, Lecture Notes in Computational Science and Engineering 79,
DOI 10.1007/978-3-642-16229-9_4,

a higher order partial differential equation. However, the numerical treatment of this new problem is inconvenient due to the interpolation conditions taken at the interpolation points.

In this paper, we convert the interpolation problem to a generalized Stokes system using variational tools. First we define an interpolation vector function by minimizing a quadratic functional on a subspace of certain divergence-free functions. Next, we show the equivalence of the resulting variational problem and a direct problem which is a fourth-order Stokes-like system. From computational point of view, the obtained direct problem seems to be more convenient than the variational one, and makes it possible to apply meshfree tools such as the well-known method of fundamental solutions (MFS, see [1]).

We restrict ourselves to 2D problems, but note that 3D problems can also be handled in a quite similar way.

2 Vectorial interpolation

The problem: Suppose that a finite set of interpolation points $x_1, ..., x_N \in \mathbf{R}^2$ is given. Moreover, suppose that a finite set of associated vectors $\mathbf{u}_1, ..., \mathbf{u}_N \in \mathbf{R}^2$ is also given. Find a (sufficiently smooth) vector function $\mathbf{u} : \Omega \to \mathbf{R}^2$ (where Ω is a bounded domain containing all the interpolation points) in such a way that both the interpolation conditions

$$\mathbf{u}(x_k) = \mathbf{u}_k \qquad (k = 1, ..., N) \tag{1}$$

and the global condition

$$\text{div}\,\mathbf{u} = 0 \qquad \text{in } \Omega \tag{2}$$

are satisfied.

This problem is of course not well-posed and is strongly underdetermined, so that additional assumptions should be taken to the vector function $\mathbf{u}$.

In practice, the problem often appears in the following context: an a priori unknown divergence-free vector function $\tilde{\mathbf{u}}$ is to be reconstructed from its values $\mathbf{u}_1, ..., \mathbf{u}_N$ taken at the interpolation points $x_1, ..., x_N$.

Componentwise interpolation. The simplest approach is to perform a componentwise interpolation independently. Applying e.g. the popular method of radial basis functions (RBFs, see e.g. [11]), the interpolation vector field $\mathbf{u} := (u, v)$ can be expressed as

$$u(x) = \sum_{j=1}^{N} \alpha_j \Phi(x - x_j), \qquad v(x) = \sum_{j=1}^{N} \beta_j \Psi(x - x_j), \tag{3}$$

where Φ, Ψ are prescribed (not necessarily different) radial basis functions. The a priori unknown coefficients α_j, β_j can be determined by solving the *interpolation equations*:

$$\sum_{j=1}^{N} \alpha_j \Phi(x_k - x_j) = u_k, \qquad \sum_{j=1}^{N} \beta_j \Psi(x_k - x_j) = v_k \qquad (k = 1, ..., N) \quad (4)$$

where $(u_k, v_k) := \mathbf{u}_k$. However, the divergence-free condition (2) is not satisfied in general.

Exactly divergence-free interpolation. It is also possible to create exactly divergence-free interpolation functions in an RBF-like form. Following Lowitzsch [12] and Fuselier [6], let us define the matrix-valued function G by $G(x) := (-\Delta I + \nabla\nabla^T)\phi(x)$, where ϕ is a properly chosen smooth RBF, I is the identity matrix and $\nabla := (D_1, D_2)$, the gradient operator, i.e.

$$G(x) := \begin{pmatrix} -D_{22}\phi(x) & D_{12}\phi(x) \\ D_{12}\phi(x) & -D_{11}\phi(x) \end{pmatrix} \quad (5)$$

Then the columns of G are always divergence-free vector functions, and the interpolation function can be expressed as

$$\mathbf{u}(x) := \sum_{j=1}^{N} G(x - x_j) \begin{pmatrix} \alpha_j \\ \beta_j \end{pmatrix}, \quad (6)$$

where the a priori unknown coefficients α_j, β_j can again be determined by solving the corresponding interpolation equations.

The generating function ϕ can be defined in various ways: it may be a Gaussian [13] or a compactly supported Wendland function [12] etc. Our approach exhibits some similarities to this idea, using the fundamental solution of a certain partial differential operator as a generating function.

2.1 Divergence-free interpolation based on the stream function

If the stream function ψ of a vector function $\mathbf{u}$ exists, then the velocity components u, v can be expressed as $u = D_2\psi$, $v = -D_1\psi$, and the divergence-free condition is automatically satisfied. In this approach, the stream function is to be approximated. With a carefully chosen radial basis function Φ, the stream function can be approximated in the following way:

$$\psi(x) \approx \sum_{j=1}^{N} \alpha_j D_2\Phi(x - x_j) - \sum_{j=1}^{N} \beta_j D_1\Phi(x - x_j)$$

Now the derivatives of ψ are to be prescribed as interpolation conditions:

$$\sum_{j=1}^{N} \alpha_j D_{22}\Phi(x_k - x_j) - \sum_{j=1}^{N} \beta_j D_{12}\Phi(x_k - x_j) = u_k$$

$$-\sum_{j=1}^{N} \alpha_j D_{12}\Phi(x_k - x_j) + \sum_{j=1}^{N} \beta_j D_{11}\Phi(x_k - x_j) = v_k$$

for $k = 1, ..., N$. In vectorial form:

$$\sum_{j=1}^{N} G(x_k - x_j) \begin{pmatrix} \alpha_j \\ \beta_j \end{pmatrix} = \mathbf{u}_k \qquad (k = 1, ..., N), \tag{7}$$

where

$$G := \begin{pmatrix} D_{22}\Phi & -D_{12}\Phi \\ -D_{12}\Phi & D_{11}\Phi \end{pmatrix}$$

(cf (5)-(6).) In general, the solvability of the above system is not assured. Moreover, numerical problems might arise which are similar to that of the method of the radial basis functions: the resulting matrix is fully populated, and may be severely ill-conditioned. Using a multi-elliptic interpolation [8], however, this computational problem can be avoided. Here the stream function ψ is supposed to satisfy the modified multi-Helmholtz equation with the exception of the interpolation points:

$$(\Delta - c^2 I)^3 \psi(x) = 0 \qquad \text{in } \Omega \setminus \{x_1, ..., x_N\}$$

supplied with the interpolation conditions

$$\text{grad } \psi(x_k) = (-v_k, u_k) \qquad (k = 1, ..., N).$$

Here the predefined constant c plays some scaling role. The problem has a unique solution in the Sobolev space $H_0^3(\Omega)$ [9]. Unfortunately, the numerical treatment of the above interpolation conditions is often inconvenient, which significantly reduces the computational advantages of the multi-elliptic interpolation approach in this case.

However, the multi-elliptic idea can be preserved, using a special variational technique. This can be considered a natural generalization of the scalar multi-elliptic interpolation method summarized in the following subsection.

2.2 Multi-elliptic interpolation, scalar problems

Here we briefly outline the main ideas of the multi-elliptic interpolation. For details, see [7], [9].

Let $\tilde{u} \in H_0^2(\Omega)$ be an arbitrary, fixed function that satisfies the scalar interpolation conditions:

$$\tilde{u}(x_k) = u_k \qquad (k = 1, ..., N).$$

Such a function $\tilde{u}$ obviously exists. Let us introduce the space

$$W := \{w \in H_0^2(\Omega) : w(x_1) = ... = w(x_N) = 0\}$$

then, due to the well-known imbedding theorems, W is a closed subspace of the Sobolev space $H_0^2(\Omega)$. Let $c \geq 0$ be a fixed scaling constant. Then the following two problems are equivalent:

Direct problem: Find a function $v \in W$ such that (in the sense of distributions):

$$(\Delta - c^2 I)^2(\tilde{u} + v) = 0 \qquad \text{in } \Omega \setminus \{x_1, ..., x_N\}. \tag{8}$$

Variational problem: Find a function $v \in W$ such that v minimizes the quadratic functional

$$F(v) := ||(\Delta - c^2 I)(\tilde{u} + v)||^2_{L_2(\Omega)} \tag{9}$$

on the closed subspace W.

Remarks:

- In a more traditional form, (8) means that the function $u := \tilde{u} + v$ is a solution of the fourth-order problem

$$(\Delta - c^2 I)^2 u = 0 \qquad \text{in } \Omega \setminus \{x_1, ..., x_N\}$$

$$u|_{\partial\Omega} = 0, \qquad \frac{\partial u}{\partial n}|_{\partial\Omega} = 0 \qquad \text{(boundary conditions)}$$

$$u(x_k) = u_k \qquad (k = 1, ..., N) \qquad \text{(interpolation conditions)}$$

- Problem (9) is clearly equivalent to the following modified problem: find a function $u \in H_0^2(\Omega)$ that minimizes the functional

$$F(u) := ||(\Delta - c^2 I)u||^2_{L_2(\Omega)}$$

  among the functions of $H_0^2(\Omega)$ which satisfy the interpolation conditions $u(x_k) = u_k$ $(k = 1, ..., N)$.
- The scaling constant c plays only a minor role in the interpolation. As shown in [7], if $c > 0$ and $\Omega = \mathbf{R}^2$, the solution of the direct problem can be expressed in an RBF-like form:

$$u(x) = \sum_{j=1}^{N} \alpha_j \Phi(x - x_j),$$

  where Φ is the fundamental solution of the modified bi-Helmholtz operator $(\Delta - c^2 I)^2$ i.e. $\Phi(x) = \frac{||x||}{4\pi c} K_1(c||x||)$. Since the function K_1 decreases rapidly, from computational point of view Φ can be regarded as if it were compactly supported. The size of the 'essential support' can be controlled by the scaling parameter c. The larger the parameter c, the smaller the 'essential support' of Φ. As a rule of thumb, c should not reach the order of magnitude of $1/h$, where h is the separation distance of the interpolation points; otherwise, numerical singularities are generated at the interpolation points.

Standard variational arguments imply that the variational problem (9) has a unique solution, which is the orthogonal projection of the function $(-\tilde{u})$ to the closed subspace W. The orthogonality is meant with respect to the scalar product $\langle v, w\rangle := \left\langle (\Delta - c^2 I)v, (\Delta - c^2 I)w\right\rangle_{L_2(\Omega)}$. Note also that in Problems (8) and (9), the modified bi-Helmholtz operator $(\Delta - c^2 I)^2$ can be replaced either by the simple biharmonic operator $\Delta\Delta$, or by the mixed Laplace-Helmholtz operator $\Delta(\Delta - c^2 I)$. The corresponding problems still have unique solutions. In the first case we obtain the simple biharmonic interpolation, while the second choice results in a quasi-harmonic interpolant provided that the scaling parameter is large enough.

In practice, however, it is worth solving the direct problem (8) instead of (9). The solution procedure can be carried out in a very economical way using quadtree cell systems generated by the interpolation points $x_1, ..., x_N$ and/or multi-level techniques. This property makes the multi-elliptic interpolation techniques competitive compared with the classical RBF-based methods. For details, see [7].

3 Multi-elliptic divergence-free interpolation, vectorial problems

Let $x_1, ...x_N \in \Omega$ be again a finite set of interpolation points scattered in the bounded domain Ω and let $\mathbf{u}_1, ...\mathbf{u}_N \in \mathbf{R}^2$ be the associated vectors (velocities in most of the applications), $\mathbf{u}_k := (u_k, v_k)$ $(k = 1, ..., N)$.

Let $\tilde{\mathbf{u}} := (\tilde{u}, \tilde{v}) \in H_0^2(\Omega) \times H_0^2(\Omega)$ be an arbitrary, fixed divergence-free vector function that satisfies the vectorial interpolation conditions:

$$\tilde{u}(x_k) = u_k, \qquad \tilde{v}(x_k) = v_k \qquad (k = 1, ..., N)$$

$$\text{div } \tilde{\mathbf{u}} = 0 \quad \text{in } \Omega.$$

Such a function $\tilde{\mathbf{u}}$ exists (moreover, it can be assumed that the components of $\tilde{\mathbf{u}}$ belong to $C_0^\infty(\Omega)$). Let us introduce the space

$$\mathbf{W} := \{\mathbf{w} \in H_0^2(\Omega) \times H_0^2(\Omega) : \mathbf{w}(x_1) = ... = \mathbf{w}(x_N) = 0, \quad \text{div } \mathbf{w} = 0 \text{ in } \Omega\}$$

Then $\mathbf{W}$ is a closed subspace of $H_0^2(\Omega) \times H_0^2(\Omega)$. The variational problem, analogously to the scalar variational problem (9), can be defined in a completely similar way:

Variational problem: Find a function $\mathbf{v} \in \mathbf{W}$ such that $\mathbf{v}$ minimizes the quadratic functional

$$F(\mathbf{v}) := ||(\Delta - c^2 I)(\tilde{\mathbf{u}} + \mathbf{v})||^2_{L_2(\Omega)\times L_2(\Omega)} \tag{10}$$

on the closed subspace $\mathbf{W}$.

This problem is clearly equivalent to the following modified problem: find a function $\mathbf{u} \in H_0^2(\Omega) \times H_0^2(\Omega)$ that minimizes the functional

$$F(\mathbf{u}) := ||(\Delta - c^2 I)\mathbf{u}||^2_{L_2(\Omega)\times L_2(\Omega)}$$

among the vector functions of $H_0^2(\Omega) \times H_0^2(\Omega)$ which satisfy the interpolation conditions $\mathbf{u}(x_k) = \mathbf{u}_k$ $(k = 1, ..., N)$ and the divergence-free condition div $\mathbf{u} = 0$ in Ω.

Problem (10) has a unique solution. Indeed, the norm of the space $H_0^2(\Omega) \times H_0^2(\Omega)$ is equivalent to the norm $||(\Delta - c^2 I)\mathbf{u}||_{L_2(\Omega)\times L_2(\Omega)}$, so that the orthogonal projection of the function $(-\tilde{\mathbf{u}})$ to the closed subspace $\mathbf{W}$ is the (unique) solution of the Problem (10), where the orthogonality is meant with respect to the scalar product $\langle \mathbf{v}, \mathbf{w} \rangle = \left\langle (\Delta - c^2 I)\mathbf{v}, (\Delta - c^2 I)\mathbf{w} \right\rangle_{L_2(\Omega)\times L_2(\Omega)}$. In practice, however, it is not a convenient task to solve the variational problem (10). Our goal is to convert the variational problem to a 'direct' problem, analogously to the problem (8). It will turn out that this direct problem is a fourth-order Stokes problem (supplied with the interpolation conditions).

3.1 The generalized Stokes problem

Temporarily omitting the pointwise interpolation conditions, Problem (10) is to minimize a quadratic functional on the closed subspace which is the kernel of the divergence operator. This is a special case of the following abstract problem. Let X be a Hilbert space, denote by X^* its dual space. Let $A \in L(X, X^*)$ be a continuous, self-adjoint, positive definite and X-elliptic (coercive) operator, i.e. $|(Ax)y| \leq M \cdot ||x|| \cdot ||y||$, $(Ax)x \geq m \cdot ||x||^2$ hold for all $x, y \in X$, with appropriate constants $M \geq 0$ and $m > 0$. Let Y be another Hilbert space. Let $B \in L(X, Y)$ be a bounded operator with closed range. Then $B^* \in L(Y^*, X^*)$ also holds, and B^* also has a closed range. Let $f \in X^*$ be an arbitrary, fixed functional.

Direct problem: Find a pair $(x, p) \in X \times Y^*$ such that

$$Ax + B^* p = f \tag{11}$$

$$Bx = 0$$

Variational problem: Minimize the quadratic functional

$$F(x) := (Ax)x - 2fx$$

on the closed subspace ker B.

Classical results imply [3] that the variational problem always has a unique solution. Moreover, if the operator B satisfies the inf-sup condition of Babuška and Brezzi, i.e.

$$\inf_{y\in Y^*, y\neq 0} \sup_{x\in X, x\neq \mathbf{0}} \frac{|(B^*y)x|}{||y||\cdot||x||} > 0,$$

then the direct problem also has a unique solution $(x,p) \in X \times Y^*$, and the vector x is the (unique) solution of the variational problem.

In the usual theory of Stokes problems, $X = H_0^1(\Omega) \times H_0^1(\Omega)$, $Y = L_{2,0}(\Omega) = \{f \in L_2(\Omega) : \int_\Omega f = 0\}$, A is the negative Laplace operator: $A = -\Delta$, while B is the (negative) divergence operator, so that B^* is the gradient operator. In this case, the direct problem has the form

$$-\Delta\mathbf{u} + \text{grad } p = \mathbf{f} \tag{12}$$

$$\text{div } \mathbf{u} = 0$$

in Ω, and $\mathbf{u}|_{\partial\Omega} = \mathbf{0}$, while the variational form of the problem is to minimize the functional

$$F(u,v) := \int_\Omega \left(||\text{grad } u||^2 + ||\text{grad } v||^2 - 2fu - 2gv\right)\, d\Omega$$

on he closed subspace of the divergence-free vector functions belonging to $H_0^1(\Omega) \times H_0^1(\Omega)$. (Here $\mathbf{u} = (u,v)$, $\mathbf{f} = (f,g) \in H^{-1}(\Omega) \times H^{-1}(\Omega)$.)

Remark: The theory is still applicable, if *homogeneous* Stokes problems are considered. Let $\tilde{\mathbf{u}} \in H^1(\Omega) \times H^1(\Omega)$ such that div $\tilde{\mathbf{u}} = 0$ in Ω. Then the corresponding direct problem is:

$$-\Delta\mathbf{u} + \text{grad } p = \mathbf{0}$$

$$\text{div } \mathbf{u} = 0 \quad \text{in } \Omega, \quad \mathbf{u}|_{\partial\Omega} = \tilde{\mathbf{u}}|_{\partial\Omega},$$

and the solution $\mathbf{u} = (u,v)$ minimizes the functional

$$F(u,v) := \int_\Omega \left(||\text{grad } u||^2 + ||\text{grad } v||^2\right)\, d\Omega \tag{13}$$

with respect to the boundary conditions and the divergence-free condition. The Lagrange multiplier is the pressure field p.

For both of the direct and the variational problems, the divergence-free condition is exactly satisfied. However, pointwise interpolation conditions cannot be prescribed, since the subspace of the functions of $H_0^1(\Omega) \times H_0^1(\Omega)$ that vanish at the interpolation points is not closed in the space $H_0^1(\Omega) \times H_0^1(\Omega)$. In other words, the pointwise interpolation conditions destroy the well-posedness of the Stokes problem. (This phenomenon is strongly related to the fact that the fundamental solution of the Stokes system has a singularity at the origin, cf [10].) This can be avoided, if, instead of (13), the functional (10) is minimized on the closed subspace $\mathbf{W}$, which results in a fourth-order generalized Stokes problem (similarly to the scalar interpolation problem).

The direct problem belonging to the variational problem (10) is as follows:

Direct problem: Find a function $\mathbf{v} \in \mathbf{W}$ and a scalar function $p \in H^{-1}(\Omega)$ such that the functions $\mathbf{u} := \tilde{\mathbf{u}} + \mathbf{v}$ and p satisfy the following system:

$$(\Delta - c^2 I)^2 \mathbf{u} + \text{grad } p = \mathbf{0} \qquad \text{in } \Omega \setminus \{x_1, ..., x_N\} \tag{14}$$

(in the sense of distributions).

Remark: The above direct problem can be reformulated in the following more traditional form: find $\mathbf{u} \in H^2(\Omega) \times H^2(\Omega)$ and $p \in H^{-1}(\Omega)$ such that

$$(\Delta - c^2 I)^2 \mathbf{u} + \text{grad } p = \mathbf{0} \qquad \text{in } \Omega \setminus \{x_1, ..., x_N\}$$

(in the sense of distributions), moreover:

$$\text{div } \mathbf{u} = \mathbf{0} \qquad \text{in } \Omega \qquad \text{(divergence-free condition)}$$

$$\mathbf{u}|_{\partial\Omega} = 0, \qquad \frac{\partial \mathbf{u}}{\partial n}|_{\partial\Omega} = 0 \qquad \text{(boundary conditions)}$$

$$\mathbf{u}(x_k) = \mathbf{u}_k \qquad (k = 1, ..., N) \qquad \text{(interpolation conditions)}$$

Now we show the equivalence of Problems (14) and (10).

Theorem 1. *If* $(\mathbf{v}, p)$ *is a solution of Problem (14), then* $\mathbf{v}$ *is a solution of Problem (10).*

Proof: Let $(\mathbf{v}, p)$ be a solution of Problem (14), and let $\mathbf{w} \in \mathbf{W}$ be arbitrary. Then (with $\mathbf{u} := \tilde{\mathbf{u}} + \mathbf{v}$):

$$F(\mathbf{v} + \mathbf{w}) = ||(\Delta - c^2 I)(\mathbf{u} + \mathbf{w})||^2_{L_2(\Omega)\times L_2(\Omega)} =$$

$$= F(\mathbf{v}) + 2\left\langle (\Delta - c^2 I)\mathbf{u}, (\Delta - c^2 I)\mathbf{w} \right\rangle_{L_2(\Omega)\times L_2(\Omega)} + ||(\Delta - c^2 I)\mathbf{w}||^2_{L_2(\Omega)\times L_2(\Omega)} =$$

$$= F(\mathbf{v}) + 2\left\langle (\Delta - c^2 I)^2 \mathbf{u}, \mathbf{w} \right\rangle_{L_2(\Omega)\times L_2(\Omega)} + ||(\Delta - c^2 I)\mathbf{w}||^2_{L_2(\Omega)\times L_2(\Omega)}$$

But $(\Delta - c^2 I)^2 \mathbf{u} = -\text{grad } p$, therefore

$$\left\langle (\Delta - c^2 I)^2 \mathbf{u}, \mathbf{w} \right\rangle_{L_2(\Omega)\times L_2(\Omega)} = -\left\langle \text{grad } p, \mathbf{w} \right\rangle_{L_2(\Omega)\times L_2(\Omega)} =$$

$$= \langle p, \text{div } \mathbf{w} \rangle_{L_2(\Omega)} = 0,$$

which implies that

$$F(\mathbf{v} + \mathbf{w}) = F(\mathbf{v}) + ||(\Delta - c^2 I)\mathbf{w}||^2_{L_2(\Omega)\times L_2(\Omega)} \geq F(\mathbf{v}),$$

i.e. $\mathbf{v}$ is a solution of Problem (10).

Thus, the divergence-free interpolation problem has been converted to a fourth-order Stokes problem.

Theorem 2. *If* $\mathbf{v}$ *is a solution of Problem (10), then there exists a functional* $p \in H^{-1}(\Omega)$ *such that* $(\mathbf{v}, p)$ *is a solution of Problem (14).*

Proof: Let $\mathbf{v}$ be a solution of Problem (10). Then, for every function $\phi \in (C_0^\infty(\Omega \setminus \{x_1, ..., x_N\}))^2$, for which div $\phi = 0$, the following inequality holds:

$$F(\mathbf{v} + \phi) \geq F(\mathbf{v})$$

Using standard variational arguments, this implies:

$$\langle (\Delta - c^2 I)^2 \mathbf{u}, \phi \rangle_{L_2(\Omega) \times L_2(\Omega)} = 0.$$

According to the theorem of de Rham (see e.g. [5]), there exists a distribution $p \in \mathcal{D}'(\Omega \setminus \{x_1, ..., x_N\})$ such that

$$\text{grad } p = -(\Delta - c^2 I)^2 \mathbf{u}.$$

Since $\mathbf{u} \in H_0^2(\Omega) \times H_0^2(\Omega)$, therefore grad $p \in H^{-2}(\Omega) \times H^{-2}(\Omega)$, i.e. $p \in H^{-1}(\Omega)$. With this p, $(\mathbf{v}, p)$ solves Problem (14).

Remark: Similarly to the scalar multi-elliptic interpolation, the choice of the scaling parameter c is not crucial (provided that c remains under the order of magnitude of $1/h$, where h is the separation distance of the interpolation points). The simplest choice is $c := 0$. Another possibility is to replace the operator $(\Delta - c^2 I)^2$ with $\Delta(\Delta - c^2 I)$. In this case, the direct problem (14) is a singularly perturbed fourth-order Stokes problem. Applying the Method of Fundamental Solutions, this approach results in a meshfree boundary-only method for the Stokes equations as pointed out in the next section.

4 Solution techniques

4.1 Uzawa's method

A usual method to solve the abstract Stokes problem (11) is the classical Uzawa algorithm:

$$x_{n+1} := A^{-1}(-B^* p_n + f)$$

$$p_{n+1} := p_n + \omega y_{n+1}$$

where $\omega > 0$ is an iteration parameter and $y_{n+1} \in Y^*$ is the functional generated by the vector $Bx_{n+1} \in Y$, i.e. $y_{n+1}\phi := \langle Bx_{n+1}, \phi \rangle$ $(\phi \in Y)$. It is known that if the operator B satisfies the inf-sup condition, then the algorithm is convergent for any sufficiently small positive iteration parameter ω.

According to the special form of the direct problem (14), an Uzawa iteration step now has the following form:

- solve the multi-elliptic equations $(\Delta - c^2 I)^2 \mathbf{u}_{n+1} = -\mathrm{grad}\, p_n$ in $\Omega \setminus \{x_1, ..., x_N\}$ supplied with homogeneous boundary conditions and the interpolation conditions $\mathbf{u}(x_k) = \mathbf{u}_k$ $(k = 1, ..., N)$.
- perform a correction by the computed divergence: $p_{n+1} := p_n + \omega \cdot \mathrm{div}((\Delta - c^2 I)\mathbf{u}_{n+1})$,

(since the operator $(c^2 I - \Delta)$ is an isomorphism between the spaces $H_0^1(\Omega)$ and $H^{-1}(\Omega)$).

Example 1. Consider the divergence-free vector field $\mathbf{u} = (u, v)$ in the unit square Ω, where

$$u(x,y) = \sin 2\pi x \cdot \sin 2\pi y, \qquad v(x,y) = \cos 2\pi x \cdot \cos 2\pi y \tag{15}$$

(the space variables are denoted by x, y, as usual.) The problem is to reconstruct this vector field from its values taken at the interpolation points $(x_1, y_1), ..., (x_N, y_N)$ scattered in the domain Ω. The interpolation function was computed by solving the generalized Stokes system (14) on a uniform 32×32 computational grid using Uzawa's method (replacing the original homogeneous Dirichlet boundary conditions with periodic boundary conditions). The applied scaling factor was set to zero. Table 1 shows the computed relative L_2-errors belonging to the different values of the number of interpolation points (N). The relative divergence (i.e. the value of the quotient $||\mathrm{div}\, \mathbf{u}||_{L_2(\Omega)} / ||\mathbf{u}||_{L_2(\Omega) \times L_2(\Omega)}$) was under 0.02% in all cases. The results show how rapidly the relative L_2-errors decrease when N increases.

Table 1. Relative L_2-errors of the computed vector field of Example 1. N is the number of the interpolation points

N	50	100	200	400
Relative L_2-error (%)	1.422	0.470	0.124	0.022

In practice, the Uzawa algorithm should be realized using multigrid tools, which significantly reduces both the computational cost and the memory requirement compared with the classical RBF-based methods. It should also be pointed out that no large, dense and ill-conditioned matrices appear when realizing the method.

4.2 The method of fundamental solutions

Another solution technique which requires no mesh or grid structure is the method of fundamental solutions [1]. This approach can also be applied to the classical Stokes problem [2], [14]. However, since the fundamental solution of the classical Stokes system has a singularity at the origin, the source points should be located outside the domain of the Stokes equations, which makes the appearing matrices extremely ill-conditioned and does not allow to take into account pointwise interpolation conditions.

The fundamental solution of the classical Stokes system (12) is a pair of a matrix function $G := \begin{pmatrix} u_1 & u_2 \\ v_1 & v_2 \end{pmatrix}$ and a vector function $\mathbf{p} := \begin{pmatrix} p_1 \\ p_2 \end{pmatrix}$ which satisfies the following pair of Stokes equations:

$$-\Delta u_1 + D_1 p_1 = \delta, \qquad -\Delta v_1 + D_2 p_1 = 0, \qquad D_1 u_1 + D_2 v_1 = 0$$

$$-\Delta u_2 + D_1 p_2 = 0, \qquad -\Delta v_2 + D_2 p_2 = \delta, \qquad D_1 u_2 + D_2 v_2 = 0$$

where δ denotes the Dirac distribution concentrated to the origin. (The fundamental solutions of the generalized Stokes system (14) can be defined in a completely similar way.) It is known (see e.g. [2], [14]) that such a fundamental solution can be expressed by the following formulas:

$$u_1(x,y) = -\frac{1}{8\pi}\left(\log(x^2+y^2) + 1 + \frac{2y^2}{x^2+y^2}\right)$$

$$u_2(x,y) = v_1(x,y) = \frac{1}{8\pi} \cdot \frac{2xy}{x^2+y^2}$$

$$v_2(x,y) = -\frac{1}{8\pi}\left(\log(x^2+y^2) + 1 + \frac{2x^2}{x^2+y^2}\right)$$

$$p_1(x,y) = \frac{1}{4\pi} \cdot \frac{2x}{x^2+y^2}, \qquad p_2(x,y) = \frac{1}{4\pi} \cdot \frac{2y}{x^2+y^2}$$

(Here the space variables are denoted by x and y, as usual.)

Straightforward calculations show that this fundamental solution can be expressed with the help of the harmonic and biharmonic fundamental solutions E_1 and E_2 in the following way:

$$G = \begin{pmatrix} -D_{22}E_2 & D_{12}E_2 \\ D_{12}E_2 & -D_{11}E_2 \end{pmatrix}, \qquad \mathbf{p} = \operatorname{grad} E_1 \tag{16}$$

(recall that, in polar coordinates: $E_1(r) = \frac{1}{2\pi}\log r$, $E_2(r) = \frac{1}{8\pi} r^2 \log r$). From the definition it is clear that all the functions u, v, p defined by

$$\begin{aligned} u(x) &\sim \sum_{j=1}^{N} a_j^{(1)} u_1(x-x_j) + \sum_{j=1}^{N} a_j^{(2)} u_2(x-x_j) \\ v(x) &\sim \sum_{j=1}^{N} a_j^{(1)} v_1(x-x_j) + \sum_{j=1}^{N} a_j^{(2)} v_2(x-x_j) \\ p(x) &\sim \sum_{j=1}^{N} a_j^{(1)} p_1(x-x_j) + \sum_{j=1}^{N} a_j^{(2)} p_2(x-x_j) \end{aligned} \tag{17}$$

satisfy the homogeneous Stokes system everywhere, with the exception of the points $x_1, ..., x_N$. Since the fundamental solution has a singularity at the origin, pointwise interpolation conditions cannot be prescribed at the interpolation points $x_1, ..., x_N$. However, the generalized Stokes system (14) avoids this difficulty (the corresponding fundamental solution is *continuous* at the origin), and the divergence-free interpolation function can be expressed in the above form: the a priori unknown coefficients $a_j^{(1)}$, $a_j^{(2)}$ can be determined by solving the system of algebraic equations

$$\sum_{j=1}^{N} a_j^{(1)} u_1(x_k - x_j) + \sum_{j=1}^{N} a_j^{(2)} u_2(x_k - x_j) = u_k \tag{18}$$

$$\sum_{j=1}^{N} a_j^{(1)} v_1(x_k - x_j) + \sum_{j=1}^{N} a_j^{(2)} v_2(x_k - x_j) = v_k$$

$(k = 1, ..., N)$, where u_1, v_1, p_1, u_2, v_2, p_2 are the components of the fundamental solution of the generalized Stokes system (14).

Now we determine this fundamental solution in some special cases. Utilizing the fact that (16) is a fundamental solution of the Stokes equations, the following two theorems can be proved by straightforward calculations.

The simplest case is when $c = 0$, i.e. (14) has the form:

$$\Delta^2 \mathbf{u} + \text{grad } p = \mathbf{0}, \qquad \text{div } \mathbf{u} = \mathbf{0} \tag{19}$$

Theorem 3. *A fundamental solution of (19) is as follows:*

$$G = \begin{pmatrix} D_{22}E_3 & -D_{12}E_3 \\ -D_{12}E_3 & D_{11}E_3 \end{pmatrix}, \qquad \mathbf{p} = \text{grad } E_1$$

where E_3 denotes the triharmonic fundamental solution:

$$E_3(r) = \frac{1}{128\pi} r^4 \log r.$$

Another special case is when one of the Laplace operators is replaced with the operator $-(I + \frac{1}{c^2}\Delta)$. Now (14) has the form:

$$-\Delta(I - \frac{1}{c^2}\Delta)\mathbf{u} + \text{grad } p = \mathbf{0}, \qquad \text{div } \mathbf{u} = \mathbf{0} \tag{20}$$

Theorem 4. *A fundamental solution of (20) is as follows:*

$$G = c^2 \cdot \begin{pmatrix} D_{22}E & -D_{12}E \\ -D_{12}E & D_{11}E \end{pmatrix}, \qquad \mathbf{p} = \text{grad } E_1$$

where E denotes the fundamental solution of the operator $\Delta^2(\Delta - c^2 I)$, i.e.

$$E(r) = -\frac{1}{2\pi c^4} \left(K_0(cr) + \log cr + \frac{(cr)^2}{4} \log cr \right),$$

and K_0 denotes the usual Bessel function of the third kind.

Remark: The system (20) can be considered a singularly perturbed approximation of the classical Stokes system, if c is large enough. This makes it possible to use the direct problem (20) and the corresponding pair of formulas (17)-(18) as a boundary meshfree method for the Stokes equation, if the interpolation points are located on the boundary of the flow domain. See [10] for details.

The general case

$$(\Delta - c^2 I)^2 \mathbf{u} + \text{grad } p = \mathbf{0}, \qquad \text{div } \mathbf{u} = \mathbf{0}, \tag{21}$$

when $c \neq 0$, cannot be deduced to the fundamental solution of the Stokes equations. However, a standard Fourier transform method is applicable, which results in the following theorem:

Theorem 5. *A fundamental solution of (21) is as follows:*

$$G = \begin{pmatrix} D_{22}E & -D_{12}E \\ -D_{12}E & D_{11}E \end{pmatrix}, \qquad \mathbf{p} = \text{grad } E_1$$

where E denotes the fundamental solution of the operator $\Delta(\Delta - c^2 I)^2$, i.e.

$$E(r) = \frac{1}{2\pi c^4} \left(K_0(cr) + \log cr + \frac{cr}{2} K_1(cr) \right),$$

and K_0, K_1 denote the usual Bessel functions of the third kind.

Note that in all of the last three cases, the matrix function G is continuous at the origin, and the divergence-free interpolation function has the same form as in (17) (using the components of the actual fundamental solution). The a priori unknown coefficients $a_j^{(1)}$, $a_j^{(2)}$ $(j = 1, ..., N)$ can be determined by solving the interpolation equations (18).

Thus, these interpolation methods can be considered special cases of the general form (5)-(6), using the fundamental solutions of the sixth-order multi-elliptic operators introduced above.

Observe that the system (18) is exactly identical to the system (7) with $\Phi = E_3$ and $\Phi = E$, respectively. However, in deriving of (18), no stream function approach was utilized.

Unfortunately, the system (18) exhibits the same computational disadvantages as the method of fundamental solutions in general: the matrix of the system is fully populated and often ill-conditioned, while in the previous approach (to solve the generalized Stokes equations directly using Uzawa's method), this problem is avoided.

Example 2. Consider again the previous Example (15) in the unit square. Now a componentwise interpolation was performed (using the thin plate spline $\Phi(r) := r^2 \log r$ as a radial basis function). At the same time, the vectorial interpolation function (17) based on the fundamental solution of (19) was also computed (i.e. $c = 0$, see G in Theorem 3). The relative L_2-norms of the errors as well as the relative divergences belonging to different numbers of interpolation points are summarized in Table 2. Here cTPS means the componentwise interpolation by the thin plate splines, and MFS refers to the method of fundamental solutions based on (19). Both methods seem to converge. Observe that the relative divergences of the componentwise interpolation also decrease when N increases. However, the interpolation vector field based on the fundamental solution of (19) is always completely divergence-free.

Table 2. Relative L_2-errors and divergences of the computed vector field of Example 2. N is the number of the interpolation points

N		10	20	50	100	200	400
Relative L_2-error (%)	(cTPS)	93.53	48.63	22.22	16.38	5.01	3.19
Relative divergence (%)	(cTPS)	258.9	255.5	147.1	115.7	72.9	54.0
Relative L_2-error (%)	(MFS)	90.22	19.18	10.16	6.10	2.72	1.63

5 Summary and conclusions

A new approach for solving divergence-free vectorial interpolation problems has been presented. The interpolation vector field is assumed to be a solution of a fourth-order generalized Stokes system, which automatically guarantees that the vector field is completely divergence-free. The interpolation conditions taken at discrete points are treated as special boundary conditions and do not destroy the well-posedness of the generalized Stokes problem.

To solve this new problem, two different techniques have been applied. Based on the fundamental solution of the generalized Stokes system, the Method of Fundamental Solutions can be built up without difficulty. Since the fundamental solution is continuous at the origin, the source points and the interpolation points are allowed to coincide. Moreover, if the interpolation points are located along the boundary of a domain, a meshfree boundary-only method for solving the Stokes problem can also be derived from the interpolation technique. However, this approach exhibits the usual disadvantages of the MFS (working with large, dense and ill-conditioned matrices).

Applying an Uzawa method and multi-level tools to the direct problem, the above phenomenon can be avoided. The results obtained by the use of a simple uniform background grid are promising. An optimal compromise would be the use of the highly economical quadtree-based multigrid solution techniques. Nevertheless, due to the higher order derivatives appearing in the generalized Stokes system, the use of more exact schemes defined in the quadtree context would be necessary, which requires some further research.

Acknowledgement: The research was partly supported by the European Union (co-financed by the European Regional Development Fund) under the project TÁMOP-4.2.2-08/1-2008-0021.

References

1. C.J.S. Alves, C.S. Chen, B. Sarler, *The Method of Fundamental Solutions for Solving Poisson Problems* (C.A. Brebbia, A. Tadeu, V. Popov, eds), Int. Series on Advances in Boundary Elements, vol. 13, WitPress, 2002, pp. 67–76.
2. C.J.S. Alves, A.L. Silvestre, *Density Results Using Stokeslets and a Method of Fundamental Solutions for the Stokes Equations*, Engineering Analysis with Boundary Elements **28** (2004), 1245–1252.
3. M. Benzi, G.H. Golub, J. Liesen, *Numerical Solution of Saddle Point Problems*, Acta Numerica (2005), 1–137.
4. F. Dudu, C. Rabut, *Vectorial Interpolation Using Radial-Basis-Like Functions*, Computers and Mathematics with Applications **43** (2002), 393–411.
5. A. Ern, J.L. Guermond, *Theory and Practice of Finite Element Method*, Applied Methematical Sciences 159, Springer, 2004.
6. E.J. Fuselier, *Improved Stability Estimates and a Characterication of the Native Space for Matrix-Valued RBFs*, Adv. Comput. Math. **29** (2008), 269–290.
7. C. Gáspár, *Multi-level Biharmonic and Bi-Helmholtz Interpolation with Application to the Boundary Element Method*, Engineering Analysis with Boundary Elements **24/7-8** (2000), 559–573.
8. C. Gáspár, *A Multi-level Solution of Scalar and Vectorial Interpolation Problems Based on Iterated Elliptic Operators*, PAMM (Proceedings in Applied Mathematics and Mechanics) **3/1** (2003), 535–536.
9. C. Gáspár, *A Meshless Polyharmonic-type Boundary Interpolation Method for Solving Boundary Integral Equations*, Engineering Analysis with Boundary Elements **28/10** (2004), 1207–1216.
10. C. Gáspár, *Several Meshless Solution Techniques for the Stokes Flow Equations*, Progress on Meshless Methods (A.J.M. Ferreira, E.J. Kansa, G.E. Fasshauer, eds.), Computational Methods in Applied Sciences, vol. 11, Springer, 2009, pp. 141–158.
11. M.A. Golberg, C.S. Chen, *A Bibliography on Radial Basis Function Approximation*, Boundary Element Communications **7/4** (1996), 155–163.
12. S. Lowitzsch, *Error Estimates for Matrix-Valued Radial Basis Function Interpolation*, Journal of Approximation Theory **137** (2005), 238–249.
13. F.J. Narcowich, J.D. Ward, *A Generalized Hermite Interpolation via Matrix-Valued Conditionally Positive Definite Functions*, Mathematics of Computation **43/208** (2004), 661–687.
14. D.L. Young, S.J. Jane, C.M. Fan, K. Murugesan, C.C. Tsai, *The Method of Fundamental Solutions for 2D and 3D Stokes Problems*, Journal of Computational Physics **211/1** (2006), 1–8.

Pressure XFEM for two-phase incompressible flows with application to 3D droplet problems

Sven Gross[1]

Hausdorff Center for Mathematics, Institute for Numerical Simulation, University of Bonn, Wegelerstr. 6, D-53115 Bonn, Germany. sgross@ins.uni-bonn.de

Summary. We consider the numerical simulation of 3D two-phase flow problems using finite element methods on adaptive multilevel tetrahedral grids and a level set approach for interface capturing. The approximation of the discontinuous pressure in standard finite element spaces yields poor results with an error of order 0.5 w.r.t. the L_2 norm. Second order approximations can be achieved by the introduction of an extended finite element space (XFEM) adding special basis functions incorporating a jump at the interface. A simple stabilization strategy for the XFEM basis is presented which also offers this optimal approximation property.

Key words: XFEM, two-phase flow, level set, surface tension, pressure approximation order

1 Introduction

Two-phase systems play an important role in chemical engineering. Two examples are extraction columns where mass transport takes place between bubbles and a surrounding liquid (liquid-liquid system), or falling films which are e.g. used for cooling by heat transfer from a thin liquid layer to the gaseous phase (liquid-gas system).

In flow simulations of such two-phase systems, special care has to be taken of the numerical treatment of the interfacial force term and the pressure space, as otherwise very large artificial spurious velocities are induced at the interface. Hence, it is often not adequate to apply methods originally designed for one-phase flow problems, but it is rather necessary to develop novel numerical approaches adapted to the special requirements of two-phase flow systems.

One example is the construction of an appropriate finite element (FE) space for the pressure approximation. The pressure is continuous in both phases, but has a jump across the interface due to surface tension. If the grid is not aligned to the interface, the approximation of such functions in standard FE spaces (including non-conformal FE) yields poor results with an error of

M. Griebel, M.A. Schweitzer (eds.), *Meshfree Methods for Partial Differential Equations V*, Lecture Notes in Computational Science and Engineering 79, DOI 10.1007/978-3-642-16229-9_5,

order 0.5 w.r.t. the L_2 norm. Second order approximations can be achieved by using an extended finite element space (XFEM) adding special basis functions incorporating a jump at the interface.

The outline of the paper is as follows. After introducing the governing equations in Section 2 the numerical methods are briefly discussed in Section 3. The construction of an enriched pressure FE space is discussed in Section 3.2. In section 4 the optimal approximation order of the proposed pressure XFEM space is discussed and a stabilization strategy for the XFEM basis is presented which also offers the optimal approximation property. Finally, numerical results for a single bubble obtained by our software package DROPS will be presented in Section 5.

2 Mathematical model

Let $\Omega_i \subset \mathbb{R}^3$, $i = 1, 2$, denote the two phases with $\Omega = \Omega_1 \cup \Omega_2$ and $\Gamma = \partial\Omega_1 \cap \partial\Omega_2$ the interface separating both phases. The level set technique [2] is used for capturing the interface, where Γ is implicitly defined by the zero level of the scalar level set function ϕ. We consider the following standard model for two-phase flows in weak formulation,

$$m(\mathbf{u}_t, \mathbf{v}) + a(\mathbf{u}, \mathbf{v}) + n(\mathbf{u}; \mathbf{u}, \mathbf{v}) + b(\mathbf{v}, p) = m(\mathbf{g}, \mathbf{v}) + f_\Gamma(\mathbf{v}) \tag{1}$$

$$b(\mathbf{u}, q) = 0 \tag{2}$$

$$(\phi_t + \mathbf{u} \cdot \nabla\phi, v)_\Omega = 0 \tag{3}$$

for all $\mathbf{v} \in \mathbf{V} := (H^1(\Omega))^3$, $q \in Q := L_{2,0}(\Omega)$, $v \in L_2(\Omega)$, comprising the Navier-Stokes equations (1)–(2) for velocity $\mathbf{u} \in \mathbf{V}$, pressure $p \in Q$ and the level set equation (3) for ϕ. The bilinear and trilinear forms are given by

$$\begin{aligned}
a(\mathbf{u}, \mathbf{v}) &= \frac{1}{2} \int_\Omega \mu(\phi) \operatorname{tr}\big(D(\mathbf{u}) D(\mathbf{v})\big) \, d\mathbf{x} \\
m(\mathbf{u}, \mathbf{v}) &= (\rho(\phi) \, \mathbf{u}, \mathbf{v})_\Omega, \\
n(\mathbf{w}; \mathbf{u}, \mathbf{v}) &= (\mathbf{w} \cdot \nabla\mathbf{u}, \mathbf{v})_\Omega, \\
b(\mathbf{v}, q) &= -(\operatorname{div} \mathbf{v}, q)_\Omega,
\end{aligned}$$

with density ρ and dynamic viscosity μ, where $(\cdot, \cdot)_\Omega$ denotes the inner product on $L_2(\Omega)$ and $D(\mathbf{u}) = \nabla\mathbf{u} + (\nabla\mathbf{u})^T$ the deformation tensor. Surface tension is modeled by the CSF term [1] in weak formulation,

$$f_\Gamma(\mathbf{v}) = \int_\Gamma \tau\kappa\mathbf{v}\mathbf{n} \, ds,$$

with κ denoting the curvature, τ the surface tension coefficient and $\mathbf{n}$ the interfacial normal.

3 Numerical methods

3.1 Overview of numerical methods

A finite element discretization on adaptive tetrahedral grids is applied for spatial discretization, with piecewise quadratic FE for the velocity space $\mathbf{V}_h$ and an extended FE space Q_h^Γ for the pressure space. The construction of Q_h^Γ is discussed in Section 3.2. For the level set function piecewise *quadratic* FE are used, which is crucial for the discretization of the CSF term, since a piecewise linear ϕ_h would not contain enough information to compute the curvature accurately. For this purpose an improved Laplace-Beltrami discretization from [4] is applied.

For time discretization a one-step θ-scheme is used. In each time step a coupled system of Navier-Stokes and level set equations has to be solved, which is treated by a fixed point approach. The linearized Oseen problems in each fixed point iteration are solved applying an inexact Uzawa method. We refer to [5] for more details.

3.2 Pressure XFEM space

For the representation of the pressure p it has to be taken into account, that the pressure is smooth in each phase Ω_i, but has a jump across Γ due to surface tension. In mathematical terms, we have

$$p \in H^m(\Omega_1 \cup \Omega_2) := \{v \in L_2(\Omega) : v|_{\Omega_i} \in H^m(\Omega_i), i = 1, 2\},$$

where m depends on the smoothness of the pressure. The use of standard FE spaces $\hat{Q}_h$ (piecewise polynomial, conformal as well as non-conformal) yields an order of 0.5 w.r.t. the L_2 norm, i.e., $\inf_{q \in \hat{Q}_h} \|p - q_h\|_0 \le c\sqrt{h}$, if the interface is *not aligned with the grid.* This is in general the case, if the interface is captured by a level set approach or a volume-of-fluid method. In the following we will construct an extended FE space, which is suitable for the approximation of such discontinuous functions.

Let Q_h be the standard FE space of piecewise linear functions and $q_1, \dots, q_n \in Q_h$ its nodal basis with $n := \dim Q_h$. Let $\mathcal{J}_\Gamma$ be the set of indices associated to the tetrahedra intersected by Γ. For each of these indices $i \in \mathcal{J}_\Gamma$, an additional basis function q_i^Γ is introduced which is discontinuous at the interface:

$$q_i^\Gamma(\mathbf{x}) := q_i(\mathbf{x}) \cdot \big(H_\Gamma(\mathbf{x}) - H_\Gamma(\mathbf{x}_i)\big), \qquad \mathbf{x} \in \Omega,$$

with $H_\Gamma(\mathbf{x}) = 0$ for $\mathbf{x} \in \Omega_1$ and $H_\Gamma(\mathbf{x}) = 1$ for $\mathbf{x} \in \Omega_2$. This so-called *Heaviside enrichment* was originally introduced and applied to fracture mechanics in [7], a related work is [6]. The pressure XFEM space Q_h^Γ is defined by the span of $\{q_i\}_{i=1}^n \cup \{q_i^\Gamma\}_{i \in \mathcal{J}_\Gamma}$. More details can be found in [3].

4 Analysis of pressure XFEM space

4.1 Approximation order of pressure XFEM space

For $v \in H^m(\Omega_1 \cup \Omega_2$ we define the corresponding Sobolev norm $\|v\|^2_{m,\Omega_1\cup\Omega_2} := \sum_{i=1,2} \|v\|^2_{m,\Omega_i}$. The following approximation result from [8] holds for $0 \le l < m \le 2$:

$$\inf_{q_h \in Q_h^\Gamma} \|p - q_h\|_{l,\Omega_1\cup\Omega_2} \le c h^{m-l} \|p\|_{m,\Omega_1\cup\Omega_2} \qquad \text{for all } p \in H^m(\Omega_1 \cup \Omega_2). \tag{4}$$

In this sense, for a pressure $p \in H^m(\Omega_1 \cup \Omega_2)$ the XFEM space Q_h^Γ possesses optimal approximation properties. E.g., for $m = 2$ we obtain second order convergence w.r.t. the L_2 norm ($l = 0$).

4.2 Stabilization of XFEM basis

Depending on the location of the interface, the interface may cut off arbitrary small parts of the tetrahedra. Consequently, the support of the corresponding extended basis functions q_i^Γ becomes very small. Numerical experiments indicate that in such situations the LBB constant of the FE pair $\mathbf{V}_h \times Q_h^\Gamma$ detoriates which has an impact on the convergence rate of iterative solvers and the stability of the discretization. Thus, on the one hand one wants to obtain a (more) stable basis of the XFEM space, on the other hand the XFEM space should have an optimal approximation property as in (4).

Let $\tilde{c} > 0$, $\alpha > 0$ be given parameters. For $j \in \mathcal{J}_\Gamma$ we consider the following condition for the corresponding extended basis function q_j^Γ:

$$\|q_j^\Gamma\|_{l,T} \le \tilde{c} h_T^{\alpha - l} \|q_j\|_{l,T} \qquad \text{for all tetrahedra } T \text{ intersecting } \Gamma. \tag{5}$$

Here $l \in \{0, 1\}$ is the degree of the Sobolev norm used for measuring the approximation error, cf. (4). This criterion quantifies the notion of a "small" contribution which can be neglected. We introduce the *reduced index set* $\tilde{\mathcal{J}}_\Gamma := \{j \in \mathcal{J}_\Gamma : (5) \text{ does } \textit{not} \text{ hold for } q_j^\Gamma\}$, resulting in the *reduced XFEM space* $\tilde{Q}_h^\Gamma$. For this reduced space an approximation property similar to the one in (4) can be shown to hold, cf. [8]:

$$\inf_{q \in \tilde{Q}_h^\Gamma} \|p - q_h\|_{l,\Omega_1\cup\Omega_2} \le c\big(h^{m-l} + h^{\alpha-l}\big) \|p\|_{m,\Omega_1\cup\Omega_2} \quad \text{for all } p \in H^m(\Omega_1 \cup \Omega_2),$$

$0 \le l < m \le 2$. Thus taking $\alpha = m$ in (5), an optimal approximation error bound is maintained.

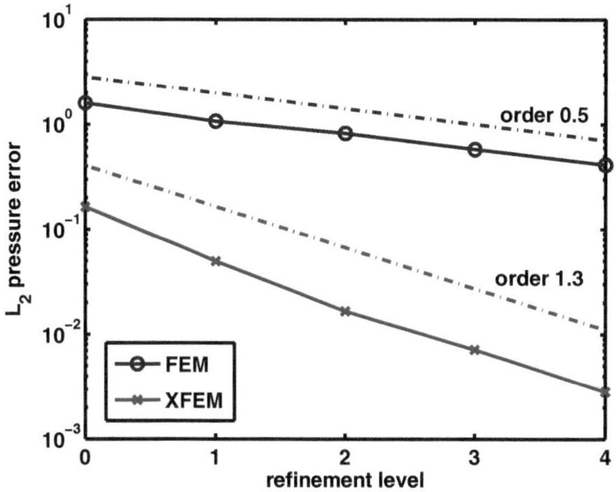

Fig. 1. Pressure error $\|p - p_h\|_{L_2(\Omega)}$ vs. refinement level for $p_h \in Q_h, Q_h^\Gamma$.

5 Numerical experiment

We consider the test case of a static bubble $\Omega_1 = \{\mathbf{x} \in \mathbb{R}^3 : \|\mathbf{x}\| < r\}$ in a cubic domain $\Omega = (-1,1)^3$. We take $r = 2/3$, $\rho = \mu = 1$, $\mathbf{g} = 0$, $\tau = 1$, i.e., surface tension is the only driving force. The analytical solution is given by $\mathbf{u} = 0$ and $p|_{\Omega_2} = C$, $p|_{\Omega_1} = C + \tau\kappa$ (with $\kappa = 2/r = 3$).

The numerical solutions $\mathbf{u}_h, p_h$ are computed for different refinement levels of the grid, and the corresponding pressure errors $\|p - p_h\|_{L_2(\Omega)}$ are given in Figure 1 for the pressure spaces Q_h and Q_h^Γ, respectively. For the standard FE space Q_h the expected order 0.5 is observed. The results for the XFEM space Q_h^Γ are much better (order > 1), however, we do not achieve second order convergence. This is due to the discretization of the CSF term, which can be shown to be (at least) first order accurate, but does not provide second order accuracy.

Figure 2 shows the pressure solutions for the standard FEM and XFEM case with refinenemt level 4. The corresponding velocity solutions are given in Figure 3. We observe large spurious velocities for the standard FEM pressure space induced by pressure oscillations at the interface. The application of the XFEM pressure space leads to an essential reduction of the spurios velocities.

References

1. J. U. Brackbill, D. B. Kothe, and C. Zemach, *A continuum method for modeling surface tension*, J. Comput. Phys. **100** (1992), 335–354.
2. Y. C. Chang, T. Y. Hou, B. Merriman, and S. Osher, *A level set formulation of Eulerian interface capturing methods for incompressible fluid flows*, J. Comput. Phys. **124** (1996), 449–464.

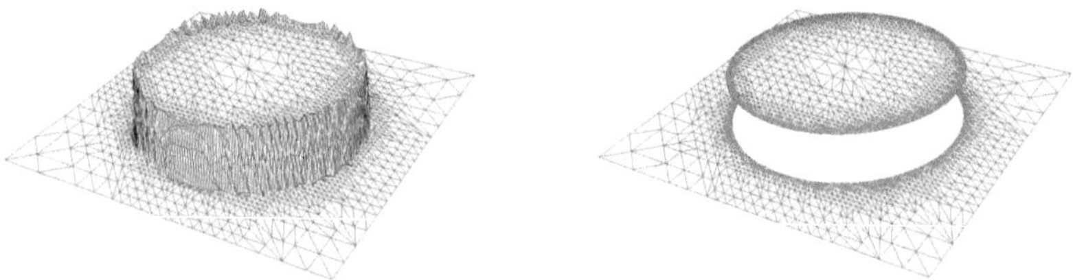

Fig. 2. FE pressure solution $p_h \in Q_h$ (left) and $p_h \in Q_h^\Gamma$ (right) for refinement level 4, visualized on slice $z = 0$.

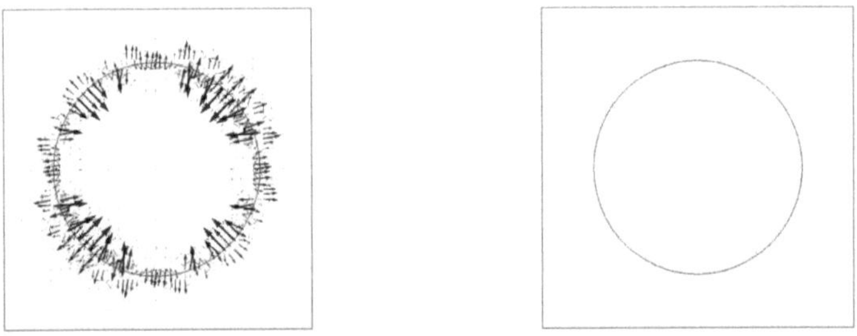

Fig. 3. FE velocity solution $\mathbf{u}_h$ for the cases $p_h \in Q_h$ (left) and $p_h \in Q_h^\Gamma$ (right) visualized on slice $z = 0$.

3. S. Groß and A. Reusken, *An extended pressure finite element space for two-phase incompressible flows with surface tension*, J. Comput. Phys. **224** (2007), 40–58.
4. ______, *Finite element discretization error analysis of a surface tension force in two-phase incompressible flows*, SIAM J. Numer. Anal. **45** (2007), no. 4, 1679–1700.
5. Sven Groß, *Numerical methods for three-dimensional incompressible two-phase flow problems*, Ph.D. thesis, RWTH Aachen, 2008.
6. A. Hansbo and P. Hansbo, *A finite element method for the simulation of strong and weak discontinuities in solid mechanics*, Comput. Methods Appl. Mech. En-

grg. **193** (2004), no. 33–35, 3523–3540.
7. N. Moës, J. Dolbow, and T. Belytschko, *A finite element method for crack growth without remeshing*, Int. J. Num. Meth. Eng. **46** (1999), 131–150.
8. A. Reusken, *Analysis of an extended pressure finite element space for two-phase incompressible flows*, Comp. Vis. Sci. **11** (2008), 293–305.

Special-relativistic Smoothed Particle Hydrodynamics: a benchmark suite

Stephan Rosswog[1]

Jacobs University Bremen, Campus Ring 1, D-28759 Bremen
s.rosswog@jacobs-university.de

Summary. In this paper we test a special-relativistic formulation of Smoothed Particle Hydrodynamics (SPH) that has been derived from the Lagrangian of an ideal fluid. Apart from its symmetry in the particle indices, the new formulation differs from earlier approaches in its artificial viscosity and in the use of special-relativistic "grad-h-terms". In this paper we benchmark the scheme in a number of demanding test problems. Maybe not too surprising for such a Lagrangian scheme, it performs close to perfectly in pure advection tests. What is more, the method produces accurate results even in highly relativistic shock problems.

Key words: Smoothed Particle Hydrodynamics, special relativity, hydrodynamics, shocks

1 Introduction

Relativity is a crucial ingredient in a variety of astrophysical phenomena. For example the jets that are expelled from the cores of active galaxies reach velocities tantalizingly close to the speed of light, and motion near a black hole is heavily influenced by space-time curvature effects. In the recent past, substantial progress has been made in the development of numerical tools to tackle relativistic gas dynamics problems, both on the special- and the general-relativistic side, for reviews see [2, 14, 20]. Most work on numerical relativistic gas dynamics has been performed in an Eulerian framework, a couple of Lagrangian smooth particle hydrodynamics (SPH) approaches do exist though.

In astrophysics, the SPH method has been very successful, mainly because of its excellent conservation properties, its natural flexibility and robustness. Moreover, its physically intuitive formulation has enabled the inclusion of various physical processes beyond gas dynamics so that many challenging multi-physics problems could be tackled. For recent reviews of the method we refer to the literature [24, 27]. Relativistic versions of the SPH method

M. Griebel, M.A. Schweitzer (eds.), *Meshfree Methods for Partial Differential Equations V*, Lecture Notes in Computational Science and Engineering 79,
DOI 10.1007/978-3-642-16229-9_6,

were first applied to special relativity and to gas flows evolving in a fixed background metric [4, 16–19, 31]. More recently, SPH has also been used in combination with approximative schemes to dynamically evolve space-time [1, 3, 8–12, 26].
In this paper we briefly summarize the main equations of a new, special-relativistic SPH formulation that has been derived from the Lagrangian of an ideal fluid. Since the details of the derivation have been outlined elsewhere, we focus here on a set of numerical benchmark tests that complement those shown in the original paper [28]. Some of them are "standard" and often used to demonstrate or compare code performance, but most of them are more violent—and therefore more challenging—versions of widespread test problems.

2 Relativistic SPH equations from a variational principle

An elegant approach to derive relativistic SPH equations based on the discretized Lagrangian of a perfect fluid was suggested in [25]. We have recently extended this approach [28, 29] by including the relativistic generalizations of what are called "grad-h-terms" in non-relativistic SPH [23, 32]. For details of the derivation we refer to the original paper [28] and a recent review on the Smooth Particle Hydrodynamics method [27].
In the following, we assume a flat space-time metric with signature (-,+,+,+) and use units in which the speed of light is equal to unity, $c = 1$. We reserve Greek letters for space-time indices from 0...3 with 0 being the temporal component, while i and j refer to spatial components and SPH particles are labeled by a, b and k.
Using the Einstein sum convention the Lagrangian of a special-relativistic perfect fluid can be written as [13]

$$L_{\rm pf,sr} = -\int T^{\mu\nu} U_\mu U_\nu \, dV, \tag{1}$$

where

$$T^{\mu\nu} = (n[1 + u(n, s)] + P) U^\mu U^\nu + P \eta^{\mu\nu} \tag{2}$$

denotes the energy momentum tensor, n is the baryon number density, u is the thermal energy per baryon, s the specific entropy, P the pressure and $U^\mu = dx^\mu / d\tau$ is the four velocity with τ being proper time. All fluid quantities are measured in the local rest frame, energies are measured in units of the baryon rest mass energy[1], $m_0 c^2$. For practical simulations we give up general covariance and perform the calculations in a chosen "computing frame" (CF). In the general case, a fluid element moves with respect to this frame, therefore,

[1]The appropriate mass m_0 obviously depends on the ratio of neutrons to protons, i.e. on the nuclear composition of the considered fluid.

the baryon number density in the CF, N, is related to the local fluid rest frame via a Lorentz contraction

$$N = \gamma n, \tag{3}$$

where γ is the Lorentz factor of the fluid element as measured in the CF. The simulation volume in the CF can be subdivided into volume elements such that each element b contains ν_b baryons and these volume elements, $\Delta V_b = \nu_b / N_b$, can be used in the SPH discretization process of a quantity f:

$$f(\boldsymbol{r}) = \sum_b f_b \frac{\nu_b}{N_b} W(|\boldsymbol{r} - \boldsymbol{r}_b|, h), \tag{4}$$

where the index labels quantities at the position of particle b, $\boldsymbol{r}_b$. Our notation does not distinguish between the approximated values (the f on the LHS) and the values at the particle positions (f_b on the RHS). The quantity h is the smoothing length that characterizes the width of the smoothing kernel W, for which we apply the cubic spline kernel that is commonly used in SPH [22,24]. Applied to the baryon number density in the CF at the position of particle a, Eq. (4) yields:

$$N_a = N(\boldsymbol{r}_a) = \sum_b \nu_b W(|\boldsymbol{r}_a - \boldsymbol{r}_b|, h_a). \tag{5}$$

This equation takes over the role of the usual density summation of non-relativistic SPH, $\rho(\boldsymbol{r}_a) = \sum_b m_b W(|\boldsymbol{r}_a - \boldsymbol{r}_b|, h)$. Since we keep the baryon numbers associated with each SPH particle, ν_b, fixed, there is no need to evolve a continuity equation and baryon number is conserved by construction. If desired, the continuity equation can be solved though, see e.g. [4]. Note that we have used a's own smoothing length in evaluating the kernel in Eq. (5). To fully exploit the natural adaptivity of a particle method, we adapt the smoothing length according to

$$h_a = \eta \left(\frac{\nu_a}{N_a} \right)^{-1/D}, \tag{6}$$

where η is a suitably chosen numerical constant, usually in the range between 1.3 and 1.5, and D is the number of spatial dimensions. Hence, similar to the non-relativistic case [23, 32], the density and the smoothing length mutually depend on each other and a self-consistent solution for both can be obtained by performing an iteration until convergence is reached.

With these prerequisites at hand, the fluid Lagrangian can be discretized [25,27]

$$L_{\mathrm{SPH,sr}} = - \sum_b \frac{\nu_b}{\gamma_b} [1 + u(n_b, s_b)]. \tag{7}$$

Using the first law of thermodynamics one finds (for a detailed derivation see Sec. 4 in [27]) for the canonical momentum per baryon

$$\boldsymbol{S}_a \equiv \frac{1}{\nu_a} \frac{\partial L_{\mathrm{SPH,sr}}}{\partial \boldsymbol{v}_a} = \gamma_a \boldsymbol{v}_a \left(1 + u_a + \frac{P_a}{n_a} \right), \tag{8}$$

which is the quantity that we evolve numerically. Its evolution equation follows from the Euler-Lagrange equations,

$$\frac{d}{dt}\frac{\partial L}{\partial \boldsymbol{v}_a} - \frac{\partial L}{\partial \boldsymbol{r}_a} = 0, \tag{9}$$

as [27]

$$\frac{d\boldsymbol{S}_a}{dt} = -\sum_b \nu_b \left(\frac{P_a}{N_a^2 \Omega_a} \nabla_a W_{ab}(h_a) + \frac{P_b}{N_b^2 \Omega_b} \nabla_a W_{ab}(h_b) \right), \tag{10}$$

where the "grad-h" correction factor

$$\Omega_b \equiv 1 - \frac{\partial h_b}{\partial N_b} \sum_k \frac{\partial W_{bk}(h_b)}{\partial h_b} \tag{11}$$

was introduced. As numerical energy variable we use the canonical energy per baryon,

$$\epsilon_a \equiv \gamma_a \left(1 + u_a + \frac{P_a}{n_a} \right) - \frac{P_a}{N_a} = \boldsymbol{v}_a \cdot \boldsymbol{S}_a + \frac{1 + u_a}{\gamma_a} \tag{12}$$

which evolves according to [27]

$$\frac{d\epsilon_a}{dt} = -\sum_b \nu_b \left(\frac{P_a \boldsymbol{v}_b}{N_a^2 \Omega_a} \cdot \nabla_a W_{ab}(h_a) + \frac{P_b \boldsymbol{v}_a}{N_b^2 \Omega_b} \cdot \nabla_a W_{ab}(h_b) \right). \tag{13}$$

As in grid-based approaches, at each time step a conversion between the numerical and the physical variables is required [4, 28].
The set of equations needs to be closed by an equation of state. In all of the following tests, we use a polytropic equation of state, $P = (\Gamma - 1)nu$, where Γ is the polytropic exponent (keep in mind our convention of measuring energies in units of $m_0 c^2$).

3 Artificial dissipation

To handle shocks, additional artificial dissipation terms need to be included. We use terms similar to [4]

$$\left(\frac{d\boldsymbol{S}_a}{dt} \right)_{\text{diss}} = -\sum_b \nu_b \Pi_{ab} \overline{\nabla_a W_{ab}} \quad \text{with} \quad \Pi_{ab} = -\frac{K v_{\text{sig}}}{\bar{N}_{ab}} (\boldsymbol{S}_a^* - \boldsymbol{S}_b^*) \cdot \hat{e}_{ab} \tag{14}$$

and

$$\left(\frac{d\epsilon_a}{dt} \right)_{\text{diss}} = -\sum_b \nu_b \boldsymbol{\Psi}_{ab} \cdot \overline{\nabla_a W_{ab}} \quad \text{with} \quad \boldsymbol{\Psi}_{ab} = -\frac{K v_{\text{sig}}}{\bar{N}_{ab}} (\epsilon_a^* - \epsilon_b^*) \hat{e}_{ab}. \tag{15}$$

Here K is a numerical constant of order unity, $v_{\rm sig}$ an appropriately chosen signal velocity, see below, $\bar{N}_{ab} = (N_a + N_b)/2$, and $\hat{e}_{ab} = (\boldsymbol{r}_a - \boldsymbol{r}_b)/|\boldsymbol{r}_a - \boldsymbol{r}_b|$ is the unit vector pointing from particle b to particle a. For the symmetrized kernel gradient we use

$$\overline{\nabla_a W_{ab}} = \frac{1}{2}\left[\nabla_a W_{ab}(h_a) + \nabla_a W_{ab}(h_b)\right]. \tag{16}$$

Note that in [4] $\nabla_a W_{ab}(h_{ab})$ was used instead of our $\overline{\nabla_a W_{ab}}$, in practice we find the differences between the two symmetrizations negligible. The stars at the variables in Eqs. (14) and (15) indicate that the projected Lorentz factors

$$\gamma_k^* = \frac{1}{\sqrt{1 - (\boldsymbol{v}_k \cdot \hat{e}_{ab})^2}} \tag{17}$$

are used instead of the normal Lorentz factor. This projection onto the line connecting particle a and b has been chosen to guarantee that the viscous dissipation is positive definite [4].
The signal velocity, $v_{\rm sig}$, is an estimate for the speed of approach of a signal sent from particle a to particle b. The idea is to have a robust estimate that does not require much computational effort. We use [28]

$$v_{\rm sig,ab} = \max(\alpha_a, \alpha_b), \tag{18}$$

where

$$\alpha_k^{\pm} = \max(0, \pm\lambda_k^{\pm}) \tag{19}$$

with $\lambda_k^{\pm}$ being the extreme local eigenvalues of the Euler equations

$$\lambda_k^{\pm} = \frac{v_k \pm c_{{\rm s},k}}{1 \pm v_k c_{{\rm s},k}} \tag{20}$$

and $c_{{\rm s},k}$ being the relativistic sound velocity of particle k. These 1D estimates can be generalized to higher spatial dimensions, see e.g. [20]. The results are not particularly sensitive to the exact form of the signal velocity, but in experiments we find that Eq. (18) yields somewhat crisper shock fronts and less smeared contact discontinuities (for the same value of K) than earlier suggestions [4].
Since we are aiming at solving the relativistic evolution equations of an *ideal* fluid, we want dissipation only where it is really needed, i.e. near shocks where entropy needs to be produced[2]. To this end, we assign an individual value of the parameter K to each SPH particle and integrate an additional differential equation to determine its value. For the details of the time-dependent viscosity parameter treatment we refer to [28].

---

[2] A description of the general reasoning behind artificial viscosity can be found, for example, in Sec. 2.7 of [27]

4 Test bench

In the following we demonstrate the performance of the above described scheme at a slew of benchmark tests. The exact solutions of the Riemann problems have been obtained by help of the RIEMANN_VT.f code provided by Marti and Müller [20]. Unless mentioned otherwise, approximately 3000 particles are shown.

4.1 Test 1: Riemann problem 1

This moderately relativistic (maximum Lorentz factor $\gamma_{\rm max} \approx 1.4$) shock tube has become a standard touch-stone for relativistic hydrodynamics codes [4, 5, 15, 20, 21, 30]. It uses a polytropic equation of state (EOS) with an exponent of $\Gamma = 5/3$ and $[P, N, v]_{\rm L} = [40/3, 10, 0]$ for the left-hand state and $[P, N, v]_{\rm R} = [10^{-6}, 1, 0]$ for the right-hand state. As shown in Fig. 1, the numerical solution at $t = 0.35$ (circles) agrees nearly perfectly with the exact one. Note in particular the absence of any spikes in u and P at the contact discontinuity (near $x \approx 0.25$), such spikes had plagued many earlier relativistic SPH formulations [17,31]. The only places where we see possibly room for improvement is the contact discontinuity which is slightly smeared out and the slight over-/undershoots at the edges of the rarefaction fan.
In order to monitor how the error in the numerical solution decreases as a function of increased resolution, we calculate

$$L_1 \equiv \frac{1}{N_{\rm part}} \sum_b^{N_{\rm part}} |v_b - v_{\rm ex}(r_b)|, \qquad (21)$$

where $N_{\rm part}$ is the number of SPH-particles, v_b the (1D) velocity of SPH-particle b and $v_{\rm ex}(r_b)$ the exact solution for the velocity at position r_b. The results for L_1 are displayed in Fig. 2. The error L_1 decreases close to $\propto N_{\rm part}^{-1}$ (actually, the best fit is $L_1 \propto N_{\rm part}^{-0.96}$), which is what is also found for Eulerian methods in tests that involve shocks. Therefore, for problems that involve shocks we consider the method first-order accurate. The order of the method for smooth flows will be determined in the context of test 6.

4.2 Test 2: Riemann problem 2

This test is a more violent version of test 1 in which we increase the initial left side pressure by a factor of 100, but leave the other properties, in particular the right-hand state, unchanged: $[P, \rho, v]_{\rm L} = [4000/3, 10, 0]$ and $[P, \rho, v]_{\rm R} = [10^{-6}, 1, 0]$. This represents a challenging test since the post-shock density is compressed into a very narrow "spike", at $t = 0.35$ near $x \approx 0.35$. A maximum Lorentz-factor of $\gamma_{\rm max} \approx 3.85$ is reached in this test.
In Fig. 3 we show the SPH results (circles) of velocity v, specific energy u, the

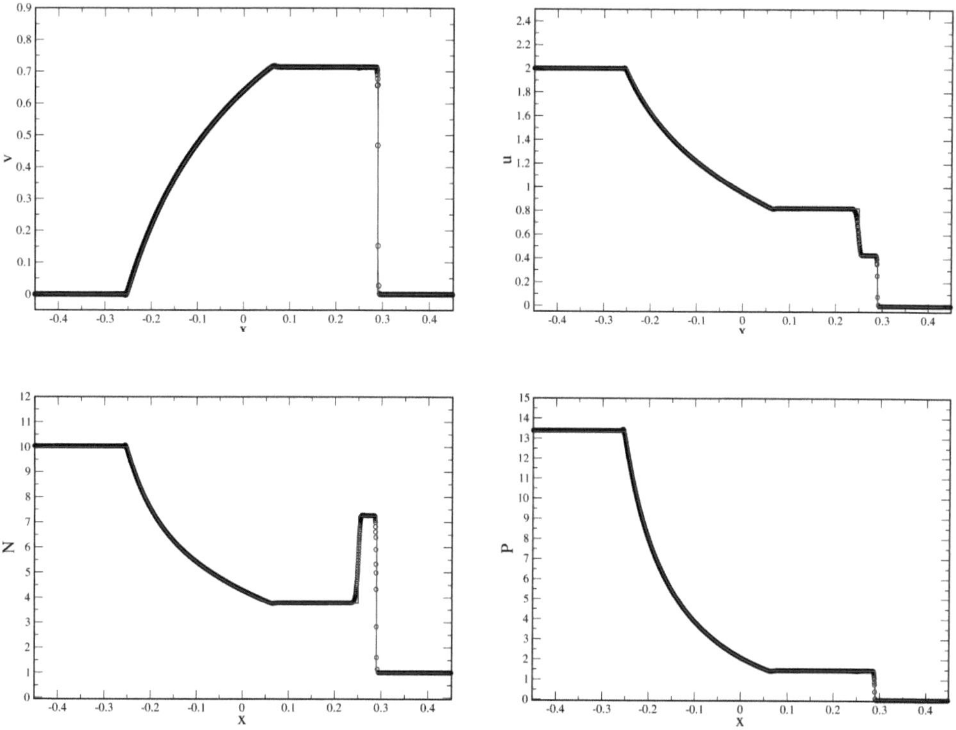

Fig. 1. Results of the relativistic shock tube of test 1 at $t = 0.35$: SPH results (circles) vs. exact solution (red line). From left to right, top to bottom: velocity (in units of c), specific energy, computing frame baryon number density and pressure.

computing frame number density N and the pressure P at $t = 0.35$ together with the exact solution of the problem (red line). Again the numerical solution is in excellent agreement with the exact one, only in the specific energy near the contact discontinuity occurs some smearing.

4.3 Test 3: Riemann problem 3

This test is an even more violent version of the previous tests. We now increase the initial left side pressure by a factor of 1000 with respect to test 1, but leave the other properties unchanged: $[P, \rho, v]_{\rm L} = [40000/3, 10, 0]$ and $[P, \rho, v]_{\rm R} = [10^{-6}, 1, 0]$. The post-shock density is now compressed into a very narrow "needle" with a width of only ≈ 0.002, the maximum Lorentz factor is 6.65.

Fig. 4 shows the SPH results (circles) of velocity v, specific energy u, the computing frame number density N and the pressure P at $t = 0.2$ together

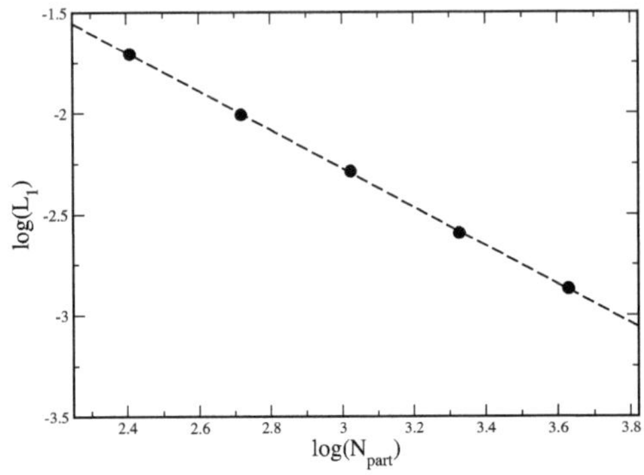

Fig. 2. Decrease of the error as defined in Eq. (21) as a function of particle number for the relativistic shock tested in Riemann problem 1. The error decreases close to $L_1 \propto N_{\rm part}^{-1}$.

with the exact solution (red line). The overall performance in this extremely challenging test is still very good. The peak velocity plateau with $v \approx 0.99$ (panel 1) is very well captured, practically no oscillations behind the shock are visible. Of course, the "needle-like" appearance of the compressed density shell (panel 3) poses a serious problem to every numerical scheme at finite resolution. At the applied resolution, the numerical peak value of N is only about half of the exact solution. Moreover, this extremely demanding test reveals an artifact of our scheme: the shock front is propagating at slightly too large a speed. This problem decreases with increasing numerical resolution and experimenting with the parameter K of Eqs. (14) and (15) shows that it is related to the form of artificial viscosity, smaller offsets occur for lower values of the viscosity parameter K. Here further improvements would be desirable.

4.4 Test 4: Sinusoidally perturbed Riemann problem

This is a more extreme version of the test suggested by [6]. It starts from an initial setup similar to a normal Riemann problem, but with the right state being sinusoidally perturbed. What makes this test challenging is that the smooth structure (sine wave) needs to be transported across the shock, i.e. kinetic energy needs to be dissipated into heat to avoid spurious post-shock oscillations, but not too much since otherwise the (physical!) sine oscillations in the post-shock state are not accurately captured. We use a polytropic exponent of $\Gamma = 5/3$ and

$$[P, N, v]^{\rm L} = [1000, 5, 0] \quad \text{and} \quad [P, N, v]^{\rm R} = [5, 2 + 0.3\sin(50x), 0]. \tag{22}$$

as initial conditions, i.e. we have increased the initial left pressure by a factor of 200 in comparison to [6]. The numerical result (circles) is shown in Fig. 5

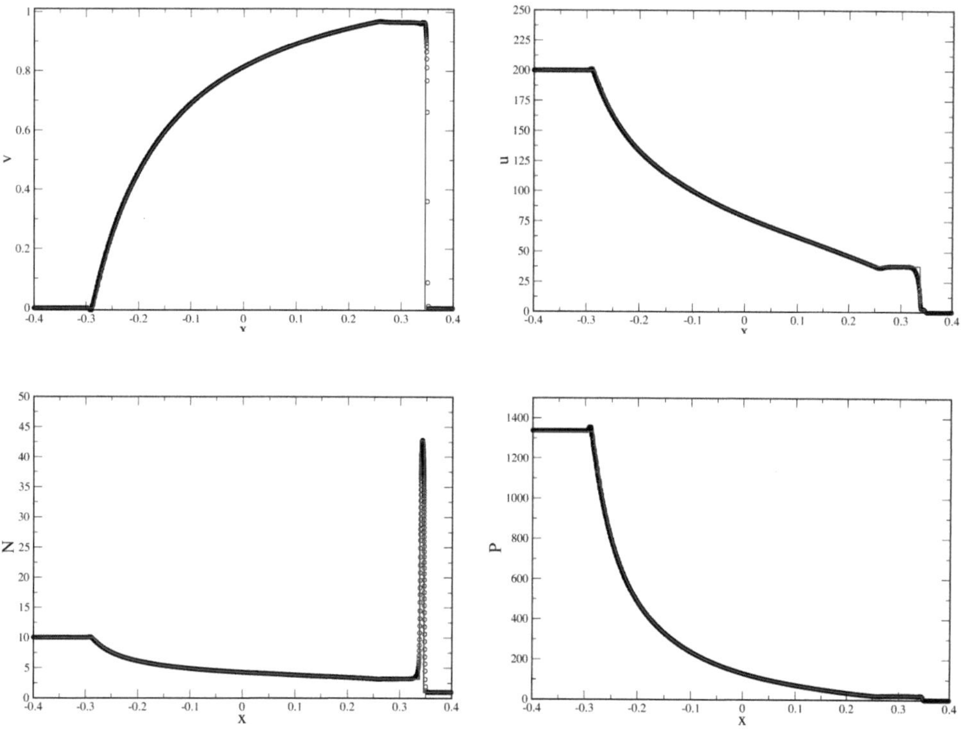

Fig. 3. Same as previous test, but the initial left hand side pressure has been increased by a factor of 100. SPH results (at $t = 0.35$) are shown as circles, the exact solution as red line. From left to right, top to bottom: velocity (in units of c), specific energy, computing frame baryon number density and pressure.

together with two exact solutions, for the right-hand side densities $N_{\rm R} = 2.3$ (solid blue) and $N_{\rm R} = 1.7$ (solid red). All the transitions are located at the correct positions, in the post-shock density shell the solution nicely oscillates between the extremes indicated by the solid lines.

4.5 Test 5: Relativistic Einfeldt rarefaction test

The initial conditions of the Einfeldt rarefaction test [7] do not exhibit discontinuities in density or pressure, but the two halfs of the computational domain move in opposite directions and thereby create a very low-density region around the initial velocity discontinuity. This low-density region poses a serious challenge for some iterative Riemann solvers, which can return negative density/pressure values in this region. Here we generalize the test to a relativistic problem in which left/right states move with velocity -0.9/+0.9 away from

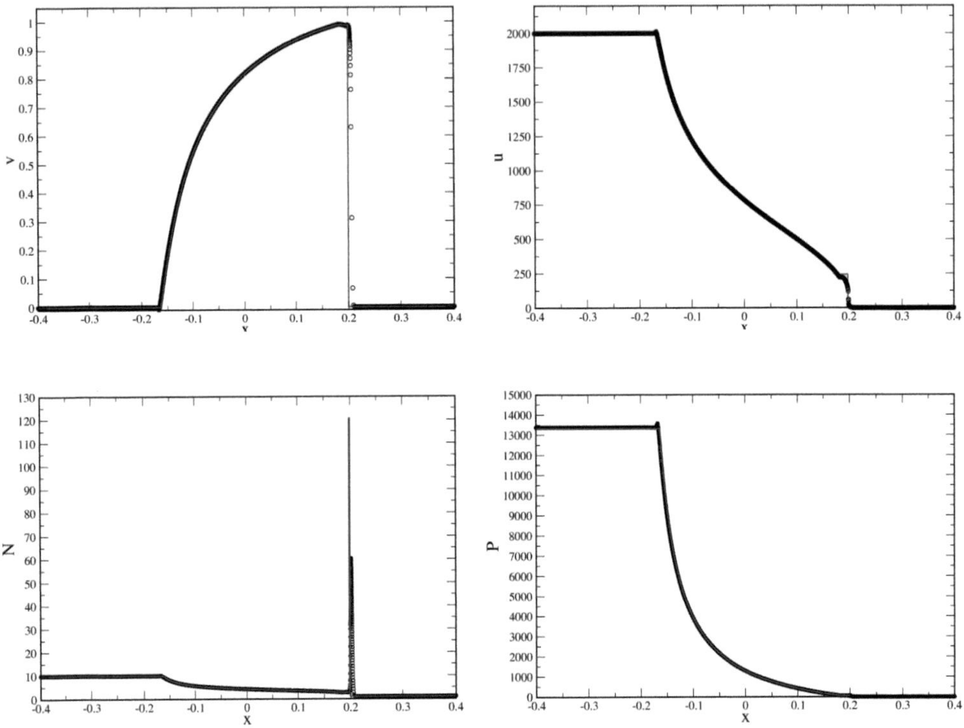

Fig. 4. Same as first shock tube test, but the initial left hand side pressure has been increased by a factor of 1000. SPH results (at $t = 0.2$) are shown as circles, the exact solution as red line. From left to right, top to bottom: velocity (in units of c), specific energy, computing frame baryon number density and pressure.

the central position. For the left and right state we use $[P, n, v]_\mathrm{L} = [1, 1, -0.9]$ and $[P, n, v]_\mathrm{R} = [1, 1, 0.9]$ and an adiabatic exponent of $\Gamma = 4/3$. Note that here we have specified the local rest frame density, n, which is related to the computing frame density by Eq. (3). The SPH solution at $t = 0.2$ is shown in Fig. 6 as circles, the exact solution is indicated by the solid red line. Small oscillations are visible near the center, mainly in v and u, and over-/undershoots occur near the edges of the rarefaction fan, but overall the numerical solution is very close to the analytical one. In its current form, the code can stably handle velocities up to 0.99999, i.e. Lorentz factors $\gamma > 200$, but at late times there are practically no more particles in the center (SPH's approximation to the emerging near-vacuum), so that it becomes increasingly difficult to resolve the central velocity plateau.

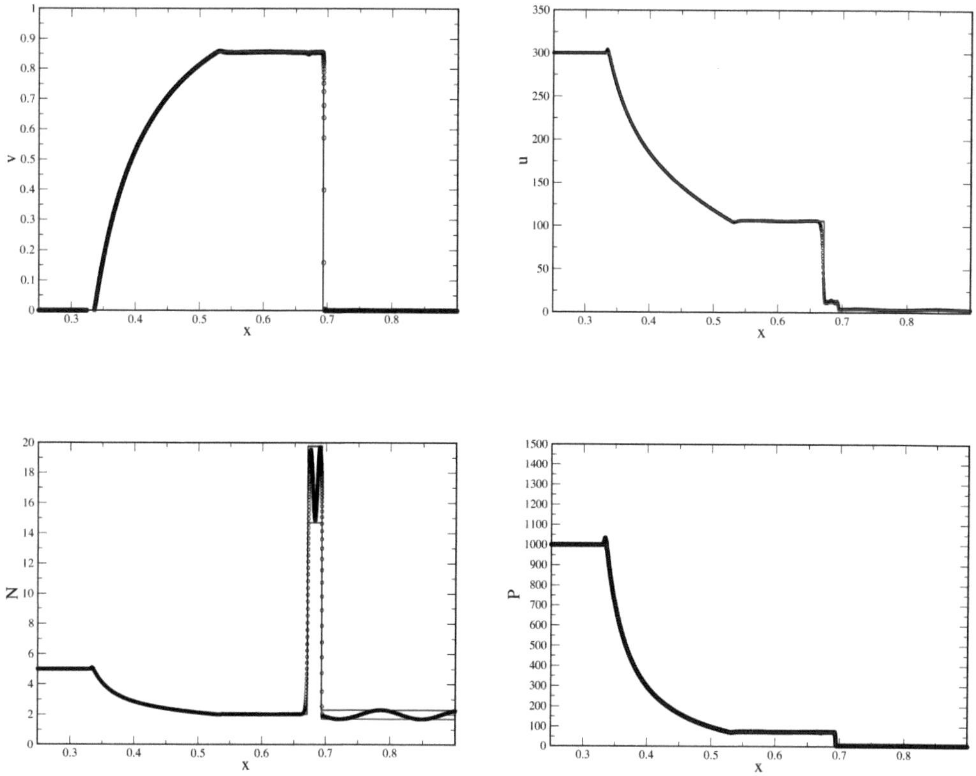

Fig. 5. Riemann problem where the right-hand side is periodically perturbed. The SPH solution is shown as circles, the exact solution for Riemann problems with constant RHS densities $N_{\rm R} = 2.3$ (blue) and $N_{\rm R} = 1.7$ (red) are overlaid as solid lines.

4.6 Test 6: Ultra-relativistic advection

In this test problem we explore the ability to accurately advect a smooth density pattern at an ultra-relativistic velocity across a periodic box. Since this test does not involve shocks we do not apply any artificial dissipation. We use only 500 equidistantly placed particles in the interval $[0, 1]$, enforce periodic boundary conditions and use a polytropic exponent of $\Gamma = 4/3$. We impose a computing frame number density $N(x) = N_0 + \frac{1}{2}\sin(2\pi x) + \frac{1}{4}\sin(4\pi x)$, a constant velocity as large as $v = 0.99999999$, corresponding to a Lorentz factor of $\gamma \approx 7071$, and instantiate a constant pressure corresponding to $P_0 = (\Gamma - 1)n_0 u_0$, where $n_0 = N_0/\gamma$ and $N_0 = 1$ and $u_0 = 1$. The specific energies are chosen so that each particle has the same pressure P_0. With

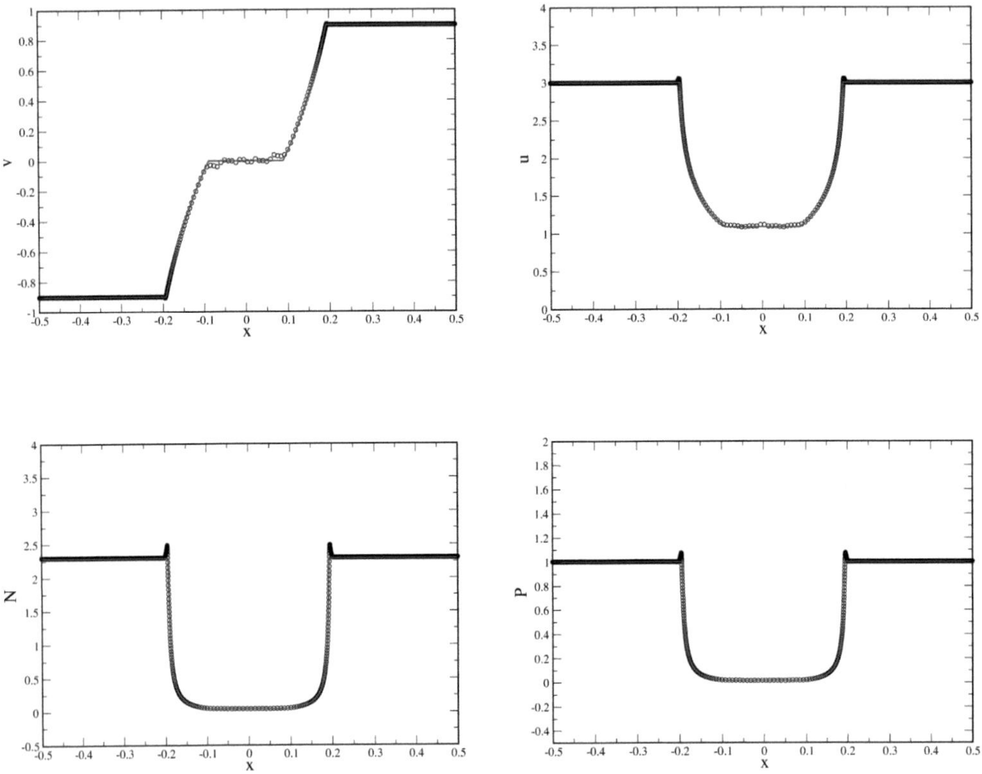

Fig. 6. Relativistic version of the Einfeldt rarefaction test. Initially the flow has constant values of $n = 1, P = 1$ everywhere, $v_L = -0.9$ and $v_R = 0.9$.

these initial conditions the specified density pattern should just be advected across the box without being changed in shape.

The numerical result after 50 times (blue circles) and 100 times (green triangles) crossing the interval is displayed in Fig. 7, left panel. The advection is essentially perfect, no deviation from the initial condition (solid, red line) is visible.

We use this test to measure the convergence of the method in the case of smooth flow (for the case involving shocks, see the discussion at the end of test 1). Since for this test the velocity is constant everywhere, we use the computing frame number density N to calculate L_1 similar to Eq. (21). We find that the error decreases very close to $L_1 \propto N^{-2}$, see Fig. 7, right panel, which is the behavior that is theoretically expected for smooth functions, the used kernel and perfectly distributed particles [22] (actually, we find as a best-fit

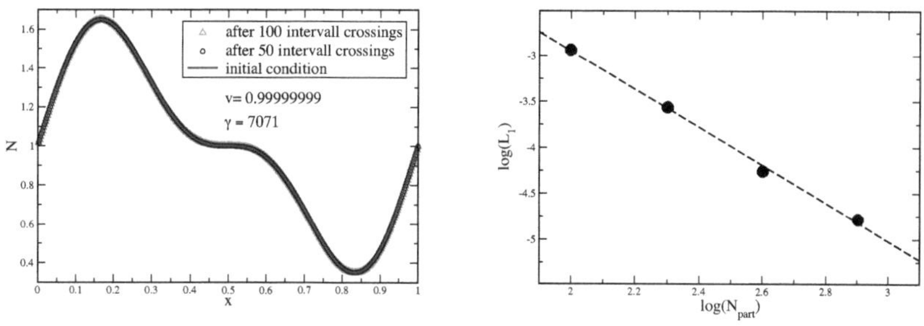

Fig. 7. Left: Ultra-relativistic advection ($v = 0.99999999$, Lorentz factor $\gamma = 7071$) of a density pattern across a periodic box. The advection is essentially perfect, the patterns after 50 (blue circles) and 100 (green triangles) times crossing the box are virtually identical to the initial condition (red line). Right: Decrease of the L_1 error as a function of resolution, for smooth flows the method is second-order accurate.

exponent -2.07). Therefore, we consider the method second-order accurate for smooth flows.

5 Conclusions

We have summarized a new special-relativistic SPH formulation that is derived from the Lagrangian of an ideal fluid [28]. As numerical variables it uses the canonical energy and momentum per baryon whose evolution equations follow stringently from the Euler-Lagrange equations. We have further applied the special-relativistic generalizations of the so-called "grad-h-terms" and a refined artificial viscosity scheme with time dependent parameters.

The main focus of this paper is the presentation of a set of challenging benchmark tests that complement those of the original paper [28]. They show the excellent advection properties of the method, but also its ability to accurately handle even very strong relativistic shocks. In the extreme shock tube test 3, where the post-shock density shell is compressed into a width of only 0.1 % of the computational domain, we find the shock front to propagate at slightly too large a pace. This artifact ceases with increasing numerical resolution, but future improvements of this point would be desirable. We have further determined the convergence rate of the method in numerical experiments and find it first-order accurate when shocks are involved and second-order accurate for smooth flows.

References

1. S. Ayal, T. Piran, R. Oechslin, M. B. Davies, and S. Rosswog, *Post-Newtonian Smoothed Particle Hydrodynamics*, ApJ **550** (2001), 846–859.
2. T. W. Baumgarte and S. L. Shapiro, *Numerical Relativity and Compact Binaries*, Phys. Rep. **376** (2003), 41–131.
3. A. Bauswein, R. Oechslin, and H. -J. Janka, *Discriminating Strange Star Mergers from Neutron Star Mergers by Gravitational-Wave Measurements*, ArXiv e-prints (2009).
4. J. E. Chow and J.J. Monaghan, *Ultrarelativistic SPH*, J. Computat. Phys. **134** (1997), 296.
5. L. Del Zanna and N. Bucciantini, *An Efficient Shock-capturing Central-type Scheme for Multidimensional Relativistic Flows. I. Hydrodynamics*, A&A **390** (2002), 1177–1186.
6. A. Dolezal and S. S. M. Wong, *Relativistic Hydrodynamics and Essentially Non-oscillatory Shock Capturing Schemes*, J. Comp. Phys. **120** (1995), 266.
7. B. Einfeldt, P. L. Roe, C. D. Munz, and B. Sjogreen, *On Godunov-type Methods Near Low Densities*, J. Comput. Phys. **92** (1991), 273–295.
8. J. A. Faber, T. W. Baumgarte, S. L. Shapiro, K. Taniguchi, and F. A. Rasio, *Dynamical Evolution of Black Hole-Neutron Star Binaries in General Relativity: Simulations of Tidal Disruption*, Phys. Rev. D **73** (2006), no. 2, 024012.
9. J. A. Faber, P. Grandclément, and F. A. Rasio, *Mergers of Irrotational Neutron Star Binaries in Conformally Flat Gravity*, Phys. Rev. D **69** (2004), no. 12, 124036.
10. J. A. Faber and F. A. Rasio, *Post-Newtonian SPH Calculations of Binary Neutron Star Coalescence: Method and First Results*, Phys. Rev. D **62** (2000), no. 6, 064012.
11. J. A. Faber and F. A. Rasio, *Post-Newtonian SPH Calculations of Binary Neutron Ntar Coalescence. III. Irrotational Systems and Gravitational Wave Spectra*, Phys. Rev. D **65** (2002), no. 8, 084042.
12. J. A. Faber, F. A. Rasio, and J. B. Manor, *Post-Newtonian Smoothed Particle Hydrodynamics Calculations of Binary Neutron Star Coalescence. II. Binary Mass Ratio, Equation of State, and Spin Dependence*, Phys. Rev. D **63** (2001), no. 4, 044012.
13. V. Fock, *Theory of Space, Time and Gravitation*, Pergamon, Oxford, 1964.
14. J. Font, *Numerical Hydrodynamics in General Relativity*, Living Rev. Relativ. **3** (2000), 2.
15. J. F. Hawley, L. L. Smarr, and J. R. Wilson, *A Numerical Study of Nonspherical Black Hole Accretion. II - Finite Differencing and Code Calibration*, ApJS **55** (1984), 211–246.
16. A. Kheyfets, W. A. Miller, and W. H. Zurek, *Covariant Smoothed Particle Hydrodynamics on a Curved Background*, Phys. Rev. D **41** (1990), 451–454.
17. P. Laguna, W. A. Miller, and W. H. Zurek, *Smoothed Particle Hydrodynamics Near a Black Hole*, ApJ **404** (1993), 678–685.
18. P.J. Mann, *A Relativistic Smoothed Particle Hydrodynamics Method Tested with the Shock Tube*, Comp. Phys. Commun. (1991).
19. P.J. Mann, *Smoothed Particle Hydrodynamics Applied to Relativistic Spherical Collapse*, J. Comput. Phys. **107** (1993), 188–198.
20. J. M. Marti and E. Müller, *Numerical Hydrodynamics in Special Relativity*, Living Rev. Relativ. **6** (2003), 7.

21. J.M. Marti and E. Müller, *Extension of the Piecewise Parabolic Method to One-Dimensional Relativistic Hydrodynamics*, J. Comp. Phys. **123** (1996), 1.
22. J. J. Monaghan, *Smoothed Particle Hydrodynamics*, Ann. Rev. Astron. Astrophys. **30** (1992), 543.
23. J. J. Monaghan, *SPH Compressible Turbulence*, MNRAS **335** (2002), 843–852.
24. J. J. Monaghan, *Smoothed Particle Hydrodynamics*, Rep. Prog. Phys. **68** (2005), 1703–1759.
25. J. J. Monaghan and D. J. Price, *Variational Principles for Relativistic Smoothed Particle Hydrodynamics*, MNRAS **328** (2001), 381–392.
26. R. Oechslin, S. Rosswog, and F.-K. Thielemann, *Conformally Flat Smoothed Particle Hydrodynamics Application to Neutron Star Mergers*, Phys. Rev. D **65** (2002), no. 10, 103005.
27. S. Rosswog, *Astrophysical Smooth Particle Hydrodynamics*, New Astron. Rev. **53** (2009), 78.
28. S. Rosswog, *Conservative, Special-relativistic Smooth Particle Hydrodynamics*, submitted to J. Comp. Phys. (2009), eprint arXiv:0907.4890.
29. S. Rosswog, *Relativistic Smooth Particle Hydrodynamics on a Given Background Space-time*, Classical Quantum Gravity, in press (2010).
30. S. Siegler, *Entwicklung und Untersuchung eines Smoothed Particle Hydrodynamics Verfahrens für relativistische Strömungen*, Ph.D. thesis, Eberhard-Karls-Universität Tübingen, 2000.
31. S. Siegler and H. Riffert, *Smoothed Particle Hydrodynamics Simulations of Ultrarelativistic Shocks with Artificial Viscosity*, ApJ **531** (2000), 1053–1066.
32. V. Springel and L. Hernquist, *Cosmological Smoothed Particle Hydrodynamics Simulations: the Entropy Equation*, MNRAS **333** (2002), 649–664.

An exact particle method for scalar conservation laws and its application to stiff reaction kinetics

Yossi Farjoun[1] and Benjamin Seibold[2]

[1] G. Millán Institute of Fluid Dynamics, Nanoscience, and Industrial Mathematics
Universidad Carlos III de Madrid
Avenida de la Universidad 30, Leganés, Spain, 28911
yfarjoun@ing.uc3m.es

[2] Department of Mathematics
Temple University
1801 North Broad Street, Philadelphia, PA 19122
seibold@temple.edu

Summary. An "exact" method for scalar one-dimensional hyperbolic conservation laws is presented. The approach is based on the evolution of shock particles, separated by local similarity solutions. The numerical solution is defined everywhere, and is as accurate as the applied ODE solver. Furthermore, the method is extended to stiff balance laws. A special correction approach yields a method that evolves detonation waves at correct velocities, without resolving their internal dynamics. The particle approach is compared to a classical finite volume method in terms of numerical accuracy, both for conservation laws and for an application in reaction kinetics.

Key words: particle, characteristic, shock, reaction kinetics

1 Introduction

In this paper, a special class of numerical methods for scalar hyperbolic conservation laws in one space dimension is presented. An important area in which such problems arise is the simulation of nonlinear flows in networks. An example is the flow of vehicular traffic on highways [16]. The flow along each network edge is described by a hyperbolic conservation law (e.g. the Lighthill-Whitham model [22] for traffic flow). The edges meet at the network vertices, where problem specific coupling conditions are imposed (such as the Coclite-Piccoli conditions [3,18] for traffic flow). Here, we focus on the evolution of the flow along a single edge. In network flows, high requirements are

M. Griebel, M.A. Schweitzer (eds.), *Meshfree Methods for Partial Differential Equations V*, Lecture Notes in Computational Science and Engineering 79,
DOI 10.1007/978-3-642-16229-9_7,

imposed on the numerical method. On the one hand, the approach must guarantee exact conservation (no cars must be lost), no spurious oscillations must occur (otherwise one may encounter negative densities), and shocks (traffic jams) should be located accurately. On the other hand, in the simulation of a large network, only very few computational resources can be attributed to each edge.

A commonly used approach is to approximate the governing conservation law by traditional finite difference [20] or finite volume methods [13]. Low order methods are generally too diffusive, thus do not admit an accurate location of shocks. High order methods, such as finite volume methods with limiters [25], or ENO [14]/WENO [23] schemes admit more accurate capturing of shocks. However, this comes at the expense of locality: stencils reach over multiple cells, which poses challenges at the network vertices. Alternative approaches are front tracking methods [17]. These do not operate on a fixed grid, but track shocks explicitly. Thus, shock are located accurately. However, smooth parts, such as rarefaction fans, are not represented very well. Another class of approaches is based on the underlying method of characteristics. An example is the CIR method [5], which updates information on grid points by tracing characteristic curves. Thus, it is a fixed-grid finite difference method. A fully characteristic approach was presented by Bukiet et al. [1] that tracks the evolution of particles. Where the solution is smooth, particles follow the characteristic curves, and where these curves collide, shocks are evolved. By construction, shocks are ideally sharp. While the tracing of the characteristics is high order accurate, the location of shocks is only first order accurate.

Another approach was presented by the authors [9–11]. In contrast to previous methods, here shocks are resolved by the merging of characteristic particles. This is made possible by the definition of a suitable interpolation, which is a similarity solution of the underlying conservation law. Hence, in [11] we suggest to call the approach *rarefaction tracking*. The method admits second order accurate location of shocks. In Sect. 2, the fundamentals of this approach, in particular the similarity interpolation, are outlined.

A generalization of the characteristic particle method presented in [10], namely *shock particles*, is introduced in Sect. 3. A shock particle is a moving discontinuity that carries two function values. If the jump height is zero, a classical characteristic particle is recovered. Even though shocks are evolved explicitly, the aforementioned similarity interpolation still plays a crucial role, thus the approach is fundamentally different from traditional shock/front tracking methods. In fact, the presented approach solves the considered hyperbolic conservation law exactly, up to the integration error of an ODE. Hence, we call it an *exact particle method*. In Sect. 4, the evolution and interaction of shock particles is shown to give rise to an actual computational method. The key idea is that the original partial differential equation is reduced to an ordinary differential equation, which then can be solved using a high order ODE solver. A comparison of the exact particle approach with a traditional finite volume method (using the package CLAWPACK [2]) is presented in Sect. 5.

The presented approach is highly accurate for hyperbolic conservation laws, and is quite amenable to extension to balance laws. Of particular interest here are stiff reaction kinetics, in which reactions happen on a faster time scale than the nonlinear advection. In Sect. 6, the problem is introduced and some properties of its solution are described. In Sect. 7, a specialized adaptation of the particle method is presented. It is based on the exact particle method introduced in Sect. 4, with a fundamental adaptation that uses the similarity interpolation to provide a certain level of subgrid resolution near the detonation wave. This method is able to track detonation waves correctly, without resolving them explicitly. Computational results for this application are presented in Sect. 8.

2 Characteristic Particles and Similarity Solution Interpolant

Consider a scalar conservation law in one space dimension

$$u_t + (f(u))_x = 0 \,, \quad u(x,0) = u_0(x) \,. \tag{1}$$

The flux function f is assumed to be twice differentiable and either convex ($f'' > 0$) or concave ($f'' < 0$) on the range of function values. We consider an approximation to the true solution of (1) by a family of functions, defined as follows.

Consider a finite number of particles. A particle is a computational node that carries a (variable) position x_i, and a (variable) function value u_i. Let the set of particles be defined by $P = \{(x_1, u_1), \ldots, (x_n, u_n)\}$ with $x_1 \leq \cdots \leq x_n$. On the interval $[x_1, x_n]$, we define the interpolant $U_P(x)$ piecewise on the intervals between neighboring particles, as follows. If $u_i = u_{i+1}$, then on the interpolant on $[x_i, x_{i+1}]$ is constant $U_P(x) = u_i$. Otherwise the interpolant on $[x_i, x_{i+1}]$ satisfies

$$\frac{x - x_i}{x_{i+1} - x_i} = \frac{f'(U_P(x)) - f'(u_i)}{f'(u_{i+1}) - f'(u_i)} \,. \tag{2}$$

This defines the inverse interpolant $x(U_P)$ explicitly on $[x_i, x_{i+1}]$. Since f is convex or concave, the interpolant $U_P(x)$ itself is uniquely defined. As shown in [10], the interpolation U_P defined above is an analytical solution of the conservation law (1), in the following sense. Consider particles moving "sideways" according to $P(t) = \{(x_1 + f'(u_1)t, u_1), \ldots, (x_n + f'(u_n)t, u_n)\}$. If at time $t = 0$, the particles satisfy $x_1 < \cdots < x_n$, then for sufficiently short times $t > 0$, the interpolant $U_{P(t)}$, defined by (2) is *the* analytical solution to the conservation law (1) at time t, starting with initial conditions $u_0 = U_P(0)$. This follows from the fact that each point $(x(t), u(t))$ on the solution moves according to the characteristic equations of (1), which are

$$\begin{cases} \dot{x} = f'(u) \\ \dot{u} = 0 \,. \end{cases} \tag{3}$$

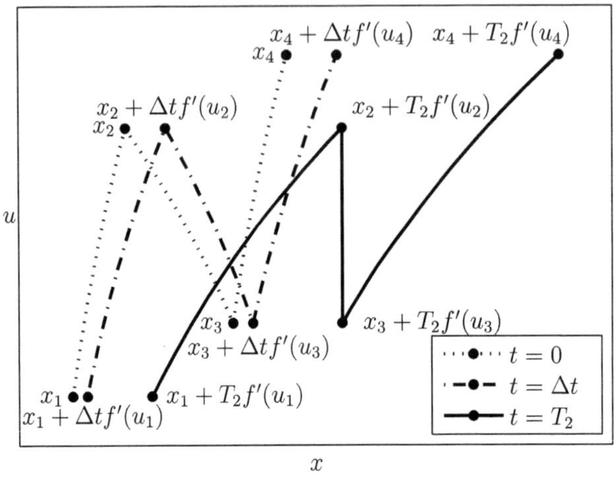

Fig. 1. An illustration of characteristic particles moving according to (3). The solution develops a shock at $t = T_2$. The dotted line is the interpolation between the particles at time t, the dash-dotted line–after a short time Δt. The solid line shows the solution at $t = T_2$ when the solution develops a shock between particles 2 and 3.

From the definition of $P(t)$ it is obvious that the particles satisfy (3). Any other point is given by the above defined interpolation. Replacing x_i by $x_i + f'(u_i)t$ in (2) and differentiating with respect to t yields that $\dot{x}(t) = f'(U_P(x))$.

The solution between neighboring particles is a similarity solution that either comes from a discontinuity (if the particles depart) or becomes a shock (if the particles approach each other). Hence, the solution $U_P(x)$ is composed of rarefaction waves and compression waves. Therefore, as described in [11], the approach can be interpreted as "rarefaction tracking", which expresses its similarity and fundamental difference to front tracking approaches [17, 19].

The interpolant U_P (2) is a solution of (1) until the time of the first collision, i.e. the moment when two neighboring particles share the same x-position. For a pair of neighboring particles (x_i, u_i) and (x_{i+1}, u_{i+1}), the time of collision is given by

$$T_i = -\frac{x_{i+1} - x_i}{f'(u_{i+1}) - f'(u_i)} . \tag{4}$$

If particles depart from each other (i.e. $f'(u_i) < f'(u_{i+1})$), one has $T_i < 0$, thus no collision happens in future time. For a set of n particles, the first time of collision is $T^* = \min(\{T_i : T_i > 0\} \cup \infty)$. The solution is continuous until that time. At $t = T^*$, a shock occurs (at x_i, between u_{i+1} and u_i), and the method of characteristics alone does not yield a correct solution further in time. An illustration of the particle movement and the development of a shock can be seen in Fig. 1.

Remark 1. We assume that one is interested in single-valued weak entropy-solutions, which possess shocks. In some applications multi-valued solutions are sought, and those can be obtained easily by continuing to move the particles according to the characteristic equations (3) beyond the occurrence of shocks.

As presented in previous papers [9–11], the use of the method of characteristics even in the presence of shocks can be made possible by a suitably designed particle management. Particles that collide are immediately merged into a single particle, with a new function value u chosen such that the total area under the interpolant U_P is preserved. After this merge, the defined interpolant is again continuous, and one can step further in time using solely characteristic particle movement (3). This approach introduces a small error around shocks. Suitable insertion of new particles near shocks (before merging) guarantees that the error remains localized near shocks.

3 Shock Particles

In this paper, we present an approach that does not introduce any errors intrinsically. While the approximation of general initial conditions by finitely many particles involves an error (see Sect. 4.1), the actual evolution under the conservation law (1) is exact—not just pointwise on the particles, but in the sense of functions. The new approach generalizes characteristic particles to *shock particles.*

3.1 Evolution of Shock Particles

A shock particle is a computational node that carries a (variable) position x_i, a (variable) left state u_i^-, and a (variable) right state u_i^+, which satisfy the Oleinik entropy condition [7] $f'(u_i^-) \geq f'(u_i^+)$). Whenever a shock particle (x_i, u_i^-, u_i^+) violates this conditions (e.g. because it is placed in the initial conditions), it is immediately replaced by two particles (x_i, u_i^-, u_i^-) and (x_i, u_i^+, u_i^+), which then depart from each other (since $f'(u_i^-) < f'(u_i^+)$). Given n shock particles $P = \{(x_1, u_1^-, u_1^+), \ldots, (x_n, u_n^-, u_n^+)\}$, the interpolant U_P on $[x_1, x_n]$ is defined piecewise: on $[x_i, x_{i+1}]$, it satisfies

$$\frac{x - x_i}{x_{i+1} - x_i} = \frac{f'(U_P(x)) - f'(u_i^+)}{f'(u_{i+1}^-) - f'(u_i^+)} \,. \tag{5}$$

The velocity of a shock particle is given by the Rankine-Hugoniot condition [7] as $\dot{x}_i = s(u_i^-, u_i^+)$, where

$$s(u, v) = \begin{cases} \frac{f(u)-f(v)}{u-v} & u \neq v \\ f'(u) & u = v \end{cases} \tag{6}$$

is the difference quotient of f, continuously extended at $u = v$.

Remark 2. When implementing this function numerically, one should avoid calculating the difference quotient when the distance $|u - v|$ is very small. For those cases one should consider the limiting value $f'(u)$ as a more accurate alternative, or even better, the next order Taylor expansion,

$$s(u, u + \epsilon) \approx f'(u) + \tfrac{1}{2}\epsilon f''(u), \tag{7}$$

can be used.

At a shock, the function values u_i^- and u_i^+ change in time as well. Their rate of change is exactly such that the interpolation (5) near the shock evolves as a smooth function should evolve under (1). Here we derive the evolution for the right value of the shock u_i^+. The argument for u_i^- works analogously. If we had $u_i^- = u_i^+$, the particle would have a function value that is constant in time $\dot{u}_i^+ = 0$, and move with velocity $\dot{x}_i = f'(u_i^+)$. The interpolant (5) with these definitions for $\dot{x}_i$ and $\dot{u}_i^+$ is the correct solution between x_i and x_{i+1}. If $u_i^- \neq u_i^+$, the shock moves at a speed different from $f'(u_i^+)$. In order to preserve the same *interpolation*, the function value u_i^+ has to evolve according to

$$\dot{u}_i^+ = (\dot{x}_i - f'(u_i^+))\frac{f'(u_{i+1}^-) - f'(u_i^+)}{x_{i+1} - x_i}\frac{1}{f''(u_i^+)} \,. \tag{8}$$

Here $\dot{x}_i - f'(u_i^+)$ is the relative velocity of the shock to a characteristic particle velocity, and $\frac{f'(u_{i+1}^-)-f'(u_i^+)}{x_{i+1}-x_i}\frac{1}{f''(u_i^+)}$ is the slope of the interpolant (5) at x_i, found by differentiating (5) with respect to x. The law of motion for a shock particle is thus

$$\begin{cases} \dot{x}_i = s(u_i^-, u_i^+) \\ \dot{u}_i^- = \left(s(u_i^-, u_i^+) - f'(u_i^-)\right) \dfrac{f'(u_{i-1}^+) - f'(u_i^-)}{x_{i-1} - x_i} \dfrac{1}{f''(u_i^-)} \\ \dot{u}_i^+ = \left(s(u_i^-, u_i^+) - f'(u_i^+)\right) \dfrac{f'(u_{i+1}^-) - f'(u_i^+)}{x_{i+1} - x_i} \dfrac{1}{f''(u_i^+)} \end{cases} \tag{9}$$

Observe that the evolution of a shock particle depends on the neighboring two particles. See Fig. 2 for an illustration of the derivation of these equations.

In the case $u_i^- = u_i^+$, we call a shock particle *characteristic*. In fact, a characteristic particle, as described in Sect. 2, is nothing else than a shock particle with jump height zero. This is motivated by

Lemma 1. *The motion of a shock particle* (9) *reduces to the motion of a characteristic particle* (3), *as* $u_i^+ - u_i^- \to 0$.

Proof. By definition (6), the first equation in (9) clearly reduces to $\dot{x}_i = f'(u_i)$. In the second and third equation, the fraction remains finite while the quantity in the parentheses converges to zero, and thus the whole expression vanishes $\dot{u}_i^- \to 0$, and $\dot{u}_i^+ \to 0$.

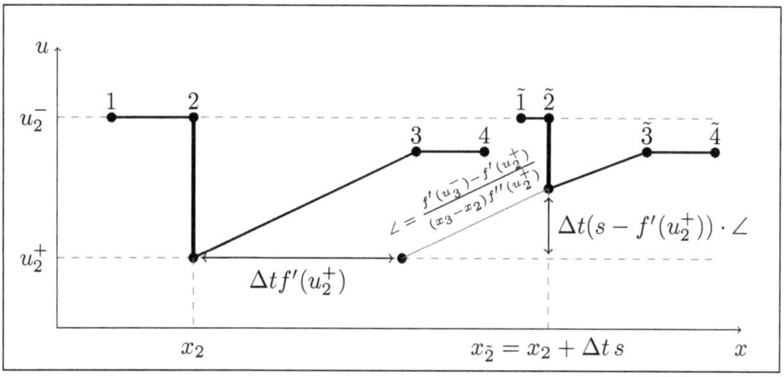

Fig. 2. The shock moves with speed s, given by the Rankine-Hugoniot condition, which is different than the characteristic speed. Thus shock particles must have varying function values. In the figure a tilde (˜) above a number, denotes the new location of a particle, and $\angle$ is the slope of the similarity solution to the left of particle 2.

Remark 3. In a numerical implementation, the right hand side (9) can almost be implemented as it stands, with the only modification that the difference quotient (6) is replaced by the characteristic speed if $\left|u_i^+ - u_i^-\right|$ is less than a sufficiently small value.

Theorem 1. *If the time evolution* (9) *of the particles* $P(t)$ *is solved exactly, then for sufficiently short times* t*, the resulting evolution of the interpolation* $U_{P(t)}$ *is the unique weak entropy solution of the conservation law* (1) *with initial conditions* $u_0 = U_{P(0)}$*, on the domain of definition* $[x_0(t), x_n(t)]$.

Proof. Due to the first equation in (9), all shocks (including those of height zero) move at their correct speeds. By construction, every shock satisfies the entropy condition. Discontinuities that violate the entropy condition immediately become rarefaction waves. The second and third equation in (9) ensure that each point on the interpolation between particles moves according to the characteristic equations (3).

3.2 Interaction of Shock Particles

After some time, neighboring shock particles may collide, i.e. share the same x-position. In this situation, the two shocks become a single shock, as the following Lemma shows.

Lemma 2. *Two neighboring shock particles* (x_i, u_i^-, u_i^+) *and* $(x_{i+1}, u_{i+1}^-, u_{i+1}^+)$ *satisfy* $u_i^+ = u_{i+1}^-$ *at their time of collision, if at least one of them is not characteristic.*

Proof. Due to (9), the difference in function values between the two shocks evolves according to

$$\begin{aligned}\frac{d}{dt}\left(u_{i+1}^- - u_i^+\right) &= \left(\frac{s(u_{i+1}^-,u_{i+1}^+)-f'(u_{i+1}^-)}{f''(u_{i+1}^-)} + \frac{f'(u_i^+)-s(u_i^-,u_i^+)}{f''(u_i^+)}\right)\frac{f'(u_{i+1}^-)-f'(u_i^+)}{x_{i+1}-x_i} \\ &= \left(\frac{s(u_{i+1}^-,u_{i+1}^+)-f'(u_{i+1}^-)}{f''(u_{i+1}^-)} + \frac{f'(u_i^+)-s(u_i^-,u_i^+)}{f''(u_i^+)}\right) f''(\xi)\frac{u_{i+1}^- - u_i^+}{x_{i+1}-x_i}\,. \end{aligned} \tag{10}$$

Here ξ is a value between $u_{i+1}^- - u_i^+$ given by the Mean Value Theorem. Due to the Oleinik entropy condition, both numerators inside the large parentheses are non-positive, and by assumption at least one is strictly negative. The signs of f'' inside and outside the large parentheses cancel each other out. Hence the right hand side always has the opposite sign than $u_{i+1}^- - u_i^+$. If we assume, by negation, that $u_{i+1}^- - u_i^+$ remains finite as $x_{i+1} - x_i \to 0$ we get a clear contradiction, and thus the difference $u_{i+1}^- - u_i^+$ goes to zero.

In the computational method, two shock particles (x_i, u_i^-, u_i^+) and $(x_{i+1}, u_{i+1}^-, u_{i+1}^+)$ that collide, i.e. $x_i = x_{i+1}$, are simply merged into a single particle (x_i, u_i^-, u_{i+1}^+). Due to Lemma 2, this merge does not change the actual solution. Hence, Thm. 1 extends to allow particle merges, assuming that the time evolution (9) is integrated exactly. If both interacting particles are characteristic, this approach automatically creates a shock.

4 An "Exact" ODE Based Method

Due to Thm. 1, the presented approach yields the exact weak entropy solution of the conservation law (1), when starting with an initial condition u_0 that can be represented by finitely many particles P_0, i.e. $u_0 = U_{P_0}$. Hence, we call this approach an *exact particle method.* In practice, two types of approximation are performed. First, a general initial function u_0 cannot be represented exactly using finitely many particles, and thus needs to be approximated. This aspect is briefly addressed in Sect. 4.1. Second, in general the time evolution (9) can not be integrated exactly. Instead, a numerical ODE solver has to be used. This aspect is addressed in Sect. 4.2.

4.1 Approximation of the Initial Conditions

Whenever the initial function u_0 can be represented exactly by an interpolation U_{P_0}, one should do so if the number of particles required is computationally acceptable. A particular advantage of the presented particle approach is that discontinuities can be represented exactly. If the initial function cannot be represented exactly, it must be approximated. It is shown in [10] that the interpolation (2) approximates piecewise smooth initial conditions with an error of $O(h^2)$, where h is the maximum distance between particles, if discontinuities are represented exactly. Furthermore, since the method does not require an equidistant placement of particles, adaptive sampling strategies should be used, such as presented in [10]. These results are based on the particles be placed exactly on the function u_0. More general approximation strategies that do not have this restriction are the subject of current research.

4.2 Integration in Time

The characteristic equation (3) can easily be integrated exactly. Therefore, characteristic particle movement incurs no integration error, and the next collision time between characteristic particles is explicitly given by (4). The particle approach presented and analyzed in [9–11] relies on these properties. The downside of those methods is an intrinsic error near shocks. In contrast, the shock-particle method presented in the current paper does not incur any errors around shocks. The downside is an error due to the integration of the ordinary differential equation (3). However, it is comparably simple to integrate systems of ODE with very high accuracy. In contrast, the construction of high order numerical approaches that approximate the PDE (1) directly (such as ENO [14]/WENO [23] or Godunov schemes with limiters [25]), is much more challenging. The numerical error analysis shown in Sect. 5 seconds this.

Another feature (besides accuracy) that the used numerical ODE solver needs to possess is *event detection*. Since at particle collisions, the system undergoes a discontinuous change (the number of particles is reduced), the ODE solver must detect such events with high accuracy. One way to do this is to use a solver that can provide a high order interpolation. Several such solvers have been derived by Dormand and Prince in [6] for the Runge-Kutta family of ODE solvers. In MATLAB, event detection is implemented in particular in `ode23.m` and `ode45.m`. As stated in [24], the latter contains an unpublished[3] variation of the interpolation presented in [6].

To enhance the performance of the adaptive ODE solver it is helpful to have an estimate about the next occurrence of a particle collision. A simple estimate may be obtained by using only the first equation of (9), thus estimating the collision time between neighboring particles by $T_i \approx -\frac{x_{i+1}-x_i}{s(u_{i+1}^-,u_{i+1}^+)-s(u_i^-,u_i^+)}$.

5 Numerical Error Analysis of the Particle Method

We investigate the order of accuracy of the presented particle method, and compare it to the benchmark PDE solver CLAWPACK [2,21], an implementation of various finite volume methods for solving hyperbolic PDE. We use

[3]Details can be found in the MATLAB file `ntrp45.m`.

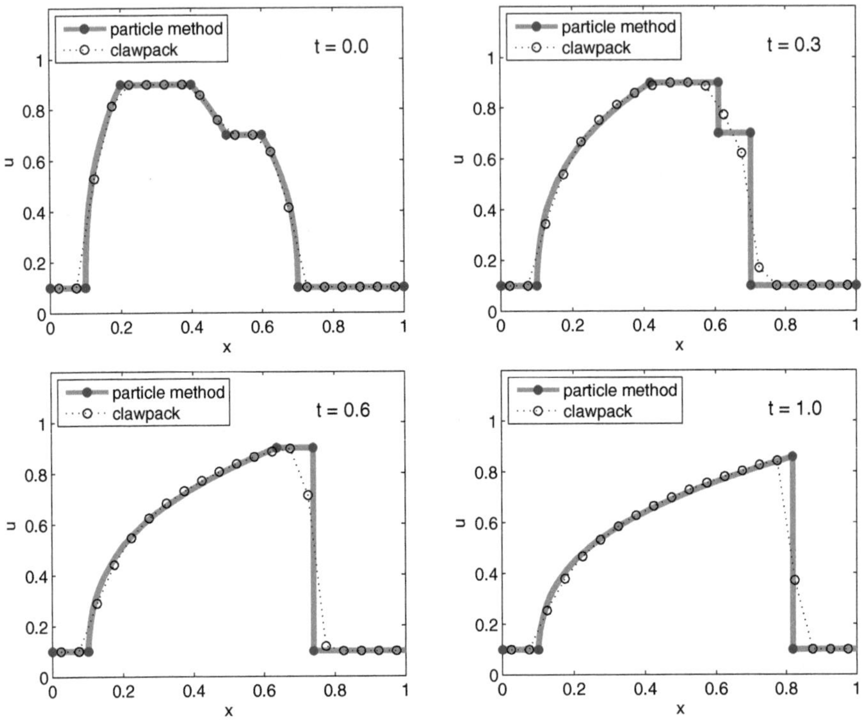

Fig. 3. Time evolution at $t \in \{0, 0.3, 0.6, 1\}$ of the solution to the conservation law (1) with $f(u) = \frac{1}{4}u^4$, both by CLAWPACK and by our particle method.

the second order scheme with limiters. The results in Fig. 4 are found using the classical "MinMod" limiter, while the results in Fig. 7 are found using the "Monotonized Centered" limiter, as the authors of CLAWPACK suggest in their treatment of this problem.

The comparison between a finite volume method and a particle method is tricky, since the two approaches are fundamentally different. First, the finite volume approach works with average function values in fixed cells, while with particles the interpolation (5) defines a function everywhere. This difference can be overcome by constructing errors in the L^1 sense from cell averages, as described in [10]. Second, the finite volume method has a fixed spacial resolution Δx, while particles move, merge, and are generally anything but equidistant. Third, in a convergence analysis of a finite volume method, the spacial resolution and the time step are chosen proportional $\Delta t = C\Delta x$. In contrast, the particle method becomes exact if $\Delta t \to 0$, assuming that the initial conditions can be represented exactly by finitely many particles.

Here, we consider an initial condition that can be represented exactly by the interpolation (5). The reason is that we do not want to measure the error in approximating general initial conditions (for this aspect, please consult [10]).

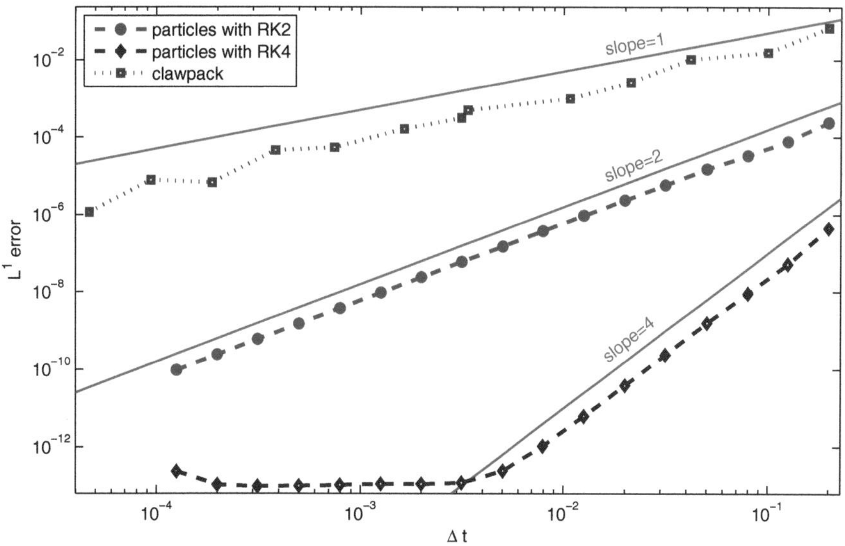

Fig. 4. Error convergence of the particle method in comparison with CLAWPACK. The dashed graphs denote the particle method, with RK2 (dots) and RK4 (diamonds) time stepping. The dotted graph represents CLAWPACK.

Instead, we want to investigate the error in the particle evolution. We consider a second order and a fourth order accurate Runge-Kutta method for the time evolution of (9). Times of particle collisions are found and resolved with the same order of accuracy. For the CLAWPACK runs, we specify a desired CFL number [4] of 0.8 and let the code choose Δt as it finds suitable. In practice, for this problem, this amounts to having $\Delta t \approx \Delta x$.

Specifically, we consider the conservation law (1) with flux function $f(u) = \frac{1}{4}u^4$, and initial function $u_0(x) = U_{P_0}(x)$, which is the interpolation (2) defined by the characteristic particles $P_0 = \{(0, 0.1), (0.1, 0.1), (0.2, 0.9), (0.4, 0.9), (0.5, 0.7), (0.6, 0.7), (0.7, 0.1), (1.0, 0.1)\}$. The time evolution of the solution is shown in Fig. 3 in four snapshots at $t \in \{0, 0.3, 0.6, 1\}$. In fact, what is shown is the solution obtained by the particle method, integrated with an accuracy that the error is not noticeably in the eye norm. For a comparison, we show the results obtained by a second order CLAWPACK method, with $\Delta x = 0.05$.

The convergence of the error for various approaches is shown in Fig. 4. We consider the $L^1([0,1])$ error (with respect to a reference solution resulting from a high resolution computation) at time $t = 1$. Note that this is possible since the approach defines a numerical solution everywhere. One can observe that the overall order of the particle method equals the order of the ODE solver used. Thus, with the standard RK4 solver, machine accuracy is obtained already for moderate time steps. Note again that we do not need to increase the number of particles to obtain convergence. This is a crucial advantage of the presented particle method. For CLAWPACK, the dotted graph in Fig. 4 shows the $L^1([0,1])$ error between the two piecewise constant functions whose cell averages agree with those of the reference solution and its finite volume approximation, respectively. The second order CLAWPACK solver yields an order of convergence only slightly better than first. This is particularly due to the presence of the shocks in the solution, and the large derivatives in the initial condition.

In general, the particle method is of the same computational complexity as classical finite volume schemes, in the sense that a computation with n particles/grid cells requires $O(n)$ time steps, and each time step requires $O(n)$ operations. The specific relative performance depends on many factors, such as the particular initial conditions (a function may be better represented by adaptively sampled particles than on a regular grid) and implementation details (while the straightforward evaluation of (9) yields a simpler code, the evaluation of (3) on characteristic particles results in a faster computation speed).

6 Stiff Reaction Kinetics

Many problems in chemical reaction kinetics can be described by advection-reaction equations, for which the reaction happens on a much faster time scale than the advection. We consider the balance law

$$u_t + (f(u))_x = \psi(u) \,, \tag{11}$$

where $0 \leq u \leq 1$ represents the density of some chemical quantity. The advection is given by the nonlinear flux function f, which is assumed convex (the case of f concave is analogous) and to be of order $O(1)$. The reactions are described by the source ψ. Here, we consider a stiff bistable reaction term

$$\psi(u) = \tfrac{1}{\tau} u(1-u)(u-\beta) \tag{12}$$

where $0 < \beta < 1$ is a fixed constant. This source term drives the values of $u < \beta$ towards 0, and values of $u > \beta$ towards 1. The reactions happen on a much faster time scale $O(\tau)$, where $\tau \ll 1$. This example is presented for instance in [15]. Since the source term ψ does not act in a discontinuity, equation (11) possesses shock solutions as the homogeneous problem (1) does. In addition, it has traveling wave solutions that connect a left state $u_L \approx 0$ with a right state $u_R \approx 1$ by a continuous function. To find those solutions, we make a traveling wave ansatz $u(x,t) = v(\xi)$, where $\xi = \frac{x-rt}{\tau}$ is the self similar variable. This transforms equation (11) into a first order ordinary differential equation for v

$$v'(\xi) = \tfrac{v(\xi)(1-v(\xi))(v(\xi)-\beta)}{f'(v(\xi))-r} \,. \tag{13}$$

For $v = \beta$, the numerator of (13) vanishes. A solution that connects a state $v_L < \beta$ to a state $v_R > \beta$ can only pass through $v = \beta$ if the denominator of

(13) vanishes as well, i.e. $r = f'(\beta)$, which yields the velocity of the traveling wave. The shape of the wave $v(\xi)$ is then found by integrating (13) using $r = f'(\beta)$. Since $u(x,t) = v(\frac{x-rt}{\tau})$, the traveling wave has a thickness (in the x-coordinate) of $O(\tau)$. This analysis of traveling wave solutions is in spirit similar to detonation waves of reacting gas dynamics [12]. The value β plays the role of a *sonic point* in detonation waves. The advection-reaction equation (11) is studied in [8]. The traveling wave solution (13) results from a balance of the advection term (which flattens the profile) and the reaction term (which sharpens the profile). Since τ is very small, these traveling waves look very similar to shocks, yet they face the opposite direction, and travel at a different velocity (if $f'(\beta) \neq f(1) - f(0)$).

In computations, the recovery of the exact shape of the traveling waves is typically not very important. However, the recovery of their correct propagation velocity is crucial. As described in [21], equation (11) can be treated in a straightforward fashion using classical finite volume approaches. However, correct propagation velocities of traveling waves are only obtained if these are numerically resolved. Thus, with equidistant grids, one is forced to use a very fine grid resolution $h = O(\tau)$, which is unnecessarily costly away from the traveling wave. This problem can be circumvented using adaptive mesh refinement techniques, however, at the expense of simplicity. An alternative approach, presented in [15], yields correct traveling wave velocities even with grid resolutions $h \gg O(\tau)$, by encoding specific information about the structure of the reaction term into a Riemann solver.

7 A Particle Method for Stiff Reaction Kinetics

Here, we present an approach based on the particle method introduced in Sect. 4 that uses the "subgrid" information provided by the interpolation (5) to yield correct propagation velocities of traveling waves, without specifically resolving them. The characteristic equations for (11) are

$$\begin{cases} \dot{x} = f'(u) \\ \dot{u} = \psi(u) \,. \end{cases} \tag{14}$$

As before, our goal is to generalize these characteristic equations to obtain an evolution for shock particles. This requires the definition of an interpolation. We use the interpolant (5), as if there were no reaction term. Clearly, this is an approximation, and the resulting method is not exact anymore. At any time t, we define the solution by shock particles $P(t) = \{(x_1, u_1^-, u_1^+), \dots, (x_n, u_n^-, u_n^+)\}$, and the interpolation $U_{P(t)}$, defined by (5). Adding the reaction term to (9), we now let the particles move according to

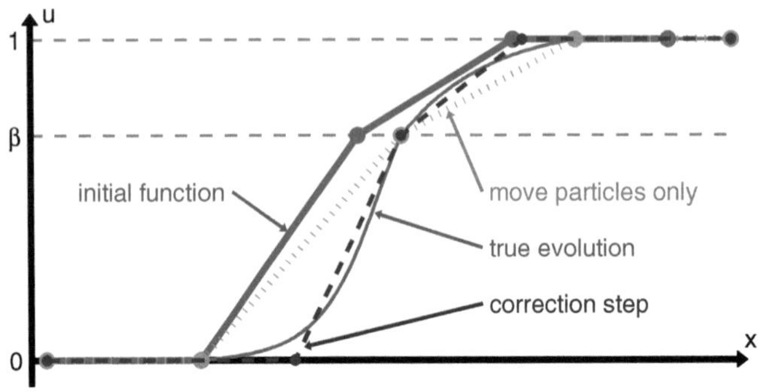

Fig. 5. Correction approach for the advection-reaction equation with dominant reaction term. The vertical dashed lines denote the three roots of the source term.

$$\begin{cases} \dot{x}_i = s(u_i^-, u_i^+) \\ \dot{u}_i^- = \left(s(u_i^-, u_i^+) - f'(u_i^-)\right) \dfrac{f'(u_{i-1}^+) - f'(u_i^-)}{x_{i-1} - x_i} \dfrac{1}{f''(u_i^-)} + \psi(u_i^-) \\ \dot{u}_i^+ = \left(s(u_i^-, u_i^+) - f'(u_i^+)\right) \dfrac{f'(u_{i+1}^-) - f'(u_i^+)}{x_{i+1} - x_i} \dfrac{1}{f''(u_i^+)} + \psi(u_i^+) \end{cases} \tag{15}$$

where the shock speed $s(u_i^-, u_i^+)$ is defined as before by (6). By construction, shocks move at their correct velocity, and for a characteristic particle $u_i^- = u_i^+$, (15) reduces to the correct characteristic evolution (14). Clearly, this approach does not remove the stiffness in time. Hence, an implicit ODE solver should be used. System (15) yields an accurate solution on the particles themselves, as well as an accurate evolution of shocks. However, traveling waves, as given by (13) are not represented well. The reason is that each particle moves very quickly towards 0 or 1. Then, the reaction term is not considered anymore, since $\psi(0) = 0$ and $\psi(1) = 0$. In order to correctly represent traveling waves, the continuous solution that goes through the sonic point β has to be considered. We do so by the following correction approach.

We assume that the bistable nature of the reaction term and the value of the unstable root β are known. Whenever the solution increases (in x) from a value $u < \beta$ to a value $u > \beta$, a special characteristic particle is placed at $u = \beta$, that moves with velocity $\dot{x} = f'(\beta)$. We call such a particle a *sonic particle*. Each particle that neighbors a sonic particle is treated in a special way. As a motivation, consider the situation shown in Fig. 5: A left state 0 is followed by a sonic particle, which is followed by a right state 1. The interpolant shown is (5) for $f(u) = \frac{1}{2}u^2$. The thick solid graph shows the initial configuration at time t. The dotted graph shows the obtained solution when evolving the particles for a time step Δt according to the method of characteristics (14). Since ψ vanishes on all particles, the source is neglected. Hence, this approach does not lead to correct traveling wave solutions. The thin solid graph shows the correct evolution of the initial function, considering the interpolation. This function cannot be represented exactly by particles and the interpolant (5), but it can be approximated by modifying the x-values of the two particles that neighbor the sonic particle, in such a way that the areas under the solution both left and right of the sonic particle are reproduced correctly. The dashed graph shows the function that results from this correction. Below, we describe the correction approach in detail.

7.1 Computational Approach

Consider a particle $(x_{i-1}, u_{i-1}^-, u_{i-1}^+)$ that is a left neighbor of a sonic particle (x_i, β). The case of a right neighbor particle is analogous. Let the interpolant on $[x_{i-1}, x_i]$ be denoted by $U(x)$, and its inverse function $X(u)$. From (5) it follows that $X'(u) = \frac{x_i - x_{i-1}}{f'(\beta) - f'(u_{i-1}^+)} f''(u)$. Using the interpolant $U(x)$, we can integrate the reaction term between the two particles. The substitution rule yields

$$\int_{x_{i-1}}^{x_i} \psi(U(x))\mathrm{d}x = \int_{u_{i-1}^+}^{\beta} \psi(u)X'(u)\mathrm{d}u = \frac{x_i - x_{i-1}}{f'(\beta) - f'(u_{i-1}^+)} \int_{u_{i-1}^+}^{\beta} \psi(u)f''(u)\mathrm{d}u \,. \tag{16}$$

This expression represents the full influence of the reaction term on the continuous solution between the value u_{i-1}^+ and the sonic value β. As derived in [10], the area under the interpolant on $[x_{i-1}, x_i]$ is given by

$$\int_{x_{i-1}}^{x_i} U(x)\mathrm{d}x = (x_i - x_{i-1})\, a(u_{i-1}^+, \beta) \,,$$

where $a(v, w) = \frac{[f'(u)u - f(u)]_v^w}{[f'(u)]_v^w}$ is a nonlinear average. Now consider a new particle $(x_{i-1} + \Delta x_{i-1}, u_{i-1}^+, u_{i-1}^+)$ be inserted between x_{i-1} and x_i. This insertion changes the area by

$$\begin{aligned}\Delta A &= \Delta x_{i-1} \left(a(u_{i-1}^+, \beta) - u_{i-1}^+\right) \\ &= \Delta x_{i-1} \frac{f'(\beta)(\beta - u_{i-1}^+) - (f(\beta) - f(u_{i-1}^+))}{f'(\beta) - f'(u_{i-1}^+)} \,.\end{aligned} \tag{17}$$

Equating the rate of area change $\frac{\Delta A}{\Delta t}$, given by (17), and expression (16), yields

$$\tfrac{\Delta x_{i-1}}{\Delta t} = c(u_{i-1}^+, \beta)\,(x_i - x_{i-1}) \,,$$

where $c(v, w) = \frac{\int_v^w \psi(u)f''(u)\mathrm{d}u}{f'(w)(w-v) - (f(w) - f(v))}$. The scaling $f = O(1)$ and $\psi = O(\frac{1}{\tau})$ implies that $c(v, w) = O(\frac{1}{\tau})$, if $w - v = O(1)$. A similar derivation for the right neighbor yields that a new particle $(x_{i+1} - \Delta x_{i+1}, u_{i+1}^-, u_{i+1}^-)$ needs to be inserted with

$$\tfrac{\Delta x_{i+1}}{\Delta t} = c(u_{i+1}^-, \beta)\,(x_i - x_{i+1}) \,.$$

Due to the bistable nature of the reaction term, one will frequently encounter a nearly constant left state, i.e. both $|u_{i-1}^+ - u_{i-1}^-|$ and $|u_{i-2}^+ - u_{i-1}^-|$ are very small. In this case, the particle $i-1$ can just be moved by Δx_{i-1}, instead of creating a new particle. Using a characteristic particle notation only, the resulting modified evolution equations are

$$\begin{cases} \dot{x}_{i-1} = f'(u_{i-1}) + c(u_{i-1},\beta)\,(x_i - x_{i-1}) \\ \dot{u}_{i-1} = \psi(u_{i-1}) \\ \dot{x}_i = f'(\beta) \\ \dot{u}_i = 0 \\ \dot{x}_{i+1} = f'(u_{i+1}) + c(u_{i+1},\beta)\,(x_{i+1} - x_i) \\ \dot{u}_{i+1} = \psi(u_{i+1})\,. \end{cases}$$

This implies that

$$\tfrac{d}{dt}\,(x_i - x_{i-1}) = (f'(\beta) - f'(u_{i-1})) - c(u_{i-1},\beta)\,(x_i - x_{i-1})\ ,$$

i.e. the distance $x_i - x_{i-1}$ converges to an equilibrium value $\frac{f'(\beta)-f'(u_{i-1})}{c(u_{i-1},\beta)}$. Since $c(u_{i-1},\beta) = O(\frac{1}{\tau})$, the equilibrium distance between the sonic particle and its neighbors is $O(\tau)$. Hence, the presented approach yields a traveling wave solution, represented by three particles that move at the correct velocity $f'(\beta)$, and whose distance from each other scales, correctly, like $O(\tau)$.

8 Numerical Results on Reaction Kinetics

For assessing our method numerically, we compare it to the benchmark PDE solver CLAWPACK [2], for the advection-reaction equation (11). We consider the reaction term (12) with $\beta = 0.8$, and choose the Burgers' flux $f(u) = \frac{1}{2}u^2$. Four different values for the reaction time scale are considered: $\tau \in \{0.1, 0.024, 0.008, 0.004\}$. The spacial resolution is $\Delta x = 0.02$. As initial condition, we use $u(x,0) = 0.9\,\exp\left(-150(x-\frac{1}{2})^4\right)$. For solving this problem using CLAWPACK we simply use the code from the CLAWPACK website [21, Chapter 17]. This code was written specifically to solve this stiff Burgers' problem.

The time evolution of the solution of (11) is shown in Fig. 6 in four snapshots at $t \in \{0.1, 0.2, 0.3, 0.4\}$. The thick grey graph shows the true solution. The solid dots denote the particle method. One can see that at $t = 0.1$ the solution is still in the transient phase, since characteristic particles are still visible on the (soon-to-be) detonation wave. At $t = 0.2$, the wave structure is almost converged. The plots at $t = 0.3$ and $t = 0.4$ show how the detonation wave catches up to the shock.

Figure 7 shows the solution at the final time $t = 0.4$, for four choices of $\tau \in \{0.1, 0.024, 0.008, 0.004\}$. The thick grey graph shows the true solution.

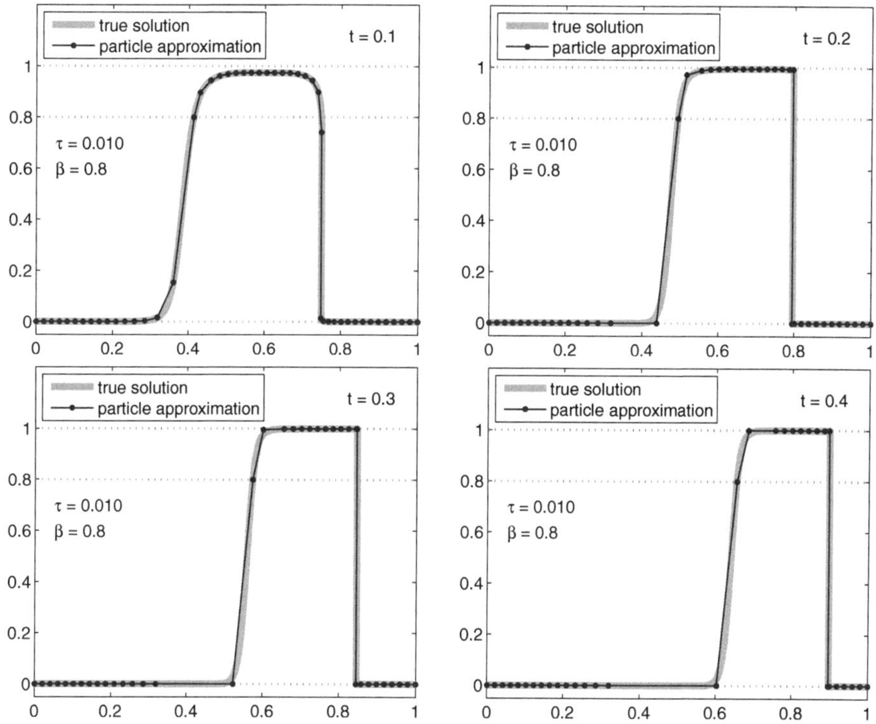

Fig. 6. Time evolution at $t \in \{0.1, 0.2, 0.3, 0.4\}$ of the advection reaction equation (11) with $\tau = 0.01$. The thick gray graph shows the true solution, while the dots denote the particle approximation.

The solid dots denote the particle method. The circles show the CLAWPACK results. Note that for $\tau = 0.1$, the solution is still in the transient phase, while for the other values, the detonation wave is comparably well established. For the selected resolution $\Delta x = 0.02$, CLAWPACK successfully captures the shock for $\tau \in \{0.1, 0.024\}$. The detonation wave for $\tau = 0.024$ is nicely represented as well. However, CLAWPACK clearly fails to resolve the shock and the detonation for $\tau = 0.004$. The intermediate value $\tau = 0.008$ is on the edge of failure. In comparison, the particle method works for all values of τ. The shock is optimally sharp, and the detonation wave moves at the correct velocity and has the correct width. The trouble that CLAWPACK is having with these equations is due to the stiff source. The problem is that the width of the shock is always $O(\Delta x)$, but this is too large when τ becomes small. The source is too active both in the detonation shock and in the regular forward-facing shock. This leads to incorrect shock speeds. Of course, CLAWPACK would resolve both fronts if a small enough grid ($\Delta x = O(\tau)$) were used, however, at a much larger computational expense.

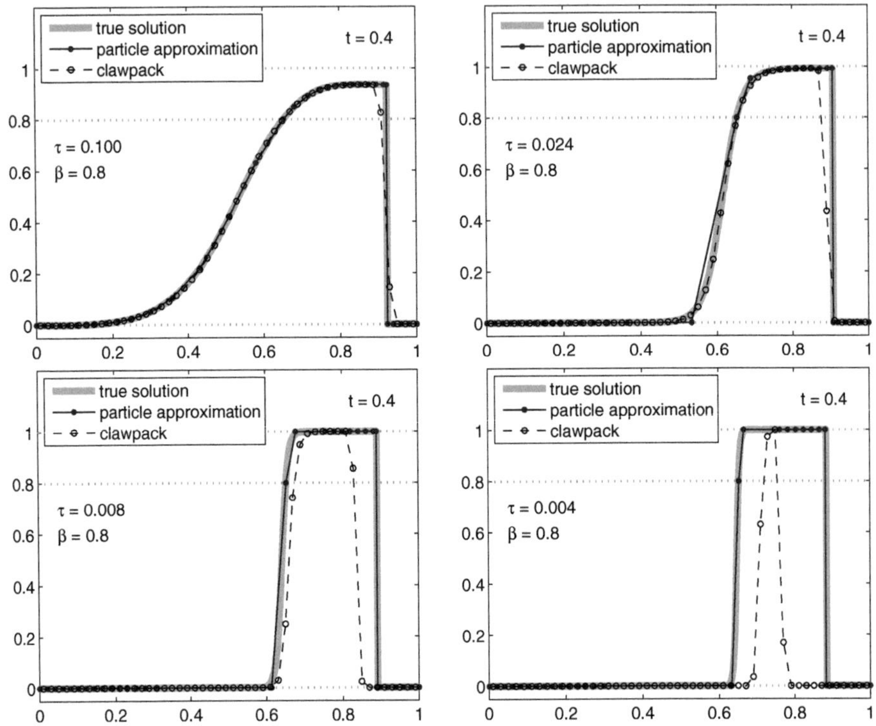

Fig. 7. Computational results for the advection reaction equation (11) with $\tau \in \{0.1, 0.024, 0.008, 0.004\}$. The thick grey graph shows the true solution, while the dots denote the particle approximation.

9 Conclusions and Outlook

We have presented a particle method that solves scalar one-dimensional hyperbolic conservation laws exactly, up to the accuracy of an ODE solver and up to errors in the approximation of the initial conditions. The numerical solution is defined everywhere. It is composed of local similarity solutions, separated by shocks. A numerical convergence analysis verified this accuracy claim for the flux function $f(u) = u^4/4$. In this example, the basic RK4 method yields solutions up to machine accuracy using a few hundred time steps. Since general initial conditions can be approximated with second order accuracy (see [10]), the overall method is at least second order accurate, even in the presence of shocks.

The method has also been extended to balance laws that describe stiff reaction kinetics. The tracking of a sonic particle in combination with a correction approach for neighboring particles yields a method that evolves detonation waves at correct velocities, without actually resolving their internal dynamics. The evolution of the sonic particle comes naturally in the considered particle method, while for classical fixed grid methods, a similar approach is much less natural. Numerical tests show that the particle method approximates the true solutions very well, even for fairly stiff systems, for which CLAWPACK fails due to an under-resolution of the wave and the shock.

The philosophy of the considered application in stiff reaction kinetics is that one can find efficient approaches for more complex problems by using the exact conservation law solver at the basis. It is the subject of current and future research to apply the same philosophy in other applications. Examples are nonlinear flows on networks. The presented particle method can be used to solve the actual evolution on each edge exactly. While an approximation has to be done at the network nodes, it is plausible to believe that this approach yields more accurate results that classical method that are far from exact on the edges themselves. Further generalizations to consider are the treatment of higher space dimensions using dimensional splitting, and systems of conservation/balance laws.

Acknowledgments

The authors would like to acknowledge the support by the National Science Foundation. Y. Farjoun was supported by NSF grant DMS–0703937, and by the Spanish Ministry of Science and Innovation under grant FIS2008-04921-C02-01. B. Seibold was partially supported by NSF grant DMS–0813648.

References

1. B. Bukiet, J. Pelesko, X. L. Li, and P. L. Sachdev, *A characteristic based numerical method for nonlinear wave equations*, Computers Math. Applic., 31 (1996), pp. 75–79.
2. *Clawpack*. Website. `http://www.clawpack.org`.
3. G. M. Coclite, M. Garavello, and B. Piccoli, *Traffic flow on a road network*, SIAM J. Math. Anal., 36 (2005), pp. 1862–1886.
4. R. Courant, K. Friedrichs, and H. Lewy, *Über die partiellen Differenzengleichungen der mathematischen Physik*, Mathematische Annalen, 100 (1928), pp. 32–74.
5. R. Courant, E. Isaacson, and M. Rees, *On the solution of nonlinear hyperbolic differential equations by finite differences*, Comm. Pure Appl. Math., 5 (1952), pp. 243–255.
6. J. R. Dormand and P. J. Prince, *Runge-Kutta triples*, Comp. Math. Appl., 12 (1986), pp. 1007–1017.
7. L. C. Evans, *Partial differential equations*, vol. 19 of Graduate Studies in Mathematics, American Mathematical Society, 1998.
8. H. Fan, S. Jin, and Z.-H. Teng, *Zero reaction limit for hyperbolic conservation laws with source terms*, J. Diff. Equations, 168 (2000), pp. 270–294.

9. Y. Farjoun and B. Seibold, *Solving one dimensional scalar conservation laws by particle management*, in Meshfree methods for Partial Differential Equations IV, M. Griebel and M. A. Schweitzer, eds., vol. 65 of Lecture Notes in Computational Science and Engineering, Springer, 2008, pp. 95–109.
10. ——, *An exactly conservative particle method for one dimensional scalar conservation laws*, J. Comput. Phys., 228 (2009), pp. 5298–5315.
11. ——, *A rarefaction-tracking method for conservation laws*, J. Eng. Math, 66 (2010), pp. 237–251.
12. W. Fickett and W. C. Davis, *Detonation*, Univ. of California Press, Berkeley, CA, 1979.
13. S. K. Godunov, *A difference scheme for the numerical computation of a discontinuous solution of the hydrodynamic equations*, Math. Sbornik, 47 (1959), pp. 271–306.
14. A. Harten, B. Engquist, S. Osher, and S. Chakravarthy, *Uniformly high order accurate essentially non-oscillatory schemes. III*, J. Comput. Phys., 71 (1987), pp. 231–303.
15. C. Helzel, R. J. LeVeque, and G. Warnecke, *A modified fractional step method for the accurate approximation of detonation waves*, SIAM J. Sci. Comput., 22 (2000), pp. 1489–1510.
16. M. Herty and A. Klar, *Modelling, simulation and optimization of traffic flow networks*, SIAM J. Sci. Comp., 25 (2003), pp. 1066–1087.
17. H. Holden, L. Holden, and R. Hegh-Krohn, *A numerical method for first order nonlinear scalar conservation laws in one dimension*, Comput. Math. Appl., 15 (1988), pp. 595–602.
18. H. Holden and N. H. Risebro, *A mathematical model of traffic flow on a network of unidirectional roads*, SIAM J. Math. Anal., 26 (1995), pp. 999–1017.
19. ——, *Front Tracking for Hyperbolic Conservation Laws*, Springer, 2002.
20. P. D. Lax and B. Wendroff, *Systems of conservation laws*, Commun. Pure Appl. Math., 13 (1960), pp. 217–237.
21. R. J. Le Veque, *Finite volume methods for hyperbolic problems*, Cambridge University Press, first ed., 2002.
22. M. J. Lighthill and G. B. Whitham, *On kinematic waves. II. A theory of traffic flow on long crowded roads*, vol. 229 of Proc. Roy. Soc. A, Piccadilly, London, 1955, pp. 317–345.
23. X.-D. Liu, S. Osher, and T. Chan, *Weighted essentially non-oscillatory schemes*, J. Comput. Phys., 115 (1994), pp. 200–212.
24. L. Shampine and M. W. Reichelt, *The MATLAB ODE suite*, SIAM J. Sci. Comput., 18 (1997), pp. 1–22.
25. B. van Leer, *Towards the ultimate conservative difference scheme II. Monotonicity and conservation combined in a second order scheme*, J. Comput. Phys., 14 (1974), pp. 361–370.

Application of Smoothed Particle Hydrodynamics to Structure Formation in Chemical Engineering

Franz Keller[1] and Ulrich Nieken[1]

Institute for Chemical Process Engineering, University of Stuttgart, Boeblinger Strasse 72, 70199 Stuttgart, Germany franz.keller@icvt.uni-stuttgart.de, ulrich.nieken@icvt.uni-stuttgart.de

Summary. In chemical engineering simulations the prediction of spatial distributions of concentrations, velocity, temperature and pressure fields in a specified environment are well established. Recently, the simulation of material structure formation gains increasing interest. In this context, the pore structure is an important material property for a large number of processes and products, ranging from heterogeneous catalysts and adsorbents to porous membranes or fibers. The goal of the present work is to describe the structure evolution and hence formation of a porous system by detailed modeling of the underlying physical and chemical processes. Presently, the development of such material relies almost completely on experimental experience, driving the need for simulation based design.

Since the described morphogenesis process is characterized by large deformation of heterogeneous material, evolving internal and external surfaces, coalescence of voids as well as fracture of material, local chemical reactions and phase changes, the treatment with classical grid-based techniques is difficult. In our opinion, meshfree methods are better suited for the stated task, and therefore (incompressible) Smoothed Particle Hydrodynamics is applied in the following work.

In the first part of the contribution, the basic chemical and physical processes are validated by simple test cases. One focus lies on modeling the visco-elastic and visco-plastic material behavior, and respective test cases are presented. Since the accurate treatment of free surfaces is decisive for the stated problem, its evolution is also validated by a test case. Lastly, a model for the inclusion of chemical reactions and phase change in the scope of pore forming is presented. In the second part, the first results of a simple pore forming process are shown to indicate the feasibility of our approach.

Key words: smoothed particle hydrodynamics, open-porous structure formation, non-Newtonian material.

M. Griebel, M.A. Schweitzer (eds.), *Meshfree Methods for Partial Differential Equations V*, Lecture Notes in Computational Science and Engineering 79,
DOI 10.1007/978-3-642-16229-9_8,

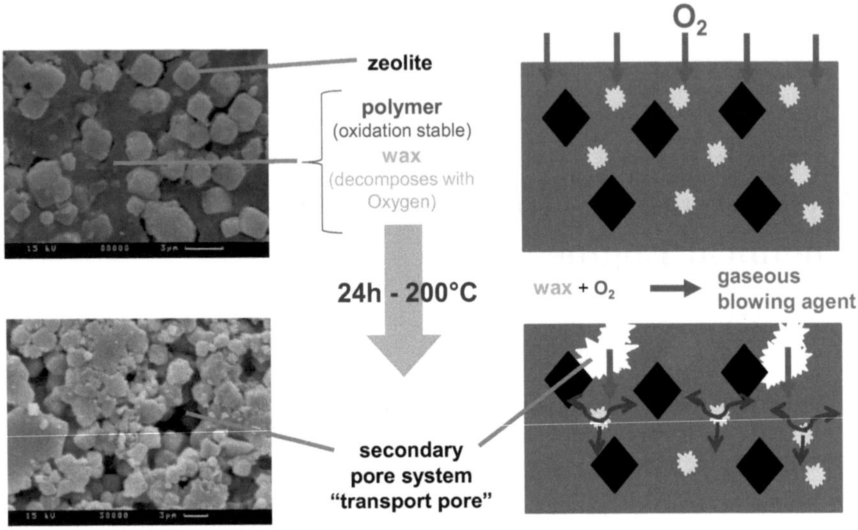

Fig. 1. Left: SEM images of adsorbent monolith before and after thermal treatment, Right: Schematic representation of pore formation process.

1 Introduction

In the field of chemical engineering modeling and simulation is mainly used to describe the macroscopic transport of matter and energy in equipments and processes. Nowadays the simulation of material structure formation with specific properties gains increasing interest. Especially the porous structure is a key parameter for many processes and products. Furthermore, open-porous materials are widely used in many applications. Examples thereof are heterogeneous catalysts and adsorbents, where a high specific internal surface with suitable access to the active sites is needed. In other applications, the pore structure is needed for the selective separation by steric effects or surface forces. A well-known example are hollow fiber membranes, which are used as artificial kidney.
For the generation of these porous structures, several manufacturing processes exist. But despite the widespread use of porous materials, these manufacturing processes rely almost completely on experimental experience and empirical correlations [14]. A model based support of the manufacturing process by prediction of the resulting material structure characteristics is desired. However these kind of simulations are difficult, since the morphogenesis is characterized by large material deformation, high density differences between pores and matrix, fracture of thin material bridges, coalescence of voids and formation of internal and external surfaces. So far only simplified models exist, making the prediction of the resulting pore structure impossible. In the following the generation of a secondary pore structure for an adsorbent monolith may serve as

an example of the formation of an open-porous structure. Before delving into the computational details, we briefly explain the manufacturing process. The regarded monolithic composites consist in general of an adsorbents and a supporting material, e.g. ceramics or polymer backbone. The adsorbent monolith considered here was developed in cooperation with the Institute for Polymer Technology of the University of Stuttgart [1], and is composed of a polymer backbone, zeolite particles as active components and a oxidizable wax. The monolith is manufactured in a two-stage process. In the first stage the monolithic structure of parallel channels of square cross-section is extruded from a suspension of polyamide, wax and a fine zeolite powder. The added wax adjusts the viscosity of the melt, thereby improving the extrusion process. However, the micro-mixing of polymer and wax during the extrusion process is far from being perfect, leaving small wax islands in the polymer matrix. After extrusion, the zeolite particles, which are the active sites, are completely embedded in the matrix. In the second manufacturing step the low molecular wax is removed by thermal decomposition at about 200° C. Together with atmospheric oxygen, the wax decomposes into a blowing agent, forming and widening pores, which result in a secondary pore system. The generation of these transport pores in the polyamide matrix is crucial for achieving high adsorption kinetics, which is essential for e.g. the application in fast pressure-swing adsorption processes, while maintaining high adsorption capacity, high mechanical stability and good embedding of zeolite particles in the polyamide matrix [2].

For modeling the presented process, which is schematically shown in Figure 1, the following processes have to be included in the detailed simulation. First, oxygen diffuses through the matrix material and, upon reaching the embedded wax islands, the oxygen reacts with wax and forms the gaseous blowing agent. During oxidation of the wax, the pressure build-up in the emerging pore by the formation of blowing agent and the resulting deformation of the surrounding heterogeneous material have to be included. The shortening of the diffusion path through the evolving pores also needs to be considered. Furthermore, the coalescence of voids needs to be taken into account as well as the evolution of internal and external surfaces. As a last step to the formation to an open-porous material, the fracture of the polymer matrix has to be modeled. To conclude, the combination of the stated processes and properties is problematic for the simulation with mesh-based methods. The evolution of external and internal surfaces, as well as large deformations of heterogeneous material and its fracture are just an excerpt of the arising difficulties. This is why most of the published simulations refer to explicitly defined structure forming processes, like the formation of close-cell foams with an initially given germs distribution in the substrate. Coupez et al. simulated the expansion of a closed cell foam structure with unstructured finite elements. For the tracking of the interface during bubble growth, Volume-of-Fluid [17] as well as Level-set methods [16] were used. Due to the additional computational costs, the simulation is limited to rather small numbers of bubbles. Furthermore, the interface has to be remeshed to prevent numerical oscillations. The bubble growth in a Newtonian fluid was also simulated by Thuerey using the Lattice Boltzmann method and an Volume-of-fluid like approach to track the interface by introducing additional interface cells [18].

As stated above, meshless methods in general are better suited for this task, since no remeshing of the domain, e.g. due to large material deformation, is needed and the relative ease of handling material fracture [15]. Additionally, by using particle methods, no explicit interface tracking or capturing technique is needed, making the simulation of heterogeneous materials and multi-phase systems possible at relative low computational costs. In the present work the particle method Smoothed Particle Hydrodynamics (SPH) is employed to simulate the overall process and therefore a brief overview of the method is given in the following section.

2 Smoothed Particle Hydrodynamics Method

As mentioned in the introduction, grid-based methods are unsuited for modeling of the morphogenesis of an open-porous material and therefore the meshless method SPH will be used in the present study. However, only a brief sketch of the concept of SPH is presented. Several review articles are available, which provide a more complete description of the practical and theoretical aspects of SPH [3], [4]. In the following a variant of SPH called incompressible SPH (ISPH) is used [5]. In order to guarantee material incompressibility, a Poisson equation for the pressure is solved. The semi-implicit time stepping is derived from the Moving Particle Semi-Implicit Method [6].

2.1 Governing equations

Subsequently, the continuity equation for an incompressible medium is depicted,

$$\frac{D\rho}{Dt} = -\rho \nabla \cdot \boldsymbol{v} \tag{1}$$

with ρ being the material density and $\boldsymbol{v}$ the velocity. The momentum conservation equation for an arbitrary material behavior reads as follows:

$$\rho \frac{D\boldsymbol{v}}{Dt} = \nabla p + \nabla \cdot \boldsymbol{\tau} + \boldsymbol{f} \tag{2}$$

where the pressure term is denoted with p, the stress tensor with $\boldsymbol{\tau}$ and external forces with $\boldsymbol{f}$.
The material balance applicable for the molecular transport of oxygen in the polymer matrix and pore system is stated below:

$$\frac{Dc_j}{Dt} = \nabla\left(\mathfrak{D}_j \nabla c_j\right) + \sum_i \nu_{ij} r_i \tag{3}$$

with c_j being the concentration of component j, $\mathfrak{D}_j$ its diffusion coefficient and r_i the reaction rate with ν_{ij} being the stoichiometric coefficient of component j in reaction i. In the next paragraph, the discretization of the governing equations via SPH as well as the solution procedure are shown.

2.2 Smoothed Particle Hydrodynamics

SPH discretization

In the following paragraph, the basic concepts of the SPH discretization are shown as well as the pseudocode of the ISPH algorithm. The basic equation of SPH is the approximation of variable A, which is a function of space, based on an integral interpolant of the form:

$$\langle A(\boldsymbol{x})\rangle = \int A(\boldsymbol{x}')\, w(\boldsymbol{x}-\boldsymbol{x}')\, d\boldsymbol{x}' \tag{4}$$

where w is the kernel and $d\boldsymbol{x}'$ is the differential volume element. By approximating the integral with a summation over the neighboring particles, the typical SPH summation interpolant is derived:

$$\langle A(\boldsymbol{x})\rangle = \sum_j m_j \frac{A_j}{\rho_j} w(\boldsymbol{x}-\boldsymbol{x}_j) \tag{5}$$

with m_j being the particle mass and ρ_j its density. As long as the kernel function is a differentiable function, derivatives can be computed easily. The derivative of variable A with respect to $\boldsymbol{x}$ is stated as follows:

$$\left\langle \frac{\partial A(\boldsymbol{x})}{\partial \boldsymbol{r}} \right\rangle = -\sum_j \frac{m_j}{\rho_j} A_j \frac{\partial w_{ij}}{\partial \boldsymbol{x}_j} \tag{6}$$

However, this form of the derivative does not vanish for constant A and several improved versions exist in the literature [3].

Kernel function

Several restrictions exist for the choice of the SPH kernel function, which can be found in the literature [3]. In the present work, a cubic-spline kernel was used. It can be seen as the "standard" kernel due to its balance in accuracy and computational efficiency [4]. The kernel is a function of particle spacing r_{ij} and the smoothing length r_e.

$$w(r_{ij}) = w_0 \begin{cases} 4\frac{r_{ij}}{r_e}^3 - 4\frac{r_{ij}}{r_e}^2 + \frac{2}{3} & 0 \le r_{ij} < \frac{r_e}{2} \\ \frac{1}{6}\left(2 - \frac{2r_{ij}}{re}\right)^3 & \frac{r_e}{2} \le r_{ij} < r_e \\ 0 & r_{ij} \ge r_e \end{cases}$$

In two dimensions the normalization factor w_0 is $\frac{15}{7\pi}$ to guarantee the kernel restriction $\int w(r_{ij})\, d\boldsymbol{x} = 1$.

Incompressible SPH algorithm

As stated in section 2, the used algorithm is based on a predictor-corrector scheme similar to the PISO-algorithm used in grid-based methods [12]. The idea was first applied to particle methods by Koshizuka et al. in the method Moving Particle Semi-implicit [6]. The approach used in this work is based on these ideas, and the pseudocode is depicted in Figure 2. In the first step, the particle velocity due to inter-particle (e.g. viscous) and external forces are calculated by means of an explicit integration step. The particles are moved according to the obtained velocity. After movement, the incompressibility conditions are not valid any more. In the second step, the pressure is calculated implicitly to ensure incompressibility. Based on the obtained pressure distribution, the particle velocities and positions are corrected.

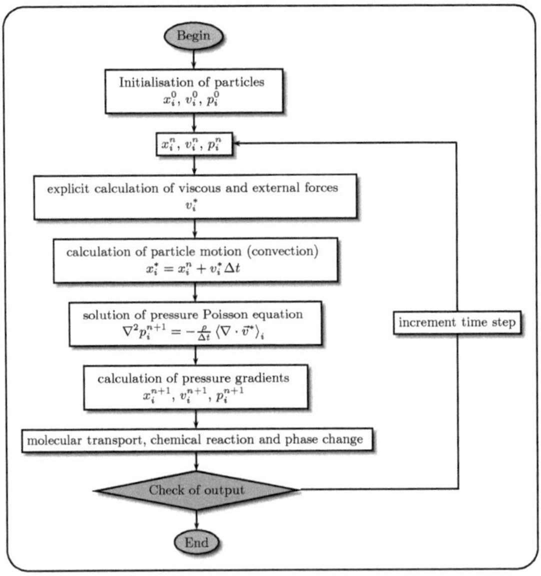

Fig. 2. Pseudocode Incompressible SPH.

3 Validation of Single Processes

As mentioned in section 1, the final pore structure is dependent on several underlying physical and chemical processes as well as material properties. In order to achieve a quantitative prediction of the resulting structure, the description of each of these processes has to be validated in advance. In the following subsections, the validation of selected processes will be presented.

Validation of material models

For the validation of the implemented material models, the Poiseuille flow was chosen as a test case. The channel walls are represented by particles fixed in space and the no-slip velocity approach of Morris in [7] is used at the solid boundary. Since the boundary particles are included in the weighting process, the wall depth should be at least as high as the kernel smoothing length. By reflecting the velocities of the near wall fluid particles, the velocity of the boundary particles v_W is obtained via the following extrapolation

$$\boldsymbol{v}_W = -\frac{d_W}{d_F}\boldsymbol{v}_F \tag{7}$$

with $\boldsymbol{v}_F$ being the fluid velocity, d_W the normal distance of the boundary particle to the wall and d_F the distance of the fluid, respectively. The boundary velocities are used for the evaluation of the velocity gradients near the wall. In the direction of flow, periodic boundary conditions are used.

Newtonian fluid

For validation purposes, the transient behavior of a Newtonian fluid is examined. Since the stress is only dependent on the velocity gradient, no complications with the chosen no-slip boundary conditions are expected. Initially, the fluid is at rest and after $t > 0\,s$ a body force of $F = 10^{-4}\,N$ acts on all fluid particles in the direction of flow. The density of the fluid particles is set to $\rho = 1\,\frac{kg}{m^3}$ and the dynamic fluid viscosity to $\eta = 10\,Pa \cdot s$. The velocity at different times is depicted over the dimensionless channel height. Satisfactory agreement between the ISPH code and the analytical solution is obtained and shown in Figure 3.

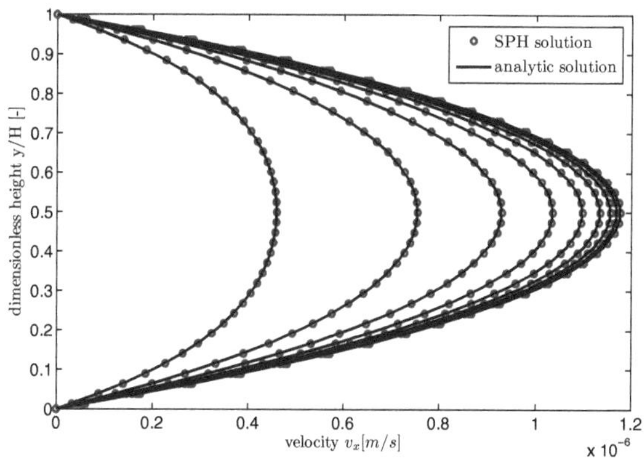

Fig. 3. Velocity profile of a Newtonian fluid at different times with time steps of $\Delta t = 5 \cdot 10^{-3}\,s$ until steady state.

Visco-elastic fluid

In order to show the limitations of the current no-slip velocity boundary model, the Poiseuille flow test case for a visco-elastic material is presented in the following paragraph. The so called Oldroyd-B-Fluid model is capable of describing the polymer behavior at least on a qualitative basis. In contrast to the Newtonian fluid, the stress tensor,

$$\tau^{\alpha\beta} = \frac{\lambda_2}{\lambda_1}\eta\left(\nabla^{\beta}v^{\alpha} + \nabla^{\alpha}v^{\beta}\right) + S^{\alpha\beta} \tag{8}$$

is a function of the velocity gradient and a time dependent elastic contribution $S^{\alpha\beta}$ is added, reading:

$$\frac{D\,S^{\alpha\beta}}{Dt} = \nabla^{\gamma}v^{\alpha}S^{\gamma\beta} + \nabla^{\gamma}v^{\beta}S^{\alpha\gamma} - \frac{1}{\lambda_1}S^{\alpha\beta} + \frac{\eta}{\lambda_1}\left(1 - \frac{\lambda_2}{\lambda_1}\right)\left(\nabla^{\beta}v^{\alpha} + \nabla^{\alpha}v^{\beta}\right) \tag{9}$$

Besides fluid viscosity, two further parameters, the relaxation time λ_1 and retardation time λ_2 have to be specified. For validation, the retardation time is set $\lambda_2 = 0$ and the model reduces to the Upper-Convected-Maxwell (UCM) model. Because of the undamped elastic contribution, the UCM model can be considered as a challenging test case. The transient velocity profiles over the channel height are depicted in Figure 4 and 5 at different times indicated in the graphs. The density of the fluid particles is set to $1\,\frac{kg}{m^3}$, the dynamic fluid viscosity to $10\,Pa \cdot s$ and the relaxation time to $\lambda_1 = 0.1\,s$, with a force of $F = 10^{-6}\,N$ acting on each particle. With increasing times the SPH solution

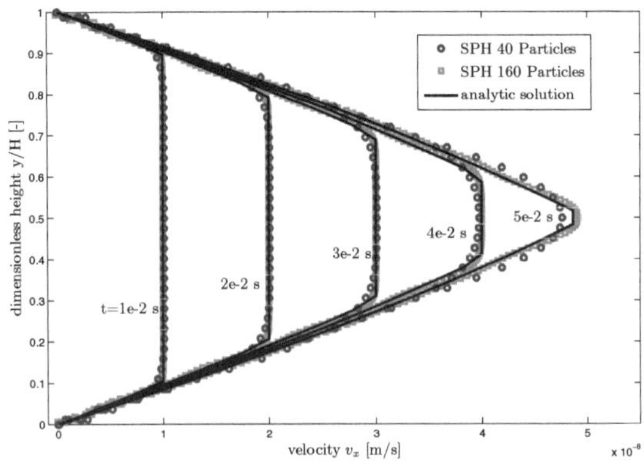

Fig. 4. Velocity profile of a UCM fluid at different times.

deviates slightly from the analytical solution. One reason for this behavior is the insufficient resolution, especially at the edges of the sharp velocity profile. As shown in Figure 4, a more accurate solution can be obtained by increasing

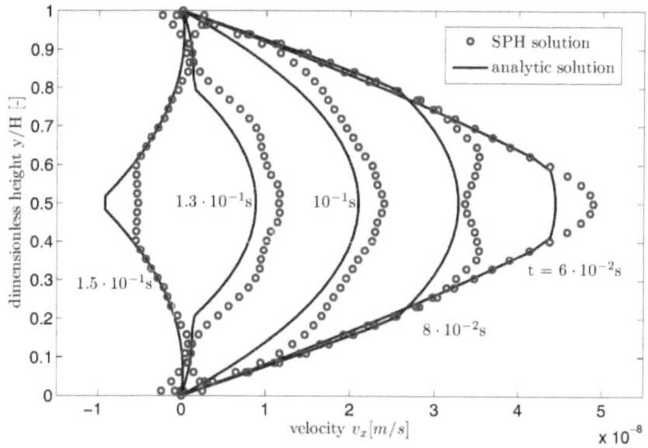

Fig. 5. Velocity profile of a UCM fluid at different times with 40 Particles.

the number of particles. Another reason for the deviation can be accounted to the treatment of the no-slip velocity near the solid wall. The technique stated in the paragraph above is not suited to handle visco-elastic flow, which can be seen especially in Figure 5, due to the time-dependence of the stress tensor. Improvements can be found in the literature [11], [10]. If looking at the transient fluid velocity in the center of the channel as displayed in Figure 6, the results are satisfactory. Again an improvement in the solution is seen by increasing the number of partilces. The L_∞-error with 40 Particles distributed over the channel height compared to the analytical solution in the steady state is 6% and for 80 Partilces 4%. Based on these results, the deviations are in

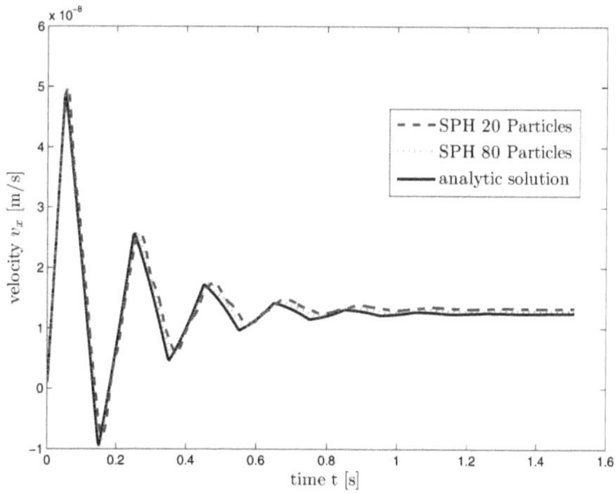

Fig. 6. Velocity of a UCM fluid in the center of the channel versus time.

large parts accounted for by the solid boundary. And since no solid boundaries occur in our application, no further improvements in the material models are needed.

To conclude, the visco-elastic material model shows satisfactory agreement in comparison with the analytical solution, while excellent consistency is obtained for the viscous case. A similar accuracy was obtained for the other material models not presented in this study (linear elastic solid, visco-plastic material).

Validation of free surfaces

As mentioned in section 1, structure formation is governed by the evolution of internal and external interfaces. For validation the dam break test case is presented and the evolution of the leading front of the collapsing water is compared to experimental data. The fluid is treated as slightly viscous ($\eta = 10^{-6}\, Pa \cdot s$) and the wall is represented by fixed particles. The depth of the wall is comparable to the smoothing length. However, only the inner boundary particles are included in the implicit density correction step. According to [13], "ghost particles" are used in the interaction of all particles identified as being on the surface. The ghost particles possess the same pressure as their origin particles, but their positions are obtained by reflection of the surface particles normal to the surface. Initially, the fluid column is at rest and for $t > 0\, s$ gravity is acting on all particles. The evolution of the water front is depicted in the following Figure 7. A stable and smooth surface is obtained by using the divergence of the velocity field as right hand side in the Poisson equation and "ghost particles" as described above. For validation, the leading edge of the water front is shown together with experimental data ([8], [9]) in Figure 8 until the front reaches the opposing wall. Satisfactory agreement has been achieved.

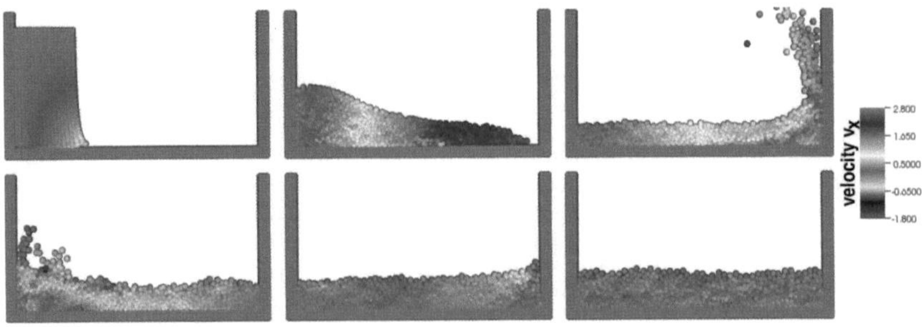

Fig. 7. Collapsing water column at $t = 5 \cdot 10^{-2}$, 0.25, 0.5, 1.15, 1.75 and $10\, s$.

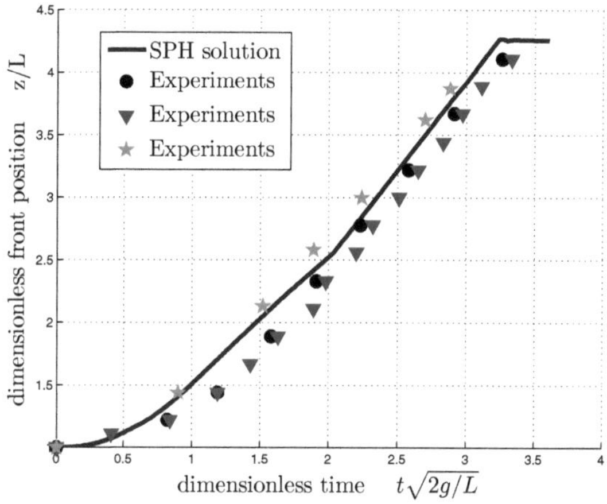

Fig. 8. Dimensionless leading edge of the collapsing water column over dimensionless time in comparison with experimental data. Experimental data taken form [8] (first two data sets) and [9] (third data set).

Validation of molecular transport

As indicated in section 1 above, the molecular transport of heat and especially of mass in the solid matrix and evolving pore system is decisive for the resulting structure. Therefore, a test case for molecular transport in homogeneous materials is presented in the following section as well as the treatment of molecular transport in multi-phase systems.
In the presented Figure 9, the cooling of a quadratic slab with an initially sinusoidal temperature distribution is compared to an analytic solution. The left and right boundary is isothermal, while the upper and lower boundary can be considered as adiabatic. The maximum temperature is set to $T_{max} = 20^oC$ with a thermal diffusivity of $\alpha = 0.1\,\frac{m^2}{s}$. As one can see, satisfactory agreement is achieved. The molecular transport in multi-phase systems, here the diffusion of oxygen in the polymer matrix and in the evolving porous system, is modeled as follows. The molecular transport in the pore system is modeled on a grid, represented by spatially fixed particles. The fixed particles are overlayed with SPH particles of the pore matrix. The fixed particles are deactivated if superimposed by matrix particles and vice versa. Molecular transport is modeled as shown above in both phases. Heat and mass transfer between the phases is modeled by linear driving forces. In the case of mass transport between gaseous and polymer phases, the exchange term $\dot{n}_j^k$ reads as follows:

$$\dot{n}_j^k = \beta\,(p_j - \mathfrak{H}_j \cdot c_j) \tag{10}$$

with p_j being the partial pressure of component j in the gas phase, $\mathfrak{H}_j$ the Henry constant of component j and c_j its concentration in the polymer phase.

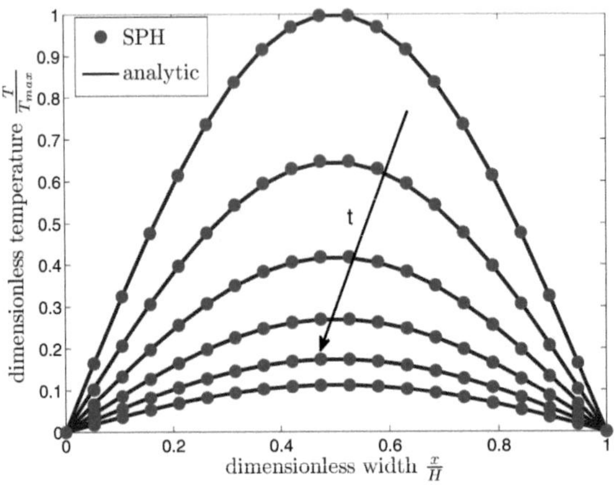

Fig. 9. Temperature distribution in quadratic slab at increasing times beginning at $t = 0\,s$ with $dt = 10^{-4}\,s$.

Furthermore, β is used as an coefficient to adjust the relaxation of the mass exchange to equilibrium.

After validating most of the relevant processes for the considered pore forming process, the inclusion of chemical reactions as well as the phase change model are presented in the context of the overall process in the following section.

4 Simulation of the overall process

After validation of the single processes at simple test cases, the individual processes are combined to model the morphogenesis of an open-porous material. Therefore, the computational domain is presented first.

Computational domain

The computational domain is shown in Figure 10 on the left. The domain consists of polymer (red particle), wax (yellow particle) and zeolite particles (black). The polymer and wax behaves like a visco-plastic Bingham material described by the Cross model as follows:

$$\eta_{eff} = \frac{\eta_0 + (K\dot{\gamma})^m \eta_\infty}{1 + (K\dot{\gamma})^m} \tag{11}$$

with η_{eff} being the effective viscosity, depending on the shear rate $\dot{\gamma}$ and consisting of the zero shear viscosity η_0 as well as infinite shear viscosity η_∞. The parameter K depicts the Cross time constant and m is known as Cross rate constant.
Zeolites are modeled as a linear elastic material, according to

$$\frac{D\,\tau^{\alpha\beta}}{D\,t} = 2\mu\left(\dot{\epsilon}^{\alpha\beta} - \delta^{\alpha\beta}\frac{1}{3}\dot{\epsilon}^{\gamma\gamma}\right) + \tau^{\alpha\gamma}\Omega^{\beta\gamma} + \Omega^{\alpha\gamma}\tau^{\gamma\beta} \tag{12}$$

with $\dot{\boldsymbol{\epsilon}}$ being the strain rate and $\boldsymbol{\Omega}$ being the rotation tensor.
Mass and heat transport are modeled as described in the paragraph above on the underlying grid depicted by blue diamonds in Figure 10.

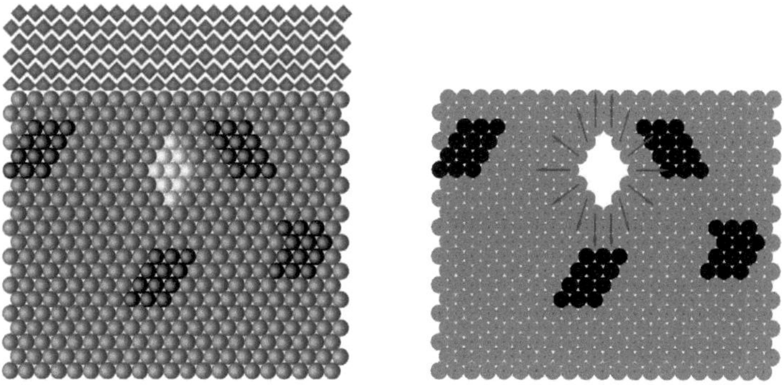

Fig. 10. Left: Computational domain of polymer-zeolite adsorber; Red: visco-plastic polymer particles; Black: zeolite particles; Yellow: visco-plastic wax particles, Blue: fixed particles for modeling molecular transport in pores and surrounding area. Right: Force vector acting on pore surface.

Chemical reaction and phase change

For modeling the overall process, the decomposition of the wax and the formation of the blowing agent has to be included into the simulation. A simple oxidation reaction is assumed to occur in the wax phase only, with the methene group ($-CH_2-$) being the representative building block. The mass balance for the wax particles is formulated as follows,

$$\frac{d\,m_i}{d\,t} = -k\,(T)\cdot c_{O_2,i}\cdot V_i\cdot MW_i \tag{13}$$

with m_i being the mass of the wax particles, $k\,(T)$ the temperature dependent reaction rate, V_i the particle volume and MW_i the molecular weight of the representative building block. Furthermore, the oxygen concentration $c_{O_2,i}$ of the regarded particle i needs to be included. As soon as the mass of the particle reaches zero, the particle is removed. The mass of the particle corresponds to the mass of the formed blowing agent. The number of moles of the formed blowing agent can be calculated and by knowing the volume of the pore, the pressure inside the pore can be obtained through the equation of state. In the next step, the normal vector on the pore surface $\boldsymbol{n}_i$ can be estimated by

$$\boldsymbol{n}_i = \frac{1}{|\boldsymbol{n}|}\sum_{j,Pore}(col_j - col_i)\frac{\partial w_{ij}^{\alpha}}{\partial dx^{\alpha}} \qquad \text{with} \quad col_i = \sum_{j,Pore} w_{ij} \tag{14}$$

with using the color function on the underlying fixed particles. The color function is evaluated for unoccupied particles and neighbors only. After knowing the pressure in the pore and the normal vector on the surface, the force vector on the matrix surface can be estimated by the summation over the pore particles with A_{ij} being the mean of the surface areas of both particles as follows:

$$\boldsymbol{F}_i = \frac{\sum_{j,pore} p_j \cdot w_{ij} \cdot A_{ij}}{\sum_{j,pore} w_{ij}} \cdot \boldsymbol{n}_i \tag{15}$$

As shown in Figure 10 on the right, an isotropic force distribution can be achieved at least for sufficiently large pores. The obtained force vector acts as an external force on the SPH particles and is included in the explicit convection step as shown in Figure 2.

Results

In the following paragraph the proof of concept for the modeling of the formation of an open-porous structure is shown for the geometry presented in Figure 10. The evolution from the compound substrate to the open-porous system is shown in the following Figure 11 from the upper left to the lower right picture. The color of the elements represent the norm of the velocity vector. In the first picture, the oxygen already diffused into the polymer matrix and the first wax particles are decomposed. In the second picture, the wax is completely decomposed and the shear rate in the upper left corner exceeds the visco-plastic yield stress due to the internal pressure on the surface. After exceeding the yield stress, the polymer begins to flow, shown in the third picture. In picture four the plastic deformation continues until the pressure compensation with the outside takes place. The final geometry is shown in subfigure five. Since the crack in the matrix can not be seen in subfigure five, the underlying grid at that time is depicted in picture six. Red diamonds denote occupied grid points, while blue ones illustrate unoccupied spaces, respectively pores. One can clearly see the pathway from the inside of the pore to the surrounding and hence the formed open-porous system.

5 Conclusion and Outlook

In the presented work the detailed simulation of the morphogenesis of a viscous substrate to an open-porous material was depicted. In a first step the decisive single processes have been validated in order to guarantee the simulation of the overall process on a quantitative level. Based on the validated single processes, the overall process was modeled and simulated as a proof of concept.
As a next step, the simulation will be extended to three dimensions and by using realistic material parameters, the validation against simple experiments will be possible.

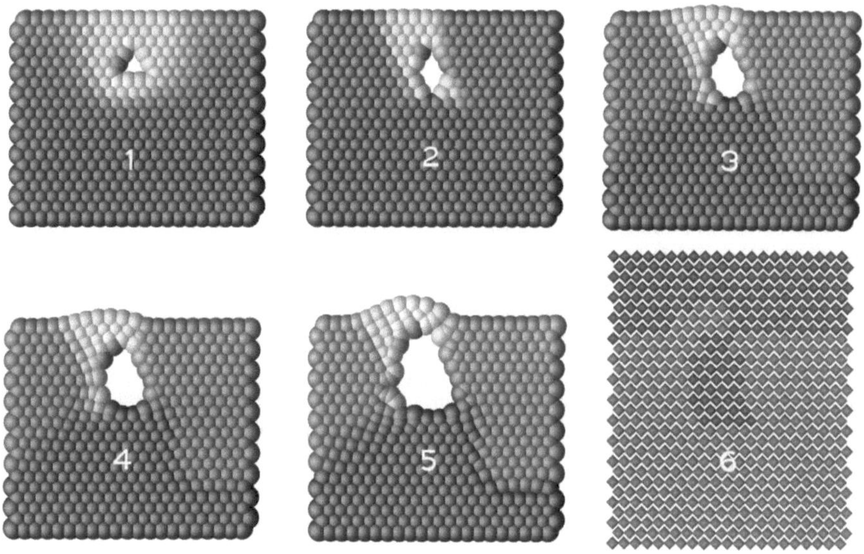

Fig. 11. Subfigure 1-5: Simulation of the overall process from the substrate compound to the open-porous material. The color coding represents the norm of the velocity vector. Subfigure 6: Sketch of the underlying grid after formation of the open-porous system (red diamonds represent occupied grid points, blue diamonds stand for void spaces).

6 Acknowledgments

The authors would like to acknowledge the grant Ni932/6-1 from the German Research Foundation (Deutsche Forschungsgemeinschaft).

References

1. H.-G. Fritz and J. Hammer, *Aufbereitung zeolitischer Formmassen und ihre Ausformung zu Adsorptionsformteilen*, Chemie Ingenieur und Technik **77** (2005), 1587-1600.
2. A. Gorbach, M. Stegmaier, J. Hammer and H.-G. Fritz, *Compact Pressure Swing Adsorption-Impact and Potential of New-type Adsorbent-Polymer Monoliths*, Adsorption **11**, (2005), 515-520.
3. J. Monaghan, *Smoothed Particle Hydrodynamics*, Reports on Progress in Physics, **68**, (2005).
4. S. Rosswog, *Astrophysical Smooth Particle Hydrodynamics*, New Astronomy Reviews **53**, (2009), 78-104.
5. S. Shao, E. Lo, *Incompressible SPH Method for Simulating Newtonian and non-Newtonian Flows with a Free Surface*, Advances in Water Resources **26**, (2003), 787-800.

6. S. Koshizuka, A. Nobe, Y. Oka, *Moving Particle Semi-implicit Method for Fragmenation of Incompressible Fluid*, Nuclear Science and Engineering **123**, (1996), 421–434.
7. J.P. Morris, P.J. Fox and Y. Zhu, *Modeling low Reynolds Number incompressible flows using SPH*, Journal of Computational Physics **136**, (1997), 214–226.
8. J.C. Martin and W.J. Moyce, *An Experimental Study of the Collapse of Liquid Columns on an rigid horizontal plane*, Philosophical Transactions of the Royal Society of London Series A, **244**, (1952), 312–324.
9. C.W. Hirt and B.D. Nichols, *Volume of Fluid Method for the Dynamics of Free Boundaries*, Journal of Computational Physics, **39**, (1981), 201–225.
10. J. Fang, R. Owens, L. Tacher and A. Parriaux, *A Numerical Study of the SPH Method for Simulating Transient Viscoelastic Free Surface Flows*, Journal of Non-Newtonian Fluid Mechanics, **139**, (2006), 68–84.
11. M. Ellero and R.I. Tanner, *SPH Simulations of Transient Viscoelastic Flows at Low Reynolds Number*, Journal of Non-Newtonian Fluid Mechanics, **132**, (2005), 61–72.
12. R.I. Issa, *Solution of the Implicit Discretized Fluid Flow Equations by Operator Splitting*, Mechanical Engineering Rept. FS/82/15, lmperical College. London, 1982.
13. B. Ataie-Ashtiani, G. Shobeyri and L. Farhadi, *Modified Incompressible SPH Method for Simulating Free Surface Problems*, Fluid Dynamics Research, **40**, 9, (2008) 637–661.
14. K. Ishizaki, M. Nanko and S. Komarneni, *Porous Materials: Process Technology and Applications*, Springer, Berlin, 1998.
15. S. Li and W.K. Liu, *Meshfree Particle Methods*, Springer, Berlin, 2004.
16. J. Bikard, J. Bruchon, T. Coupez and B. Vergnes, *Numerical Prediction of the Foam Structure of Polymeric Materials by Direct 3D Simulation of their Expansion by Chemical Reaction based on a Multidomain Method*, Journal of Material Science, **40**,(2005), 5875–5881.
17. J. Bruchon, A. Fortin, M. Bousmina and K. Benmoussa, *Direct 2D Simulation of Small Gas Bubble Clusters: From the Expansion Step to the Equilibrium State*, International Journal for Numerical Methods in Fluids, **54**, (2007), 73–101.
18. N. Thuerey, *Phyiscally based Animation of Free Surface Flows with the Lattice Boltzmann Method*, PhD thesis University of Erlangen-Nuernberg,(2007)

Numerical validation of a constraints-based multiscale simulation method for solids

Konstantin Fackeldey[1], Dorian Krause[2], and Rolf Krause[3]

[1] Konstantin Fackeldey, Associate Member of Institute of Computational Science, Universita della Svizzera Italiana `konstantin.fackeldey@usi.ch`
[2] Dorian Krause, Institute of Computational Science, Universita della Svizzera Italiana `dorian.krause@usi.ch`
[3] Rolf Krause, Institute of Computational Science, Universita della Svizzera Italiana `rolf.krause@usi.ch`

Summary. We present numerical validation studies for a concurrent multiscale method designed to combine molecular dynamics and finite element analysis targeting the simulation of solids. The method is based on an overlapping domain-decomposition and uses weak matching constraints to enforce matching between the finite element displacement field and the projection of the molecular dynamics displacement field on the mesh. A comparison between our method and the well-known bridging domain method by Xiao and Belytschko [22] is presented. As part of our validation study we discuss applicability of the method to the simulation of fracture propagation and show results.

Key words: molecular dynamics, multiscale, finite elements, weak coupling

1 Introduction

For an efficient and accurate modeling of material properties it seems favorable to separate the modeling problem into at least two different scales and models, such that the accuracy of a fine scale model can be combined with the advantages of a computational efficient model. In previous work [13, 17] we have developed a multiscale method for the coupling of molecular dynamics as a micro- (or fine-scale) model and continuum based finite elements as the macro- (coarse-scale) model. Concurrent multiscale methods of this type can be applied e.g. to the study of fracture processes where non-smooth displacement fields in the crack tip region can only be represented in a fine-scale model without additional modeling efforts. In this paper we present new results from validation studies for the method and compare the method with the state-of-the-art bridging domain method from literature. As part of our validation studies we present an application to the above mentioned field of fracture mechanics.

M. Griebel, M.A. Schweitzer (eds.), *Meshfree Methods for Partial Differential Equations V*, Lecture Notes in Computational Science and Engineering 79, DOI 10.1007/978-3-642-16229-9_9,

2 Coupling with projection-based constraints

In this section we present the two scales that are involved in the concurrent multiscale simulations aimed at and discuss our approach to manage the information transfer between these scales.

2.1 Molecular Dynamics

To simulate material behavior on the atomistic level where critical phenomena connected to nonsmooth effects and oscillating behavior, such as growing crack fronts or line dislocations in a crystalline material, might occur, molecular dynamics (MD) [1,5] is a widely used approach. A system of N particles (for simplicity with unit masses) evolves according to the Newtonian equations

$$\ddot{\mathbf{q}}_i = -\nabla_{\mathbf{q}_i} V + \mathbf{f}_i^{\text{ext}}, \quad \text{for } i = 1, \dots, N, \tag{1}$$

with the interaction potential $V = V(\mathbf{q}_1, \dots, \mathbf{q}_N)$. For reasons of computational efficiency, a popular choice for the potential V in the qualitative study of solids is the Lennard-Jones potential

$$V(\mathbf{q}_1, \dots, \mathbf{q}_N) = \sum_{i=1,\, j>i}^{N} v\left(|\mathbf{q}_i - \mathbf{q}_j|\right), \quad v(r) = 4\left(r^{-12} - r^{-6}\right).$$

It is well known that the applicability of MD studies to larger systems of interest is often prohibited by the intrinsic restrictions on time- and length-scale. Resolving atomistic vibrations using an explicit time integration scheme requires time steps on the order of femtoseconds. Moreover, the number of nuclei even in nano- or micro scale specimens is very high. Therefore simulations are often carried out with smaller sized systems (resulting in finite-size effects, cf. [8]) and under extreme conditions (e.g. highly amplified external strains or forces).

2.2 Multiscale Coupling

To overcome the above mentioned restrictions of atomistic simulation techniques, many researchers have investigated approaches to accelerate MD.

Hierarchical coarse-graining, where atoms are grouped together into beads which interact by a coarse-grained force field, have been applied successfully in protein and polymer sciences. Speedups of three- to four orders of magnitude have been reported in the literature [9]. This computational gain comes from the reduction of degrees of freedom as well as the larger time steps which can be used in coarse-grained simulation.

Concurrent coupling approaches on the other hand try to retain the high accuracy of MD simulations in a localized region (e.g. around a crack tip or material defect) and couple the dynamics of the system to a coarse-grained (typically finite element based) model which provides boundary conditions for the MD system and can be used to sample in the regions were elastic deformations are smooth. For a recent topical review of some concurrent coupling methods we refer to [11]. Let us remark that most coupling methods found in the literature are limited to the zero-temperature case where the atomistic simulation is carried out using constant-NVE molecular dynamics. The same limitation applies to our method described below.

The coarse-grained model we employ is a finite element model based on a non-linear stress-strain relation given by homogenization of the atomistic interaction using the Cauchy-Born relation [12]. Denoting by $\mathbf{U}_I$ the displacement at the mesh node I and by M_I the lumped mass matrix on the finite element mesh, the governing equations for the nodal displacement values is

$$\ddot{\mathbf{U}}_I = M_I^{-1}\left(\mathbf{F}_I + \mathbf{F}_I^{\text{ext}}\right), \tag{2}$$

which is similar to the MD equations (1). For details we refer to [15]. Denoting by θ_I the Lagrangian finite element basis function at node I the force $\mathbf{F}_I$ can be written as

$$\mathbf{F}_I = \int \sum\nolimits_{k,\ell} \mathbf{P}_{k\ell} \frac{\partial \theta_I}{\partial \mathbf{q}_\ell} \mathbf{e}_k \, \varrho \, d\mathbf{q} \,,$$

with the density ϱ. Therefore, force computation requires element wise quadrature and the evaluation of the stress $\mathbf{P}$ by measuring atomistic stress in a representative lattice at each quadrature point using the Cauchy-Born rule. Fortunately, despite the strong nonlinearity of the integrand, in our numerical experiments we have found low-order Gauss integration to be sufficient for accurate force evaluation.

Nevertheless, to maximize the gain in computational efficiency we target coarse finite element meshes for which the mesh size is orders of magnitude larger than the mean nearest neighbor distance in the atomistic configuration. The rest of this section is concerned with the construction of a robust transfer method to couple these heterogeneous systems.

2.3 A method for weak coupling conditions

In [13] a new method for coupling molecular dynamics and (coarse) finite element dynamics using constraints was developed. The method assumes an overlapping decomposition of the simulated domain Ω into overlapping parts Ω_{M} and Ω_{F} where we use MD and finite elements resp. to describe the dynamics of the solid. In the handshake (or bridging) region

$$\Omega_{\text{B}} = \Omega_{\text{M}} \cap \Omega_{\text{F}}$$

both models coexist. To match the micro- and macro-model (i.e. MD and finite elements) we use local constraints

$$\mathbf{0} = \mathbf{g}(\mathbf{u}, \mathbf{U}) \text{ in } \Omega_{\mathrm{B}} .$$

The purely displacement based formulation of the coarse finite element model requires us to introduce also an atomistic displacement field $\mathbf{u}$. To do so, we choose a reference configuration $\{\mathbf{q}_i^0\}$ which typically coincides with the starting positions of the atoms and define

$$\mathbf{u}_i = \mathbf{q}_i - \mathbf{q}_i^0 .$$

Finding a good reference configuration can be a subtle task itself and might require energy minimization (cf. [14]). Fortunately, in most simulations the vibration of atoms around lattice sites in a standard lattice (such as FCC or BCC) are small compared to the motion due to external forces such that $\mathbf{q}_i^0$ can be chosen as a lattice site.

The constrains $\mathbf{g}$ introduced in [13] are of the form

$$\mathbf{g} = \Pi \mathbf{u} - \mathbf{U} \tag{3}$$

where Π is a projection operation from the space of atomistic displacement fields in Ω_{B} to the finite element space $\mathcal{S}_{\mathrm{B}}$ over the domain Ω_{B}. In view of a scale-decomposition (cf. [16])

$$\mathbf{u} \sim \overline{\mathbf{u}} + \mathbf{u}' = \Pi \mathbf{u} + (\mathrm{id} - \Pi)\, \mathbf{u} \tag{4}$$

these constraints enforce pointwise matching between the coarse atomistic displacements $\overline{\mathbf{u}}$ and the finite element displacement field $\mathbf{U}$. On the other hand the fine fluctuation field $\mathbf{u}'$ which lies in the kernel of Π and which is orthogonal to $\mathcal{S}_{\mathrm{B}}$ is not affected by the constraints since we can add an arbitrary fine fluctuation field $\mathbf{v}'$ to $\mathbf{u}$ without affecting the value of $\mathbf{g}$ in (3). Note that in (4) we do not claim equality between $\mathbf{u}$ and its scale decomposition. In general we need to apply a linear operator (such as an interpolation operator) to the right hand side to give this equation a mathematical precise meaning since $\mathbf{u}$ and $\Pi \mathbf{u}$ lie in different spaces.

In [13] the projection Π is computed as the composition of a function-space embedding followed by an L^2 projection [17]. In this so called "weak coupling framework" the projection Π is computed from the condition

$$\left(\Pi \textstyle\sum_i \mathbf{u}_i \varphi_i, \mathbf{V}\right)_{L^2(\Omega_{\mathrm{B}}, \varrho\, d\mathbf{q})} = \left(\textstyle\sum_i \mathbf{u}_i \varphi_i, \mathbf{V}\right)_{L^2(\Omega_{\mathrm{B}}, \varrho\, d\mathbf{q})}$$

for all $\mathbf{V} \in \mathcal{S}_{\mathrm{B}}$. Here, ϱ denotes the density of the finite element system and $\varphi_i \in L^2(\Omega_{\mathrm{B}})$ is a partition of unity. We note that

$$\textstyle\sum_i \mathbf{u}_i \varphi_i$$

is a function which approximates (or – depending on φ_i – interpolates) the atomistic displacement field in the handshake region. The partition of unity basis functions can be computed via mesh-based or mesh-less approximation techniques, e.g. Shepard's method [17]. The advantage of this approach is the higher flexibility in the choosing the weights in the projection Π. Moreover, since $\mathbf{u}$ and $\Pi\mathbf{u}$ can be treated as elements of the function space $L^2(\Omega_\mathrm{B})$ we can give a rigorous meaning to the scale decomposition (4) in this space. On the other hand the construction of φ_i is highly non-trivial and a prohibitive task in higher dimensions since it involves costly cut-detection and quadrature.

The high computational cost involved in assembling the L^2 project in the weak coupling framework motivates the search for approximations. In our current studies have investigated the (mass-weighted) least squares projection

$$\Pi\mathbf{u} = \operatorname{argmin}_{\mathbf{V}\in\mathcal{S}_\mathrm{B}} \frac{1}{2}\sum\nolimits_i \left|\mathbf{u}_i - \mathbf{V}(\mathbf{q}_i^0)\right|^2$$

as used by Wagner and Liu in the Bridging Scale method [16].

In our results we observed little difference between the two approaches for constructing Π. In fact, heuristically, we can understand the L^2 projection as a higher order least squares projection (and therefore the least squares projection as an approximation to the L^2 projection). This can be seen by applying lowest order midpoint quadrature to the integral in the algebraic representation of the L^2 projection:

$$\int \theta_I \varphi_i \varrho \, d\mathbf{q} \approx \varrho(\mathbf{q}_i^0) V_i \cdot \varphi_i(\mathbf{q}_i^0)\theta_I(\mathbf{q}_i^0) = \theta_I(\mathbf{q}_i^0)\,,$$

where V_i denotes the volume of the support of θ_i. Here we have used the mass normalization (which implies $\varrho(\mathbf{q}_i^0)V_i = 1$) and we have assumed that $\varphi_i(\mathbf{q}_i^0) = 1$ which is fulfilled, e.g., when the supports of partition of unity functions φ_i are pairwise disjoint. Note however that the approximation is not accurate (2.3) near the boundaries of the coupling zone.

The coupled equations of motion for the MD and finite element system can be derived from a Hamiltonian which is the weighted sum of the Hamiltonians h of MD and H of the finite elements plus a contribution by the constraints [22]. We choose a weighting function $\alpha : \Omega \to \mathbb{R}$ which is 1 on Ω_M and 0 on Ω_F. The multiscale Hamiltonian is then defined as

$$H_\mathrm{tot}(\mathbf{u},\mathbf{p},\mathbf{U},\mathbf{P};\lambda) = \alpha h(\mathbf{u},\mathbf{p}) + (1-\alpha)\,H(\mathbf{U},\mathbf{P}) + \lambda\cdot\mathbf{g}(\mathbf{u},\mathbf{U}) \qquad (5)$$

Let us note that the multiplication of a Hamiltonian (defined on the phase-space) by a spatial varying function is to be understand symbolically and requires a definition. We refer to [13] for a discussion of this issue.

The Hamiltonian equations derived from (5) are a set of differential algebraic equations for the unknowns $\mathbf{u}_i$ and $\mathbf{U}_I$. Linearity of the constraints allow us to apply well known time integration techniques (e.g. a RATTLE integrator [18]). These require the solution of one or more (in this case linear) systems per time-step to compute the Lagrange multipliers λ.

The above presented method employs a scale separation idea to construct constraints which are designed for the stable transfer of displacements between different scales with orders of magnitude in difference in resolution. The projection operator Π therein ensures that only displacement fields which are exactly representable on the coarse mesh will contribute to the residual $\mathbf{g}$ so that the fine scale displacement field is not affected. We refer to Subsection 3.1 for a discussion of the advantages of this approach in comparison to pointwise matching conditions.

2.4 Damping in zero-temperature simulations

To ensure that the fine fluctuation field $\mathbf{u}'$ (which is not affected by the constraints in the handshake region Ω_{B}) is not reflected at the boundary of the molecular dynamics domain, we employ a tailored perfectly matched layer (PML) method. In the original PML method [19, 20] a modified force term

$$\mathbf{F}_i^* = \mathbf{F}_i - 2D_i\left(\mathbf{v}_i + D_i\mathbf{u}_i\right)$$

is used to damp the displacement field $\mathbf{u}$ in the surrounding of the MD simulation domain, i.e. $D_i = D(\mathbf{q}_i^0) \neq 0$ only outside of Ω_{M}. Exploiting the linearity of the PML force terms a tailored damping technique has been proposed in [13]. Herein, the force is defined by

$$\mathbf{F}_i^* = \mathbf{F}_i - 2D_i\Big((\mathcal{F}\mathbf{v})_i + D_i(\mathcal{F}\mathbf{u})_i\Big) \tag{6}$$

with the fine fluctuation filter $\mathcal{F} = \mathbf{1} - (\text{interpolation}) \circ \Pi$. Using this damping approach allows us to remove the “extension” of the MD domain and apply the damping directly in the handshake region, i.e. $D_i \neq 0$ only in Ω_{B}. This reduces the computational cost of the damping since no additional atoms need to be introduced. However, the matrix-vector multiplication with the fine fluctuation filter can be expensive due to the large bandwidth of $\mathcal{F}$.

Let us point out that relying on this damping technique to remove reflection and spurious energy accumulation in the MD domain is likely to be the biggest issue in applying our algorithm to finite temperature (i.e., constant-NVT molecular dynamics) simulation. However, in principle Langevin-type non-reflecting boundary conditions (e.g. [21]) can be applied in replacement.

3 Numerical Validation

In this section we present numerical results for our multiscale simulation method. In the literature a set of (as far as possible) standardized benchmarks have been used to validate dynamical multiscale simulation techniques. The most popular benchmark problems are wave propagation problems [26, 27] which allow for estimating the amount of reflection as waves propagate from the fine into the coarser domain. We have used these benchmarks to validate our multiscale coupling method along with its parallel implementation. Further, to validate the method using a real-case example we have chosen a test problem from fracture mechanics.

3.1 Comparison with pointwise constraints

Initially, one motivation for the development of the constraints (3) was given by observations of weaknesses of the bridging domain method by Belytschko and Xiao [22]. Hence a comparison of the performance is an interesting test case. Here, we use a one-dimensional wave propagation benchmark. It should be emphasized that the nature of molecular dynamics prohibits direct extrapolation of these one-dimensional results to higher dimensions.

The approach of the bridging domain method employs pointwise constraints

$$\mathbf{0} = \mathbf{g}_i = \mathbf{u}_i - \mathbf{U}(\mathbf{q}_i^0) \text{ for } \mathbf{q}_i^0 \in \Omega_{\mathrm{B}} .$$

This method is known to work well in dynamic simulations only with a specially adapted time integration scheme. This is already apparent from the definition of $\mathbf{g}$: Since the constraints enforce pointwise matching between the atomistic and finite element displacement field at every lattice site, the atomistic displacement in Ω_{B} is required to be element wise affine. This sudden change in the dimension of space of atomistic displacements at the interface between Ω_{M} and Ω_{B} results in severe reflections of the fine fluctuation contributions $\mathbf{u}'$ of $\mathbf{u}$. This is in contrast to the constraints (3) which do not reduce the resolution of the atomistic system in the handshake region.

In dynamical simulations using the bridging domain method (e.g. [23]) usually a modified RATTLE algorithm is employed which omits the displacement corrections and therefore only enforces the secondary constraints $\dot{\mathbf{g}} = \mathbf{0}$ exactly. Moreover, within the linear system to determine the Lagrange forces a lumped multiplier matrix is used. It can be shown heuristically [24] that this yields the desired dissipation of the fine fluctuation field $\mathbf{u}'$ in the handshake region Ω_{B} by means of velocity rescaling.

On the one hand this argument suggests – as can we observe in practice – that the bridging domain method performs badly when combined with structure-preserving symplectic integrators, such as RATTLE, which enforce also the primary constraints in the displacements. The constraints (3) on the other hand do not require approximations in the time integration since the separation of the scale transfer and the dissipation of the fine-fluctuation field was one of the particular goals of the design of the constraints.

In our numerical experiment, we consider a one-dimensional Lennard-Jones atomic chain with nearest neighbor interaction and a coarse grained

Cauchy-Born mesh as depicted in Fig. 1. In the line with the numerical studies in [27] we conduct an initial amplitude

$$u_i = A \cdot \sin(kq_i)e^{(-kq_i/\sigma)^2} \quad \text{with } A = 0.015\,, \sigma = 3$$

and wavenumber $k = 2\pi/\lambda$ with varying wavelength $\lambda = 2, 4, \dots, 60$. We propagate the system with $\tau = 0.05$ for 2,000 timesteps and measure the maximal reflection coefficient r as defined in Eq. (7). The results for mesh size $5 \cdot 2^{1/6}$ and $10 \cdot 2^{1/6}$ are shown in Fig. 2. The figure shows the measured amount of reflection for the new projection-based constraints and the classical bridging domain method (with lumped multiplier matrix). For both systems a RATTLE time integration scheme was applied. For the computation of the weak constraints (3) we use an L^2 projection employing Shepard's method to create the partition of unity. The support of each φ_i has diameter $1.5 \cdot 2^{1/6}$.

We see that with our new coupling method the amount of reflection is more than an order of magnitude lower than the reflection rate measured for the bridging domain method. Moreover, we see that the reflection rate of the new method is less dependent on the mesh size.

In comparison with the results of Anciaux et al. [27] we find that due to the separation of the information transfer between micro- and macro-scale and the dissipation of the fine-fluctuation field, the proposed coupling method is able to reduce reflection rates even with the symplectic structure-preserving RATTLE integrator. It should however be pointed out that our method has a higher computational demand than the bridging domain approach as two linear systems need to be solved per timestep in contrast to a simple rescaling in the lumped bridging domain method.

3.2 Energy and reflection measurements

We consider the propagation of a radial wave

$$\mathbf{u}_i = \frac{A}{A - e^{-25}} \left(A \exp\left[-|\mathbf{q}_i|^2/\sigma^2\right] - e^{-25}\right) \Big(1 + b\cos\left(8\pi|\mathbf{q}_i|/\sigma\right)\Big) \frac{\mathbf{q}_i}{|\mathbf{q}_i|}$$

with $A = 0.15$, $\sigma = 15$ and $b = 0.1$, from a square shaped molecular dynamics domain into a surrounding finite element region, cf. Fig. 3. The MD domain Ω_{M} of size $\approx 475 \times 479$ contains 209.546 atoms in a hexagonal lattice with lattice constant $2^{1/6}$. We use a nearest neighbor Lennard Jones potential (i.e. cut-off radius equal to 1.5σ) with normalized $\sigma = 1$ and $\varepsilon = 1$. The atomistic domain is surrounded by a finite element layer of thickness 230 and 260 with the handshake width 30 and 60 resp. The finite element mesh consists of 9.184 and 9.728 quadrilaterals resp. Each finite element in Ω_{B} contains ≈ 50 atoms. We use Cauchy-Born consecutive equation with density $\varrho = \sqrt{3} \cdot 2^{-2/3}$ in the finite element domain. It is important to notice that by the choice of nearest-neighbor interaction, the system is free of lattice vibrations enabling

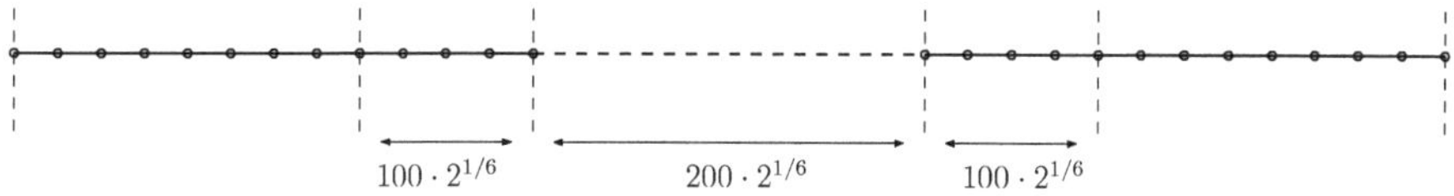

Fig. 1. Geometry of the numerical experiment.

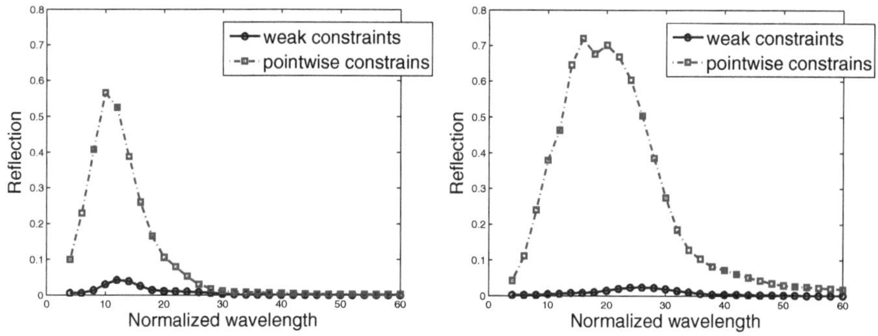

Fig. 2. Comparison of the reflection rate, defined as the maximum in time of the reflection coefficients, for the new coupling method based on weak constraints and the (lumped) Bridging domain method using a RATTLE integrator. *Left:* Mesh size $5 \cdot 2^{1/6}$. *Right:* Mesh size $10 \cdot 2^{1/6}$.

us to visualize the wave as depicted in Fig. 3. For time integration we use a RATTLE scheme with $\tau = 0.005$. The weighting function α interpolates linearly between 1 and 0.01 in the handshake region. We employ a purely frictional damping (i.e. omitting the stiffness change in Eq. (6)) with quadratic damping coefficients D_i as in [13]. The projection Π is given by the L^2 projection with the partition of unity functions φ_i evaluated by means of Shepard's method. The simulation runs on 24 MD and 8 finite element processors, resp. The arising multiplier system is easily solved in parallel with a few iterations of the conjugated gradient solver implemented in the Trilinos library [25]. The finite element computation is based on UG [2]. For the MD computation we employ a modified version of Tremolo [5–7].

The time evolution of the total (weighted) energy in the atomistic domain is shown in Fig. 4. We see a drop of the energy to only about 7 percent of the initial energy. This is in line with results reported in the literature (e.g. [26]) especially considering the fact that (due to the applied weighting) we cannot correct the energy by subtracting the energy values from a purely atomistic simulation as in [26]. We find that the modified PML method is able to efficiently reduce reflections near the corners of Ω_{M}, though the original PML is known to have problems handling corners. Also we see that a handshake width

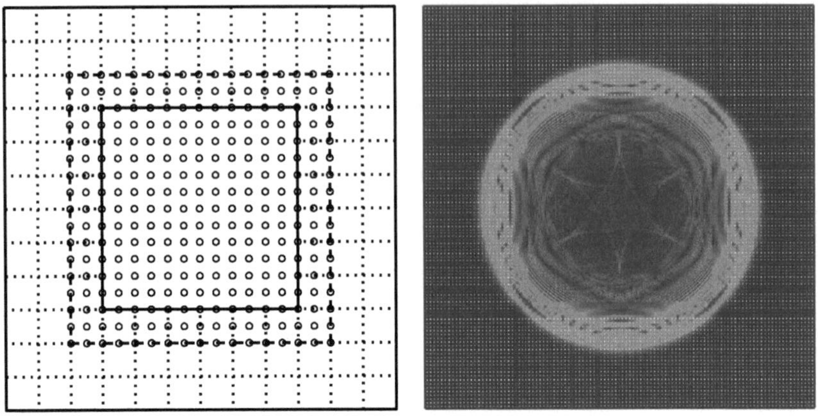

Fig. 3. *Left:* Geometry sketch. *Right:* Displacement field after 6500 timesteps as the wave passes the handshake region (handshake width 30).

of 30 (i.e. 4 element widths) is sufficient for the damping method to reduce reflections and a doubled handshake region yields only minor improvements.

We notice an increase in energy, especially between the 5000th and 6000th timestep, which happens to be the time where the maximal atom deflection passes the boundary of the molecular dynamics domain. This increase seems to be related to the weighting which results in small (effective) masses of atoms near the boundary and therefore to an amplified instability.

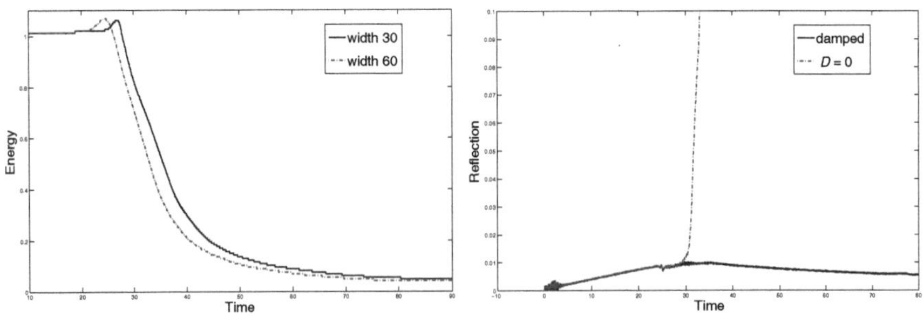

Fig. 4. *Left:* Normalized energy and reflections as a function of time. The total weighted energy in the atomistic domain for different handshake sizes are depicted in the left picture. We can see a drop of the initial energy to less than 7 percent of the initial energy. *Right:* Reflection coefficient for the coupled equation with (solid) and without (dotted). This plot shows the efficiency of the damping method.

Another measure for reflection that has been proposed in [27] is the reflection coefficient

$$r = \frac{K^{\mathrm{MS}} - k^{\mathrm{MD}}}{k_0^{\mathrm{MD}}} . \tag{7}$$

Here K^{MS} denotes the kinetic energy in the multiscale in the pure atomistic domain $\Omega_{\mathrm{M}} \setminus \Omega_{\mathrm{B}}$ and k^{MD} denotes the kinetic energy in the same domain taken from a reference pure MD simulation. The energy is normalized by k_0^{MD} which is the kinetic energy at a reference point after stabilization of the kinetic energy (and before the wave leaves the domain). As can be seen from Fig. 4 the damping method bounds the reflection coefficient to ~ 1 percent. In contrast without damping, we measure a reflection coefficient of up to 70 percent.

3.3 Mode-I fracture simulation

The simulation of crack propagation is a challenging task due to the multiple linked scales which determine the material toughness and the fracture process [10]. Hence, multiscale simulation methods are a good match as they allow to resolve the critical material behavior in the vicinity of the crack tip using molecular dynamics while retaining the computational efficiency of standard finite element techniques in the majority of the computational domain. Several multiscale simulations have been reported in the literature. Abraham et al. [3] developed a multiscale method for coupling electronic structure calculations (tight binding), molecular dynamics and finite elements in a concurrent simulation and applied their method to the simulation of fracture processes in silicon. Liu et al. [4] have applied their bridging scale multiscale method to the simulation of intersonic crack propagation previously performed with pure atomistic simulations by Abraham and Gao.

In this subsection we present preliminary results from the validation of our multiscale method for the simulation of fracture. We conducted two-dimensional simulations of a mode-I fracture using our multiscale method. Our focus here is to determine the effect of the choice of the fine scale region on the simulation result. We have performed simulations for three different geometries with 44,579, 62,390 and 80,195 atoms each (cf. Fig. 5). In all cases the finite element domain was used to pad the geometry to the same width. The same surface forces (± 0.25 at the left and right surface) have been applied. We have measured the maximal BDT stress [28] (sampled every 10th timestep and averaged over 10 samples) in the atomistic region. Our results show a good agreement between the stress profile for the first and third geometry with 44,579 and 80,195 atoms respectively. Although Fig. 5 shows that both simulations feature a different crack path, the crack tip velocity is comparable.

We note that for the simulation with geometry 2 a different stress profile and a different crack tip velocity is measured. To understand whether this is caused by the instability of the system or whether this is also connected to the multiscale approach presented here we plan to investigate the statistical behavior of the system using more simulation samples.

Irrespectively, our results are encouraging. The first geometry contains only about half the number of atoms of the third geometry, but still simulations on both geometries feature the same quantitative behavior. This suggests that we can gain one or two orders of magnitude reduction in the number of degrees of freedom compared to fully atomistic simulations. On the other hand, while the hyperelastic material behavior is localized around the crack tip region the crack path is in general unknown. Since adaptivity of the fine scale domain Ω_F is not well understood by now, this yields a striking lower bound for the size of the fine scale domain and hence for the number of degrees of freedom. Therefore, more research is needed to gain the desired efficiency of multiscale methods for real-world applications.

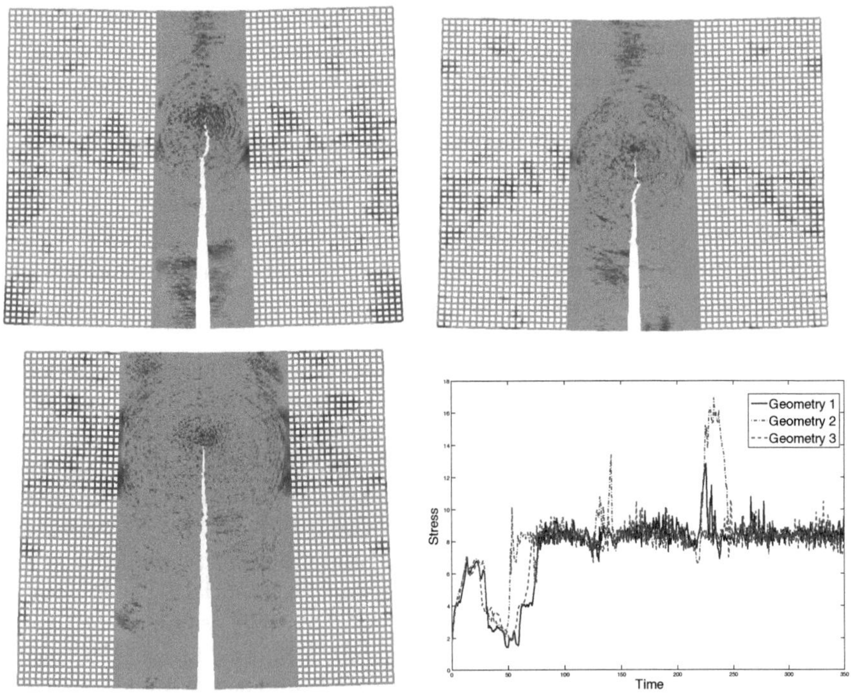

Fig. 5. Velocity distribution after 55,000 timesteps in the simulation of a mode-I crack using a variable sized fine scale domain with 44,579, 62,390 and 80,195 atoms resp. (left to right). Measurements of the maximal stress in the system show that the results of simulation 1 and simulation 3 are in good agreement . The behavior crack in geometry 2 however shows an abrupt change after $\approx 45,000$ timesteps.

4 Conclusion

We have presented numerical validation studies for a recently developed multiscale method for the coupling of molecular dynamics and finite elements for the simulation of crystalline solids. The methods is based on weak constraints which ensure matching between the micro- and macro-scale displacements in an overlapping region. The constraints are designed to ignore high-fluctuation information in the fine model. A modified perfectly matched layer method has been developed to cope with the fine-fluctuation field in zero-temperature simulations.

Numerical results prove the efficiency of the damping method and show advantages of the averaging constraints in comparison with the classical point wise approaches.

We have started to conduct numerical studies of the dependence of mode-I fracture on the fine-scale domain size. Based on our experiences in these simulations we have commented on challenges in bringing multiscale simulation to practice.

References

1. M. P. Allen and D. J. Tildesley, *Computer Simulation of Liquids*, Oxford Science Publications, 1987.
2. P. Bastian and K. Birken and K. Johannsen and S. Lang and N. Neuss and H. Rentz-Reichert and C. Wieners, *UG - A Flexible Software Toolbox for Solving Partial Differential Equations*, Comp. Vis. Science 1 (1997), pp. 27–40.
3. J. Q. Broughton and F. F. Abraham and N. Bernstein and E. Kaxira, *Concurrent coupling of length scales: Methodology and application*, Phys. Rev. B 60 (1999), pp. 2391–2403.
4. D. E. Farrell and H. S. Park and W. K. Liu, *Implementation aspects of the bridging scale method and application to intersonic crack propagation*, Int. J. Numer. Meth. Engng. 60 (2007), pp. 583–605.
5. M. Griebel, S. Knapek and G. Zumbusch, *Numerical Simulation in Molecular Dynamics*, Springer, 2007.
6. M. Griebel and J. Hamaekers, *Molecular dynamics simulations of the mechanical properties of polyethylene-carbon nanotube composites*, Handbook of Theoretical Comp. Nanotechnology 9 (2004), pp. 409–454.
7. M. Griebel and J. Hamaekers, *Molecular dynamics simulations of the elastic moduli of polymer-carbon nanotube composites*, Comp. Meth. Appl. Engng. 193 (2006), pp. 1773–1788.
8. B. L. Holian and R. Ravelo, *Fracture simulation using large-scale molecular dynamics*, Phys. Rev. B 51 (1995), pp. 11275–11288.
9. J.-S. Chen, H. Teng and A. Nakano, *Wavelet-based multi-scale coarse graining approach for DNA molecules*, Finite Element Anal. and Design 43 (2007), pp. 346–360.
10. A. Needleman and E. Van der Giessen, *Micromechanics of Fracture: Connecting Physics to Engineering*, MRS Bulletin 26 (2001), pp. 211–214.

11. R. E. Miller and E. B. Tadmor, *A unified framework and performance benchmark of fourteen multiscale atomistic/continuum coupling methods*, Modelling Simul. Sci. Engng. 17 (2009), pp. 053001–053052.
12. W. E and P. Ming, *Cauchy-Born Rule and Stability of Crystalline Solids: Static Problems*, Arch. Rational Mech. Anal. 183 (2007), pp. 241–297.
13. K. Fackeldey, D. Krause, R. Krause and C. Lenzen, *Coupling Molecular Dynamics and Continua with Weak Constraints*, Submitted to SIAM MMS.
14. D. Thomas *A Generic Approach to Multiscale Coupling – Concepts and Applications*, Diploma thesis, Institute for Numerical Simulation Bonn, 2008.
15. T. Belytschko, W. K. Liu and B. Moran, *Nonlinear Finite Elements for Continua and Structures*, Wiley, 2006.
16. G. J. Wagner and W. K. Liu, *Coupling of atomistic and continuum simulations using a bridging scale decomposition*, J. Comp. Phys. 190 (2003), pp. 249–274.
17. K. Fackeldey and R. Krause, *Multiscale Coupling in Function Space - Weak Coupling between Molecular Dynamics and Continuum Mechanics*, Int. J. Numer. Meth. Engng. 79 (2009), pp. 1517–1535.
18. E. Hairer, C. Lubich and G. Wanner, *Geometric Numerical Integration. Structure-Preserving Algorithms for Ordinary Differential Equations*, 2. ed, Springer (2006).
19. A. C. To and S. Li, *Perfectly matched multiscale simulation*, Phys. Rev. B 72 (2005), pp. 035414–035422.
20. S. Li, X. Liu, A. Agrawal and A. C. To, *Perfectly matched multiscale simulations for discrete lattices: Extension to multiple dimensions*, Phys. Rev. B 74 (2006), pp. 045418–045432.
21. X. Li and W. E, *Variational boundary conditions for molecular dynamics simulations of crystalline solids at finite temperature: Treatment of the thermal bath*, Phys. Rev. B 76 (2007), pp. 10078–10093.
22. S. P. Xiao and T. Belytschko, *A bridging domain method for coupling continua with molecular dynamics*, Comp. Meth. Appl. Engng. 193 (2004), pp. 1645–1669.
23. G. Anciaux and O. Coulaud and J. Roman, *High Performance Multiscale Simulation or Crack Propagation*, Proceedings of the 2006 International Conference Workshops on Parallel Processing, pp. 473–480.
24. K. Fackeldey, D. Krause and R. Krause, *A Note on the Dissipative Effect of Lumping in the Bridging Domain Method*, Private Notes.
25. Michael A. Heroux et al., *An overview of the Trilinos project*, ACM Trans. Math. Softw. 31 (2005), pp. 397–423.
26. H. S. Park, E. G. Karpov, W. K. Liu and P. A. Klein, *The bridging scale for two-dimensional atomistic/continuum coupling*, Philosophical Magazine 85 (2005), pp. 79–113.
27. G. Anciaux, O. Coulaud, J. Roman and G. Zerah, *Ghost force reduction and spectral analysis of the 1D bridging method*, Technical Report INRIA (2008).
28. M. Zhou, *A new look at the atomic level virial stress: on continuum-molecular system equivalence*, Proc. R. Soc. London A 459 (2003), pp. 2347–2392.

Coupling of the Navier-Stokes and the Boltzmann equations with a meshfree particle and kinetic particle methods for a micro cavity

Sudarshan Tiwari and Axel Klar

Department of Mathematics, TU Kaiserslautern, 67663 Kaiserslautern, Germany
tiwari@mathematik.uni-kl.de and klar@mathematik.uni-kl.de

Summary. We present a coupling procedure of a meshfree particle method to solve the Navier-Stokes equations and a kinetic particle method, a variant of the Direct Simulation Monte Carlo(DSMC) method, to solve the Boltzmann equation. A $2D$ micro cavity problem has been simulated for different Knudsen numbers. An adaptive domain decomposition approach has been implemented with the help of a continuum breakdown criterion. The solutions from the Navier-Stokes equations and the coupling algorithm are compared with the ones from the Boltzmann equation. Moreover, it is shown that for larger Knudsen numbers, where the Navier-Stokes equations fail to predict the correct flow behaviors, its stationary solutions are still good candidate to initialize a Boltzmann solver. The CPU time for the coupling code is upto 5 times faster than the CPU time for the code solving Boltzmann equation for the same accuracy of the solutions.

Key words: meshfree method, DSMC, micro fluidics, coupling Boltzmann and Navier-Stokes

1 Introduction

The coupling of the Boltzmann and the Euler/Navier-Stokes equations is initiated from simulations of hypersonic flows around a space vehicle during the re-entry phase, where it experiences several flow regimes that are characterized by the Knudsen number, $Kn = \lambda/H$, where λ is the mean free path and H is the characteristic length of the domain. The degree of rarefaction of a gas can be measured through the Knudsen number. With the help of the Knudsen number one can characterized different flow regimes. For examples, for $Kn < 0.001$, the flow is in the continuum regime, where the Navier-Stokes equations with no-slip boundary conditions are solved. For $0.001 < Kn < 0.1$, the flow is in slip regime, where the Navier-Stokes equations with velocity-slip and temperature jump conditions are solved [1]. For $Kn > 0.1$ kinetic type

M. Griebel, M.A. Schweitzer (eds.), *Meshfree Methods for Partial Differential Equations V*, Lecture Notes in Computational Science and Engineering 79,
DOI 10.1007/978-3-642-16229-9_10,

approach, based on the Boltzmann equation is required. We note that the kinetic approach is valid in the whole range of rarefaction of a gas.

The general assumption of gaseous flows in macro scaled domains are not always applicable for flows in micro-sized domain. Thus the Navier-Stokes equations are no longer considered to be valid when the characteristic length is within the micron range [12]. In this paper, we have considered the slip regime and we have coupled the Boltzmann and the Navier-Stokes equations with no slip boundary conditions wherever it is possible.

Usually particle methods, like DSMC [5] and its variants (see, for example, [2, 21]) are used for simulations of the Boltzmann equation. However, for smaller Knudsen numbers DSMC type particle methods are becoming increasingly expensive since the cell size must be smaller than the mean free path. On the other hand, in the continuum regime one can solve the Euler or the Navier-Stokes equations. However, the continuum equations are not valid everywhere in the computational domain, for example, in shock and boundary layers, etc. This leads to domain decomposition approaches including continuum and kinetic domains. For these approaches it is first necessary to define the domains of validity and then to solve the equations in their respective domains. Several criteria have been suggested for the breakdown of the continuum equations [6, 15, 19, 27]. Many works have been reported in the development of hybrid solvers for macro scale domains, see for example [9, 11, 16–19, 27, 29, 30, 35]. In recent years the gas dynamics has been a active research area in micro and nano sized domains. Some works on coupling of the Boltzmann and the Navier-Stokes equations are reported also for flow problems on small scale domains [1, 25].

Most of these papers deal with the coupling of a particle method for the Boltzmann equation and with a Finite Volume or Finite Element code for the fluid dynamic equations. The more natural choice (and more straightforward to implement also for complicated applications) is to choose kinetic particle methods for both equations [29]. This simplifies the treatment of interface conditions between the two domains considerably. In particular, this is important if the decomposition process results in complicated (time dependent) domains for the Boltzmann and the Navier-Stokes equations. In [29] the Euler equations have been solved by a kinetic particle method. However, this method is not the optimal one in terms of CPU time, since it is - for the same spatial and temporal grids - as expensive as the particle method for the Boltzmann equation.

In our earlier work we have reported the hybrid method for the Boltzmann and the Navier-Stokes equations, where we have solved the compressible Navier-Stokes equations by a meshfree particle method and the Boltzmann equation by a kinetic particle method [30]. The problem had been considered a $1D$ time dependent problem. In this paper we present the extension of the work, reported in [30], into a $2D$ stationary micro flow. We use two types of particle methods, for example a meshfree Lagrangian particle method to solve the compressible Navier-Stokes equations [28, 32] and a mesh-based kinetic Lagrangian particle method to solve the Boltzmann equation [2, 5, 21]. The kinetic particle move with their kinetic velocities. They carry only the kinetic velocities and positions with them, where as meshfree particles are numerical interpolation points, which move with fluid velocities and carry all necessary fluid information, like density, velocity, pressure, etc. with them. Differential operators at an arbitrary particle position are approximated from its neighboring cloud of particles. Meshfree methods are in particular, suitable for the coupling of the Boltzmann and fluid dynamic equations since they allow for an treatment of arbitrary shape of the interface between the two regimes [29].

The particle methods for both the Boltzmann and the Navier-Stokes equations utilize a grid on which particles move. We use different grid spacings and time steps in both cases. In general, the Boltzmann grid size is chosen smaller then the mean free path and the Navier-Stokes grid size (i.e. the distance between Navier-Stokes particles) is chosen several times larger than the mean free path. An adaptive grid refinement technique used here is similar to that of above mentioned earlier works [29, 30]. To determine the domains of validity for the Boltzmann and the Navier-Stokes equations we use the breakdown criterion suggested in [27]. It can be computed as a function of the stress tensor and the heat flux vector, which in turn can be computed from the Navier-Stokes solver. The numerical example we consider in this paper is a $2D$ micro cavity flow, where we apply Boltzmann, Navier-Stokes and hybrid solvers for $Kn = 0.05, 0.02, 0.01$. It is observed that for a large Knudsen number, like $Kn = 0.05$, the solutions of the Navier-Stokes solver deviate from the solutions of the Boltzmann solver. However, the solutions from the coupling algorithm are close to the ones from the Boltzmann solver. This indicates that one can avoid the unnecessary use of the Boltzmann solver in the entire domain even for larger Knudsen numbers. Moreover, we note that for larger Knudsen numbers, where the validity of the Navier-Stokes equations are questionable, its stationary solutions are still good candidates to initialize the Boltzmann solver.

The paper is organized as follows. In section 2 we present the mathematical models. In section 3 the numerical methods for the Boltzmann and the Navier-Stokes equations are described. The description of the hybrid method is explained in section 4. Finally, some numerical tests are presented in section 5.

2 Governing equations

The Boltzmann equation describes the time evolution of a distribution function $f(t, x, v)$ for particles of velocity $v \in \Re^3$ at point $x \in D \subset \Re^s (s = 1, 2, 3)$ and time $t \in \Re_+$. It is given by

$$\frac{\partial f}{\partial t} + v \cdot \nabla_x f = Q(f, f), \tag{1}$$

where

$$Q(f,f)=\int_{\Re^3}\int_{S^2}\beta(|v-w|,\eta)[f(v')f(w')-f(v)f(w)]d\omega(\eta)dw$$

with

$$v'=T_{v,w}(\eta)=v-\eta<\eta,v-w>,\ w'=T_{w,v}(\eta).$$

Here, β denotes the collision cross section, η is the unit normal vector on the sphere and $<,>$ is the scalar product. For the sake of simplicity, we have not used any bold letters for vector quantities, like x, v, w, etc. Writing the equations in dimensionless form one observes that Q is of the order $\mathcal{O}(\frac{1}{Kn})$. The local mean free path $\lambda=\lambda(x,t)$ is given by

$$\lambda=\frac{kT}{\sqrt{2}\pi p d^2}, \tag{2}$$

where k is the Boltzmann constant, $T=T(x,t)$ the temperature, $p=p(x,t)$ the pressure and d is the diameter of molecules. For more details we refer to [8, 23].

For Kn tending to zero one can show that the Boltzmann distribution function f tends to the local Maxwellian [7]

$$f_M(t,x,v)=\frac{\rho}{(2\pi RT)^{3/2}}\,e^{-\frac{|v-U|^2}{2RT}}, \tag{3}$$

where $\rho=\rho(x,t)$ is the density, $U=U(x,t)$ the mean velocity and R is the gas constant. The parameters of the Maxwellian ρ, U, T solve the compressible Euler equations. This can be verified from the asymptotic expansion of f in Kn, where the zeroth order approximation gives the local Maxwellian distribution and the first order approximation [4] gives the Chapman-Enskog distribution

$$f_{CE}(t,x,v)=f_M(v;x,t)\left[1+\phi(v;x,t)\right], \tag{4}$$

with

$$\phi(t,x,v)=\frac{2}{5}\frac{q\cdot c}{\rho(RT)^2}\left(\frac{|c|^2}{2RT}-\frac{5}{2}\right)-\frac{1}{2}\frac{\tau:c\otimes c}{\rho(RT)^2}, \tag{5}$$

where $c=v-U$. Here, $\phi=\mathcal{O}(Kn)$ and the parameters ρ, U, T, q, τ satisfy the compressible Navier-Stokes equations

$$\begin{aligned}\frac{\partial\rho}{\partial t}+\nabla\cdot(\rho U)&=0\\ \frac{\partial(\rho U)}{\partial t}+\nabla\cdot(\rho U\otimes U+pI-\tau)&=0\\ \frac{\partial(\rho E)}{\partial t}+\nabla\cdot[(\rho E+p)U-\tau\cdot U-q]&=0,\end{aligned} \tag{6}$$

where $E=|U|^2/2+e$ is the total energy and e is the internal energy, the stress tensor τ and heat flux vector q are of order Kn and given by

$$\tau_{ij} = \mu \left(\frac{\partial U_i}{\partial x_j} + \frac{\partial U_j}{\partial x_i} - \frac{2}{3} \nabla \cdot U \, \delta_{ij} \right), \qquad q = -\kappa \nabla T. \tag{7}$$

The dynamic viscosity $\mu = \mu(x,t)$ and the heat conductivity $\kappa = \kappa(x,t)$ for a monatomic gas of hard sphere molecules are of order Kn. They are given, see [5], by

$$\mu = \frac{5}{16 d^2} \sqrt{\frac{mkT}{\pi}}, \qquad \kappa = \frac{15k}{4m} \mu, \tag{8}$$

where m is the molecular mass. In this paper we have considered a monatomic gas of hard sphere.

3 Numerical methods

As we have already mentioned that, we apply Lagrangian particle methods of different characters for both types of equations. The Boltzmann equation is solved by a DSMC type Monte Carlo method, whereas the Navier-Stokes equations are treated with a meshfree particle method, which is called the Finite Pointset Method (FPM). For the Boltzmann solver a computational domain is divided into rectangular cells. The cell or mesh is used for the Boltzmann solver to choose the collision pairs and sampling the macroscopic flow properties such as mean velocity, pressure, density and temperature, etc.

3.1 Particle Method for the Boltzmann equation

For solving the Boltzmann equation we have used a variant of the DSMC method [5], developed in [2, 3, 21]. The method is based on the time splitting of the Boltzmann equation. Introducing fractional steps one solves first the free transport equation (the collisionless Boltzmann equation) for one time step. During the free flow, boundary and interface conditions are taken into account. In a second step (the collision step) the spatially homogeneous Boltzmann equation without the transport term is solved. To simulate this equation by a particle method an explicit Euler step is performed. The result is then used in the next time step as the new initial condition for the free flow. To solve the homogeneous Boltzmann equation the key point is to find an efficient particle approximation of the product distribution functions in the Boltzmann collision operator given only an approximation of the distribution function itself. To guarantee positivity of the distribution function during the collision step a restriction of the time step proportional to the Knudsen number is needed. That means that the method becomes exceedingly expensive for small Knudsen numbers.

3.2 Meshfree particle method for the Navier-Stokes equations

We solve the Navier-Stokes equations by a meshfree Lagrangian particle method similar to the smoothed particle hydrodynamics (SPH) method [13].

Our approach differs with the SPH while approximating the spatial derivatives. We approximate the spatial derivatives at an arbitrary particle from its surrounding clouds of points with the help of the least squares method. We express the compressible Navier-Stokes equations in the primitive variables according to the Lagrangian form as

$$\frac{Dx}{Dt} = U \tag{9}$$

$$\frac{D\rho}{Dt} = -\rho \nabla \cdot U \tag{10}$$

$$\frac{DU}{Dt} = -\frac{1}{\rho}\nabla p + \frac{1}{\rho}\nabla \cdot (\tau) \tag{11}$$

$$\frac{DT}{Dt} = \frac{\gamma - 1}{\rho R}\left[-p\nabla \cdot U + (\tau \cdot \nabla) \cdot U + \nabla(\kappa \nabla T)\right], \tag{12}$$

where D/Dt we denote the material derivative. γ is the ratio of specific heats. Moreover, we consider the equation of state

$$p = \rho RT. \tag{13}$$

The equations (10-12) are to be solved with the appropriate initial and boundary conditions which are specified in the section where numerical tests are performed.

We first fill a computational domain by a finite number of particles or grids and assign all fluid quantities on them. Then we approximate the spatial derivatives on the right hand side of (10 - 12) at every particle position from its surrounding neighbors with the help of the least squares method. We have reported about the meshfree approximation of spatial derivatives in earlier issues, see [31–33]. Due to the restriction of the space, we do not repeat the same details in this paper. The resulting equations reduce to a time dependent system of ordinary differential equations. This system can be solved by a simple integration scheme. One can use the explicit Euler scheme, but it requires very small time step. Here a two steps Runge-Kutta method is proposed which is sufficient for the test cases considered in this paper [24].

4 Hybrid method

Our main goal is to couple a mesh based particle method with a meshfree particle method in their domains of validity with the help of some breakdown criteria for the Navier-Stokes equations.

We start by defining regular cells as usual for DSMC simulations. In the cell centers we store the macroscopic quantities obtained by averaging over the particles contained in the cell. As far as the Navier-Stokes equations are concerned we use the cell centers as starting positions for the corresponding particles. Then, we prescribe the initial conditions on each Navier-Stokes particle and solve the Navier-Stokes equations in the whole domain until the steady state. Next, depending on a breakdown criterion, the cell may either remain a Navier-Stokes cell or it may be defined as a Boltzmann cell which may need to be subdivided and new particles (gas molecules) have to be generated according to the requirements of the DSMC method. We use a breakdown criterion suggested in [27], where the distribution function $f(t,x,v)$ is written as a deviation from the local Maxwellian $f_M(t,x,v)$ according to (4). The size of the deviation ϕ is then estimated with the norm $\|\phi\|^2 = \int_{\Re^3} \frac{f_M}{\rho}\phi^2 dv$. A local equilibrium can be assumed if $\|\phi\| << 1$. Using a first order expansion in Kn one obtains the explicit expression (5) for ϕ and

$$\|\phi\| = \frac{1}{\rho RT}\left[\frac{2}{5}\frac{|q|^2}{RT} + \frac{1}{2}|\tau|^2\right]^{1/2}. \tag{14}$$

From the point of view of solutions of the Navier Stokes equations the quantities q and τ are given by (7). From the point of view of the Boltzmann equations we have

$$q = \frac{1}{2}\int_{\Re^3} c|c|^2 f dv, \quad \tau = \int_{\Re^3} c \cdot c^T f dv - \rho RTI.$$

We define a cell as Navier-Stokes cell if $\|\phi\| \leq \epsilon$, otherwise it is a Boltzmann one. The main difficulty is to find the cutoff value ϵ for the breakdown criteria. Since q and τ are of the order of the Knudsen number Kn, the cutoff value can be defined as a function of Kn as in [30]. However, it depends on problems considered, see [9, 11, 16–19, 27, 29, 30, 35] and there exists no global cutoff value. In this paper, we have used $\epsilon = 0.004$. Varying the ϵ give different domain decompositions.

In this paper we have considered a driven micro cavity problem. In Fig. 1 we see the domain decomposition of the Boltzmann and Navier-Stokes domains for the Knudsen number 0.05, 0.02 and 0.01. As we observe that the Boltzmann domain becomes smaller when the Knudsen number decreases.

4.1 Adaptive grid refinement

The spatial grid size for the Boltzmann solver has to be chosen as the order of the mean free path λ. However, the grid size for the Navier-Stokes solver is independent of λ. This means that we usually need to subdivide a Boltzmann cell into smaller units to achieve the desired numerical accuracy. In the following, we refer to these units as Boltzmann sub cells. Let N be the number of cells in each direction of a square cavity. The particle distance for the Navier-Stokes solver is given by $dx_{NS} = 1/N$. The number of Boltzmann sub cells per Boltzmann cell is defined by

$$L = \max\left[\text{Int}(\frac{dx_{NS}}{\lambda}) + 1,\ 2\right]. \tag{15}$$

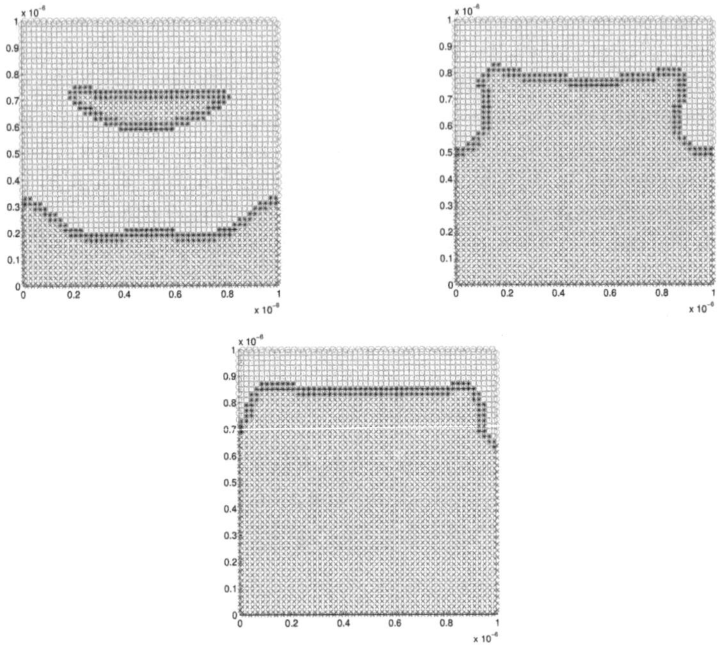

Fig. 1. Domain decompositions: Kn = 0.05 (top left), Kn = 0.02 (top right) and Kn = 0.01 (bottom), green(o) = Boltzmann cell , red(x) = Navier-Stokes cells and blue(*)=interface cells for Boltzmann boundary conditions

This means, we define at least two Boltzmann sub cells per Boltzmann cell(coarser cell). The Boltzmann grid size dx_B is defined by

$$dx_B = \frac{dx_{NS}}{L}. \tag{16}$$

In Fig.2 we have demonstrated the different grid sizes corresponding to the domain decomposition in Fig.1 for $Kn = 0.05$. The finer cells are the Boltzmann and the coarser are the Navier-Stokes ones.

4.2 Selection of time steps

The time step used for the DSMC method, Δt_B, should be less than the mean collision time, so that the time step is taken as

$$\Delta t_B = 0.9 \, \frac{dx_B}{c_m}, \tag{17}$$

where $c_m = \sqrt{2RT_0}$ is the most probable molecular speed [5], and T_0 is the reference temperature.

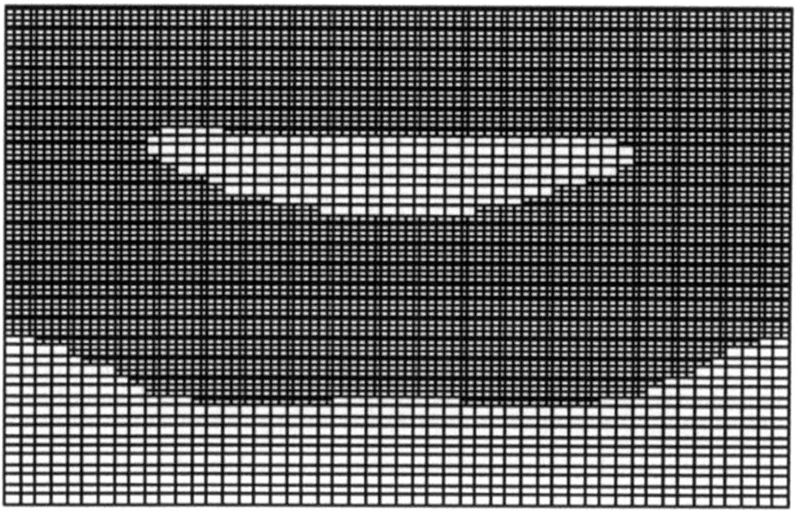

Fig. 2. Finner cells are Boltzmann and coarser cells are Navier-Stokes cells for $Kn = 0.05$

For the Navier-Stokes solver we choose the time step, which is consistent suggested in [26], as

$$\Delta t_{NS} = \frac{dx_{NS}^2}{|U_{max}|dx_{NS} + 0.7\ \nu}, \tag{18}$$

where $\nu = \max[\mu/\rho, \kappa(\gamma - 1)/(\rho R)]$. The number of iterations I for the Boltzmann solver per Navier-Stokes iteration is performed from the relation

$$I = Int[\frac{\Delta t_{NS}}{\Delta t_B}] + 1. \tag{19}$$

4.3 Coupling condition

In this paper we have considered the stationary problem. Therefore, we solve first the Navier-Stokes equations and reach its steady state. Then we apply the breakdown criterion (14) and obtain the domain decomposition. If a cell is predicted to be a Boltzmann cell, we use the macroscopic quantities ρ, U, T from the Navier-Stokes solver and generate the Boltzmann particles according to the Maxwellian distribution. One can also generate the Chapman-Enskog distribution using the acceptance-rejection method suggested by Gracia and Alder [14]. In this paper we did not observed much difference by using the Maxwellian or the Chapman-Enskog distribution. The reason would be that, we are looking for a stationary solution and take several sampling to reduce the statistical noise from the Monte Carlo simulation.

In order to apply the boundary conditions for the Boltzmann equation, we have to define the boundary cells (or interface cells) between the Boltzmann and Navier-Stokes domains. The interface cells for the Boltzmann domain are, in fact, the Navier-Stokes cells, which are adjacent to the Boltzmann domain. Hence, the boundary conditions for the Boltzmann equation at the interface are obtained by generating again the Boltzmann particles according to a Maxwellian distribution from the Navier-Stokes values at the interface. Since the interface between two domains can be zig-zagging, we need to consider double layers of the interface boundary for the Boltzmann domain to guarantee the continuous of the flux from the corner. If the interface is straight, one layer of interface cell is sufficient. Then we perform the Boltzmann simulation for I time steps. In all I time iterations of the Boltzmann simulation we compute the macroscopic quantities in each coarser cell and then take the averaging over I steps. As in all DSMC codes there are some statistical fluctuations in the Boltzmann data. These fluctuating data destabilize the Navier-Stokes solver. Therefore, we need a smoothing operator. Here we have used the Shephard interpolation. For example, for the density at cell center x, the Shephard interpolation is defined as

$$\tilde{\rho} = \frac{\sum_{i=1}^{m} W_i \rho_i}{\sum_{i=1}^{m} W_i}, \tag{20}$$

where m is the number of neighboring cell centers x_i, which are Boltzmann cells and W_i is the weight function depends on the distance between central and the neighbor points. Similarly, we smooth U and T and then update μ and κ.

The boundary conditions for the Navier-Stokes equations are applied as follows. Near interface a Navier-Stokes cell consists of many Boltzmann cells as its neighbor. In this case we consider all neighboring Boltzmann and Navier-Stokes cells as neighbor and approximate the spatial derivatives from the least squares method. Instead of using the Dirichlet boundary condition at the Boltzmann interface cell, we find this approach is stable and sufficient.

As we have already stated that we initialize generating coarse regular cells in the domain. The centers of these cells are the positions of the Navier-Stokes particles. Since Navier-Stokes particles move with their fluid velocities, they will be redistributed arbitrarily. In this case it is meaningful to reproject all Navier-Stokes solutions to the old positions and then reconsider the old positions as the current positions of particles. This is called particle remeshing and is used by several authors, see [10] and other references therein. The fluid quantities at the old positions are approximated from the neighbor particles at the new positions with the help of the least squares method. To reconstruct the flow fields accurately one needs at least a second order least squares approximation. The solutions obtained from the remeshing particle method are consistent with the moving particle method as well as with corresponding analytical solutions [28, 34].

4.4 Coupling Algorithm

Summarizing the above, we present the following coupling algorithm.

1. Generate cells with cell size dx_{NS} and define cell centers as Navier-Stokes particles in the whole domain. Prescribe initial value on Navier-Stokes particles.
2. Solve the Navier-Stokes equations in all cells until the steady state.
3. Compute $\|\phi\|$ from (14) in all cells and define

$$\text{cell} = \begin{cases} \text{Navier-Stokes}, & \text{if} \quad \|\phi\| \leq \epsilon \\ \text{Boltzmann}, & \text{otherwise} \end{cases}$$

4. Generate particles in Boltzmann cells according to the local Maxwellian distribution with parameters as solutions of the Navier-Stokes equations
5. Refine the Boltzmann cells by $dx_B = dx_{NS}/L$.
6. Do for i = 1, I (in all Boltzmann sub cells)
   a) Generate particles on interface cells of Boltzmann domain according to the local Maxwellian distribution
   b) free flow of Boltzmann particles over a time step Δt_B
   c) if a Boltzmann particle hits solid wall apply gas surface interaction, if it eaves Boltzmann domain and enters to NS domain, delete it
   d) sort particles into Boltzmann sub cells and perform intermolecular collisions
   e) sample macroscopic quantities on coarser Boltzmann cells and average over I

   end do
7. Smooth the macroscopic quantities in the Boltzmann cells near to interface
8. Solve the NS equations in all NS cells for a time step Δt_{NS}. Boundary conditions from the Boltzmann cells are taken from the values of the Boltzmann cells, which fall under the neighbor list of NS cells.
9. Reproject Navier-Stokes particles onto old positions
10. goto 6 and repeat until the final time.

5 Numerical examples

The flow in a cavity driven by the velocity on the top has become a popular example for testing and comparing numerical methods. In recent years the micro cavity flows has been a popular test case for the micro flows, see [20,22] and the references therein. The solutions predicted by the Navier-Stokes equations and the coupled algorithm are compared with the macroscopic quantities obtained from the Boltzmann solver.

We consider a square cavity of size $[0, H] \times [0, H]$, where $H = 1 \times 10^{-6}$. The velocity U on the top side of the cavity is chosen as $U = (u_{top}, 0)$ and zero velocities are chosen on the other walls in all cases. The Mach number, defined as

$Ma = u_{top}/\sqrt{2RT_0} = 0.1132$, where $T_0 = 300K$. The temperature boundary conditions on all walls are $T = T_0$. Furthermore, we used the following parameters. The gas is chosen as Argon with a molecular mass $m = 6.63 \times 10^{-26} kg$. The Boltzmann constant $k = 1.38 \times 10^{-23} JK^{-1}$, the molecular diameter $d = 3.68 \times 10^{-10} m$, the ratio of specific heats $\gamma = 5/3$. These parameters give the gas constant $R = 208 JkgK^{-1}$. Here we consider three different test cases for varying the initial pressure $P_0 = 125100, 312750, 625500$. The corresponding Knudsen number are $0.05, 0.02$ and 0.01, respectively.

The boundary condition for the Boltzmann solver a Maxwellian scattering with perfect accommodation is used, this means, if a particle hits the boundary it will be reflected with the wall temperature and the wall velocity.

We choose the coarser grids 50×50 for all cases ($N = 50$). The Boltzmann grid sizes are chosen according to the initial mean free path from the relation (15). The initialization of number of Boltzmann particles is proportional to the initial density ρ_0. For the Knudsen numbers 0.05 and 0.02, we have generated the initial number of particle $n_0 = 200$ per coarser cell. For the Knudsen number $= 0.01$ we have considered $n_0 = 400$ per coarser cell.

Similarly, in the coupling code, if the cell becomes a Boltzmann cell, or in the interface boundary cells of the Boltzmann domain, the number of particles is given by

$$Int\left(\frac{\rho(t) n_0}{\rho_0 \; L}\right).$$

This guarantees the number of particles per cell in the Boltzmann and hybrid cells are approximately same such that the CPU time can be compared for both solvers.

In all three cases, we have considered the final time is equal to $6.5 \times 10^{-6}\ s$, where the steady state is guaranteed.

In Fig. 3 we have presented the plots of the steady state velocity fields from all three solvers for $Kn = 0.05$. Because of the discontinuity of the boundary velocity at the top corners, the solution of the Navier-Stokes equations is in some way “singular” at these points (the vorticity becomes infinite). As a consequence, it is difficult to compare solutions since the approximation of the flow close to the singularity affect the accuracy of the solution. This is particularly true when the mesh is refined so that computational points are drawn near to the corners. The grid which we have considered is not finer enough, however, we can see some disturbance on the left corner on the bottom of Fig. 3. In this range of the Knudsen number, the Navier-Stokes cells are almost of the same order of the Boltzmann cells. Therefore, considering the finner grid for the Navier-Stokes solver is not the purpose of this paper.

In Fig.4, we have plotted the x-velocity components along the vertical central line from all three solvers for $Kn = 0.05, 0.02$ and 0.01. One can observed that the velocity from Navier-Stokes solver over predict compared to the one from the Boltzmann solver for both $Kn = 0.05$ and 0.02, where as the coupling solution has good agreement with the Boltzmann solution. However,

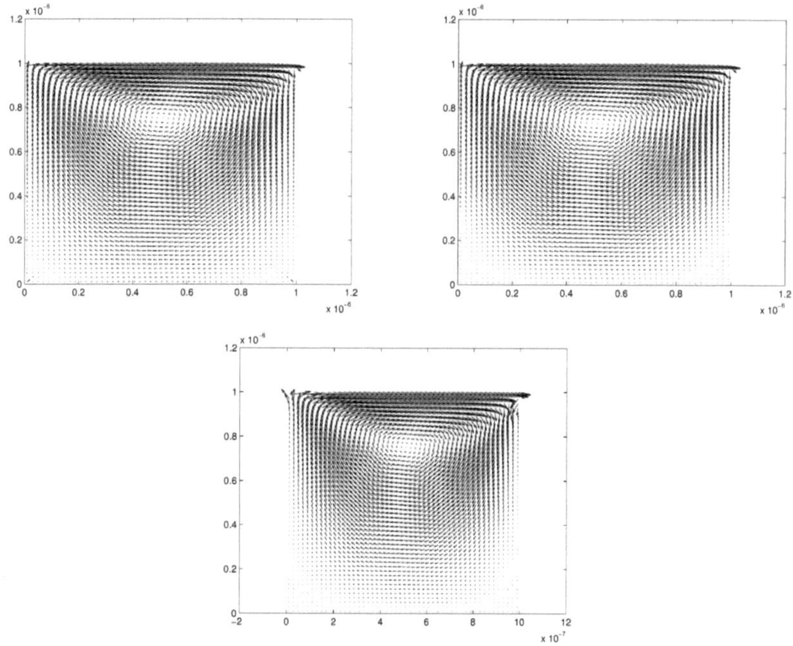

Fig. 3. Velocity fields from Boltzmann solver(top left), Coupled solver(top right) and Navier-Stokes solver (bottom) for the Knudsen number = 0.05

for $Kn = 0.05$ we observed that the velocity from the coupling solver is under predicted compared to the solution from the Boltzmann solver. This is due to the smoothing effect in the Boltzmann domain near to the interface. Some investigations are required to reduce this smoothing effect.

Similarly, in Fig.5, the y-velocity components are plotted along the horizontal central lines. Here also one can observe the same behaviors as in the x-velocity components along the vertical central lines.

For the Knudsen number 0.01 it is almost continuum regime, all solutions are close to each other, plotted on the bottoms of Figs. 4 and 5.

5.1 CPU time

As we already mentioned that we consider the final time is equal to 6.5×10^{-6} s. For the Navier-Stokes solver, we stopped the simulation after 2000 time steps. For the hybrid solver, we compute the breakdown criterion after 2000 time steps and perform domain decomposition and solve respective equations in their domains of validity. We further compute another 8000 time steps such that the Boltzmann equation recovers the flow field in the entire domain. Then we start sampling the macroscopic quantities after 10000 time steps and

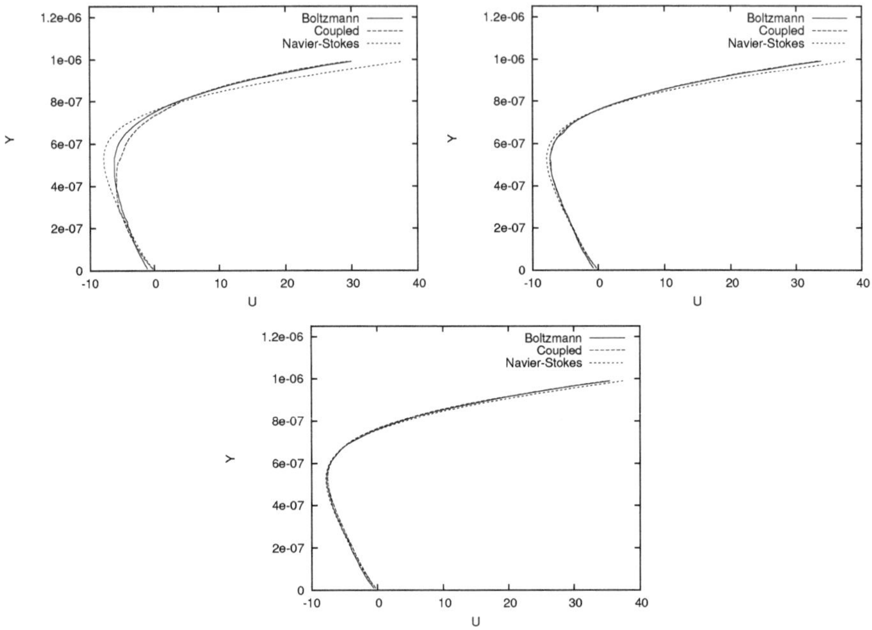

Fig. 4. x-velocity component along the vertical center line. Kn = 0.05(top left), Kn = 0.02 (top right) and Kn = 0.01 (bottom)

taken the mean over the number of sampling. To compare the CPU time, the sampling is performed also after 10000 time step for the Boltzmann solver.

The computation was carried out on a single processor Intel Xeon E5420 (2.5 GHz). In Table 1 the CPU times are shown for the the Boltzmann and hybrid solvers. The CPU times for the Navier-Stokes solver are not presented since we continue the hybrid solver after the Navier-Stokes equations reach steady state, therefore, it is negligible compared to the other two. For $Kn = 0.05$ hybrid solver is slightly faster than the Boltzmann one, however, for $Kn = 0.02$ and $Kn = 0.01$ hybrid solver is more than two and five times faster than the Boltzmann solver. We remark that in the case of hybrid solver, the CPU time for $Kn = 0.02$ is slightly smaller than the CPU time for $Kn = 0.01$. The reason is that the size of Boltzmann domains for both cases are almost same, however, the number of inner Boltzmann iterations for $Kn = 0.01$ is double of the case for $Kn = 0.02$.

6 Conclusion

We have presented a coupling procedure of meshfree particle method and mesh based kinetic particle method for a stationary micro cavity flow problem. The Boltzmann equation is simulated by a variant of the DSMC method

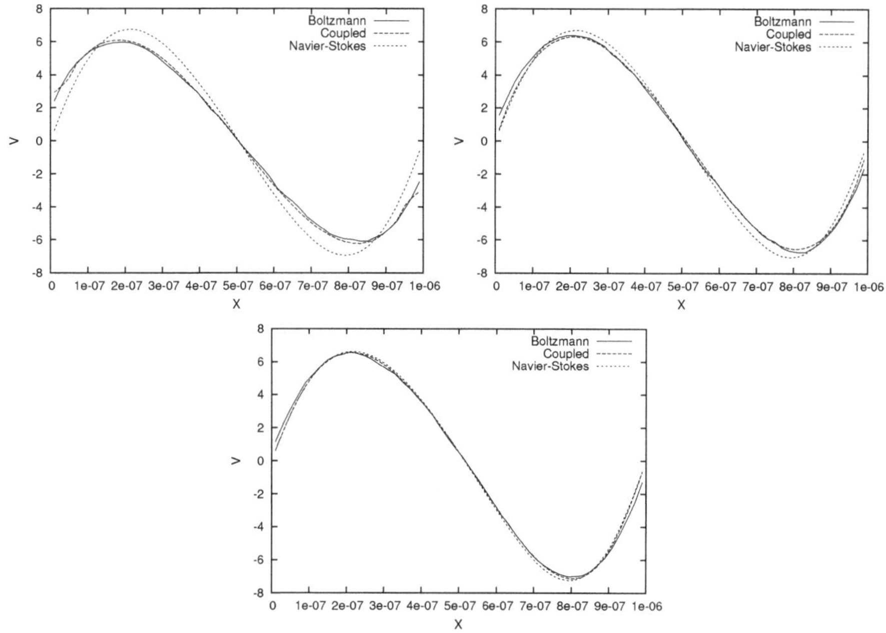

Fig. 5. y-velocity component along the horizontal center line. Kn = 0.05(top left), Kn = 0.02 (top right) and Kn = 0.01 (bottom)

Table 1. Comparison of CPU time for Boltzmann and hybrid solvers

Kn	Boltzmann	Hybrid
0.05	1507 min	1230 min
0.02	1997 min	823 min
0.01	4243 min	853 min

and the Navier-Stokes equations are solved by a meshfree Lagrangian particle method. For the coupling solver, we started with the Navier-Stokes equations and reach the steady state and then apply a breakdown criterion to decompose the domains. The coupling procedure shows that, for larger Knudsen numbers, where the Navier-Stokes equations fail to predict the correct flow behavior, its stationary solutions are the good candidate to initialize the Boltzmann solver. The coupling between the two domains is done by sampling macroscopic field quantities from the particle ensembles in one direction and by creating particle ensembles from a Maxwell distribution with suitable parameters in the other direction. A smoothing operator has been used in the Boltzmann domain near to the interface in order to pass the smooth information to the Navier-Stokes domain. A very satisfactory agreement between the solutions of the coupling and the Boltzmann codes was found. However, the coupling code is faster than the Boltzmann code. Future work will be concentrating on simulations of multiphase flows in nano devices.

Acknowledgment: This work was supported by the German Research Foundation (DFG), KL 1105/17-1. We would like to thank the (DFG) for the financial support.

References

1. O. Aktas, N. R. Aluru, A Combibed Continuum/DSMC Technique for Multiscale Analysis of Microfluidic Filters, J. Comput. Phys., 178 (2002) 342-372.
2. H. Babovsky, A convergence proof for Nanbu's Boltzmann simulation scheme, Eur. J. Mech., 8:41, 1989.
3. H. Babovsky, R. Illner, A convergence proof for Nanbu's simulation method for the full Boltzmann equation. SIAM J. of Num. Anal., 26:45, 1989.
4. C. Bardos, F. Golse, D. Levermore, Fluid dynamic limits of kinetic equations. JSP 63 (1991) 323-344.
5. G. A. Bird, Molecular Gas Dynamics and Direct Simulation of Gas Flows, Oxford University Press, New York, 1994.
6. I. D. Boyd, G. Chen, C. V. Candler, Predicting failure of the continuum continuum fluid equations in transitional hypersonic flows, Phys. Fluids, 7 (1995) 210.
7. R. Caflish. The fluid dynamical limit of the nonlinear Boltzmann equation. CPAM, 33:651, 1980.
8. C. Cercignani, R. Illner, M. Pulvirenti, *The Mathematical Theory of Dilute Gases.* Springer, 1994.
9. S. Chen, Weinan E, Y. Liu, C.-W. Shu, A discontinuous Galerkin implementation of a domain decomposition method for kinetic hydrodynamic coupling multiscale problems in gas dynamics and device simulations. J. Comput. Phys., 225 (2007) 1314-1330.
10. A. K. Chaniotis, D. Poulikakos, P. Koumoutsakos, Remeshed Smoothed Particle Hydrodynamics for the Simulation of Viscous and Heat Conducting Flows, J. Comput. Phys. 182 (2002) 67-90.
11. P. Degond, G. Dimarco, L. Mieussens, A moving interface method for dynamic kinetic-fluid coupling, J. Comput. Phys. 227 (2007) 1176-1208.
12. M. Gad-el Hak, The Fluid Mechanics of Microdevices - The Freeman Scholar Lecture, ASME J. Fluids Engs., 121(403), 5-33, 1999.
13. R. A. Gingold, J. J. Monaghan, Smoothed Particle Hydrodynamics: theory and application to non-spherical stars, Mon. Not. Roy. Astron. Soc. 181 (1977) 375-389.
14. A. L. Gracia, B. J. Alder, Generation of the Chapmann-Enskog Distribution, J. Comput. Phys. 140 (1998) 66-70.
15. A. L. Gracia, J. B. Bell, W. Y. Crutchfield, B. J. Adler, Adaptive mesh and algorithm refinement using direct simulation Monte Carlo, J. Comput. Phys., 154 (1999) 134.
16. A. Klar, Domain Decomposition for Kinetic Problems with Noneqilibrium States, Eur. J. Mech., B/Fluids, 15, 2, (1996) 203-216.

17. V.I. Kolobov, R.R. Arslanbekov, V.V. Aristov, A.A. Frolova, S.A. Zabelok, Unified solver for rarefied and continuum flows with adaptive mesh and algorithm refinement, J. Comput. Phys. 223 (2) (2007) 589-608.
18. P. Le Tallec, F. Mallinger, Coupling Boltzmann and Navier-Stokes equations by half fluxes, J. Comput. Phys. 136, (1997) 51-67.
19. D. Levermore, W. J. Morokoff, B. T. Nadiga, Moment realizability and the validity of the Navier-Stokes equations for rarefied gas dynamics, Phys. Fluids 10 (12) (1998).
20. S. Mizzi, D. R. Emerson, S. K. Stefanov, R. W. Barber, J. M. Reese, Effects of Rarefaction on Cavity Flow in the slip regime, J. Comp. Theor. Nanoscience, Vol. 4, No. 4, 817-622, 2007.
21. H. Neunzert and J. Struckmeier, Particle methods for the Boltzmann equation. Acta Numerica, page 417, 1995.
22. S. Narris, D. Valougeorgis, The driven cavity flow over the whole range of the Knudsen number, Phys. of Fluids, 17, 097106, 2005.
23. Y. Sone, Molecular Gas Dynamics, Theory, Techniques and Applications, Birkhaueser, 2007.
24. J. STOER AND R. BULIRSCH, *Introduction to numerical analysis*; Spring-Verlag, New York (1980)
25. Q. Sun, I. D. Boyd, G. V. Candler, A hybrid continuum/particle approach for modeling subsonic, rarefied gas flows, J. Comput Phys. 194 (2004) 256-277.
26. J. C. Tannehill, T. L. Holst, J. V. Rakich, Numerical computation of two-dimensional viscous blunt body flows with an impinging shock; AIAA Paper 75-0154, AIAA 13th Aerospace Sciences Meeting, 1975.
27. S. Tiwari, Coupling of the Boltzmann and Euler equations with automatic domain decomposition, J. Comput. Phys. 144 (1998) 710-726.
28. S. Tiwari, A LSQ-SPH Approach for Solving Compressible Viscous Flows, International Serier of Numerical Mathematics, 141, Freistüler and Warnecke (Eds), Birkhäuser, 2001.
29. S. Tiwari, A. Klar, An adaptive domain decomposition procedure for Boltzmann and Euler equations, J. Comput. Appl. Math. 90, 233, 1998.
30. S. Tiwari, A. Klar, S. Hardt, A particle-particle hybrid method for kinetic and continuum equations, J. Comput. Phys., 228, 7109-7124, 2009.
31. S. Tiwari and J. Kuhnert, *Finite pointset method based on the projection method for simulations of the incompressible Navier-Stokes equations*, (M. Griebel and M. A. Schweitzer, eds.), Lecture Notes in Computational Science and Engineering, vol. 26, Springer, 2002, pp. 373–387.
32. S. Tiwari and J. Kuhnert, *A numerical scheme for solving incompressible and low Mach number flows by Finite Pointset Method*, (M. Griebel and M. A. Schweitzer, eds.), Lecture Notes in Computational Science and Engineering, vol. 43, Springer, 2005, pp. 191–206.
33. S. Tiwari et al, *A Meshfree Method for Simulations of Interactions between Fluids and Flexible Structures*, (M. Griebel and M. A. Schweitzer, eds.), Lecture Notes in Computational Science and Engineering, vol. 57, Springer, 2006, pp. 249–264.
34. S. Tiwari, S. Manservisi, Modeling Incompressible Navier-Stokes flows by least squares approximation. The Nepali Math. Sci. Report, Vol. 20, No. 1 &2, 2002.
35. H. S. Wijesinghe, N. G. Hadjiconstantinou, A discussion of hybrid atomistic-continuum methods for multiscale hydrodynamics, Int. J. Multiscale Comput. Eng. 2 (2004) 189.

Accuracy and Robustness of Kinetic Meshfree Method

Konark Arora[1] and Suresh M. Deshpande[2]

[1] CFD Division, DOCD, DRDL, Hyderabad - 500058 konark.arora@gmail.com
[2] EMU, JNCASR, Jakkur, Bangalore - 560064 smd@jncasr.ac.in

Summary. Meshfree methods are gaining popularity over the conventional CFD methods for computation of inviscid and viscous compressible flows past complex configurations. The main reason for the growth of popularity of these methods is their ability to work on any point distribution. These methods donot require the grid for flow simulation, which is an essential requirement for all other conventional CFD methods. However these methods are limited by the requirement of a good connectivity around a node. Here, a very robust form of the meshfree method called Weighted Least Squares Kinetic Upwind Method using Eigendirections (WLSKUM-ED) has been used to avoid the problem of code divergence due to the bad connectivity. In WLSKUM-ED, the weights are calculated to diagonalize the least squares matrix $A(w)$ such that the x and y directions become the eigen directions along which the higher dimensional least squares formulae reduce to the corresponding one dimensional formulae. Here an effort has been made to explain the enhanced robustness of the WLSKUM-ED meshfree method over the conventional LSKUM meshfree method. The accuracy of the kinetic meshfree method for the Euler equations has been enhanced by use of entropy variables and inner iterations in the defect correction method. It is observed that the use of entropy variables and inner iterations in the defect correction method helps in obtaining the formasl order of accuracy in case of a non-uniform point distribution.

Key words: LSKUM, WLSKUM-ED, Eigendirections, SVD, LED, Rank deficiency, Entropy variables.

1 Introduction

Meshfree methods are gaining popularity over the conventional CFD methods for computation of inviscid and viscous compressible flows past complex configurations. The main reason for the growth of popularity of these methods is their ability to work on any point distribution. These methods do not require the grid for flow simulation, which is an essential requirement for all other conventional CFD methods. But they do require point distribution or a cloud

M. Griebel, M.A. Schweitzer (eds.), *Meshfree Methods for Partial Differential Equations V*, Lecture Notes in Computational Science and Engineering 79,
DOI 10.1007/978-3-642-16229-9_11,

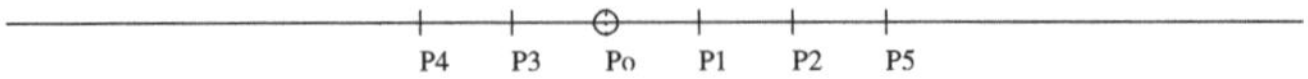

Fig. 1. Point distribution for 1D least squares formula

of points. To fully exploit the advantage of meshfree methods, it is necessary to enhance their robustness and accuracy so that they can work efficiently and effectively on all types of point distributions. In this paper, we have addressed the issue of accuracy and robustness of kinetic meshfree method.

2 Least Squares Meshfree Method

The basic idea behind the least squares meshfree method is the discretization of the spatial derivatives using the least squares approach [1,11,17]. The spatial derivatives are obtained by minimizing the sum of the squares of the error (deviation) in the truncated Taylor's series, leading to a system of linear algebraic equations whose solution gives us the formulae for the spatial derivatives. Consider in 1D, a distribution of points shown in Fig.1. It is desired to get the derivative of a function $F(x)$ at point P_o shown in Fig.1. We expand F_i around point P_o in terms of Taylors series :

$$F_i = F_o + (x_i - x_o)F_{xo} + O(\Delta x)^2 \tag{1}$$

Define

$$\Delta x_i = x_i - x_o, \qquad \Delta F_i = F_i - F_o$$

Neglecting the higher order terms in Eq. 1, we define the deviation e_i by

$$e_i = (\Delta F_i - \Delta x_i F_{xo}) \tag{2}$$

The sum of the squares of deviation e_i at point P_o is given by

$$E = \sum_{i=1}^{p} e_i^2 = \sum_{i=1}^{p} (\Delta F_i - \Delta x_i F_{xo})^2 \tag{3}$$

where p is the number of points in the stencil or connectivity. Minimizing E in Eq.(3) with respect to F_{xo} and simplifying, we get the first order accurate least squares (LS) formula for the derivative in one dimension as

$$F_{xo}^{(1)} = \frac{\sum\limits_{i=1}^{p} \Delta x_i \Delta F_i}{\sum\limits_{i=1}^{p} \Delta x_i^2} \tag{4}$$

The least squares matrix A for 2-D and 3-D cases respectively is given by

$$A = \begin{bmatrix} \sum \Delta x_i{}^2 & \sum \Delta x_i \Delta y_i \\ \sum \Delta x_i \Delta y_i & \sum \Delta y_i{}^2 \end{bmatrix} \qquad A = \begin{bmatrix} \sum \Delta x_i{}^2 & \sum \Delta x_i \Delta y_i & \sum \Delta x_i \Delta z_i \\ \sum \Delta x_i \Delta y_i & \sum \Delta y_i{}^2 & \sum \Delta y_i \Delta z_i \\ \sum \Delta x_i \Delta z_i & \sum \Delta y_i \Delta z_i & \sum \Delta z_i{}^2 a \end{bmatrix}$$

It is observed that the least squares (LS) matrix A so obtained is a geometric matrix whose elements are functions of the coordinate differentials of the nodes in the connectivity. The bad connectivity at a node makes the corresponding LS matrix rank deficient matrix [6], which results in the loss of accuracy or even bad connectivity related code divergence. The LS matrix for a node with good connectivity is found to be a full rank matrix. Thus to enhance the robustness of the least squares meshfree solver, it is required that the LS matrix should always be a full rank matrix. There are various methods of improving the rank deficiency of the LS matrix. Praveen [15,16] has enhanced the connectivity of the nodes to improve the rank deficiency of the LS matrix. We have made use of the weighted least squares approach. The formulae for the derivatives using the weighted least squares (WLS) method are obtained in an identical manner [2]. For WLS, it is required to minimize the weighted sum of the squares of the deviation to get the formulae for the derivatives. The 1-D formula for the derivative using the WLS method is given as

$$F_{x_o}^{(1)} = \frac{\sum\limits_{i=1}^{p} w_i \Delta x_i \Delta F_i}{\sum\limits_{i=1}^{p} w_i \Delta x_i{}^2} \tag{5}$$

The weighted LS formulae for the spatial derivatives in 2-D is

$$F_{x_o}^{(1)} = \frac{\sum w_i \Delta y_i{}^2 \sum w_i \Delta x_i \Delta F_i - \sum w_i \Delta x_i \Delta y_i \sum w_i \Delta y_i \Delta F_i}{\sum w_i \Delta x_i{}^2 \sum w_i \Delta y_i{}^2 - (\sum w_i \Delta x_i \Delta y_i)^2}$$

The weights are suitably chosen so that the (X,Y) direction along which the upwinding is done becomes one of the eigendirections of the LS matrix A along which the multidimensional LS formulae reduce to the corresponding one dimensional formulae [2–4].

$$F_{x_o}^{(1)} = \frac{\sum w_i \Delta y_i{}^2 \sum w_i \Delta x_i \Delta F_i}{\sum w_i \Delta x_i{}^2 \sum w_i \Delta y_i{}^2} = \frac{\sum w_i \Delta x_i \Delta F_i}{\sum w_i \Delta x_i{}^2} \tag{6}$$

because by choice

$$\sum w_i \Delta x_i \Delta y_i = 0$$

Further, the weights chosen must be positive so that the Local Extremum Diminishing (LED) property is satisfied. A very simple and novel way to calculate the positive weights, utilizing the coordinate differentials of the neighbouring nodes in the connectivity in 2-D and 3-D, has been developed for the purpose [2].

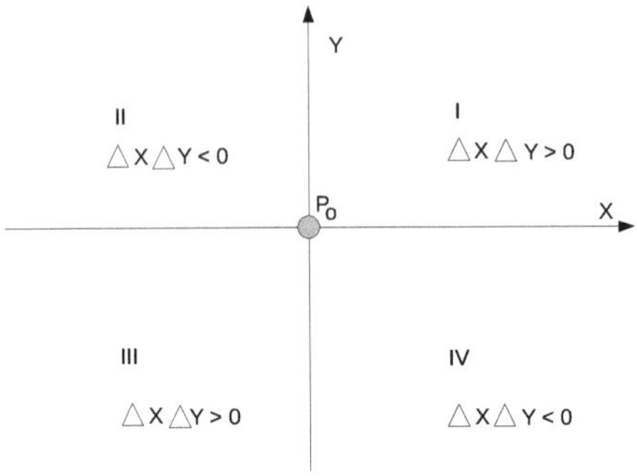

Fig. 2. Quadrants for the split stencil of a node

3 Method of calculation of Weights in 2-D

Fig. 2 shows the four quadrants around the point P_o in which the neighbours of the point P_o are located. We observe that the product of the coordinate differentials $(\Delta x \Delta y)$ is always positive in the first and the third quadrant, while it is always negative in the second and the fourth quadrant. The upwinding in the least squares method is done by stencil splitting. The split stencil of the point P_o in case of stencil splitting will always contain two quadrants : one with the positive product of the coordinate differentials while the other with the negative product of the coordinate differentials. Suppose, we want to find the weights for the LS formula using the points in the left stencil only (consisting of second ane third quadrants only). Making use of the above observation, we can easily find the positive weights which satisfy the condition

$$\Sigma_{II+III}(w_i \Delta x_i \Delta y_i) = 0$$

It must be kept in mind that none of the quadrants of the connectivity stencil should be empty. Let w_{II} be the weight assigned to the points lying in the quadrant II of the stencil while w_{III} be the weight assigned to the points lying in the quadrant III of the stencil. We then get

$$w_{III}(\Sigma \Delta x_i \Delta y_i)_{III} + w_{II}(\Sigma \Delta x_i \Delta y_i)_{II} = 0 \tag{7}$$

Introducing the notation for the cross products:

$$C_{xy}^{III} = (\Sigma \Delta x_i \Delta y_i)_{III} \text{ and } C_{xy}^{II} = (\Sigma \Delta x_i \Delta y_i)_{II}$$

we obtain

$$w_{II} C_{xy}^{II} + w_{III} C_{xy}^{III} = 0 \tag{8}$$

Here : $C_{xy}^{II} < 0$ and $C_{xy}^{III} > 0$

$$\frac{w_{II}}{w_{III}} = -\frac{C_{xy}^{III}}{C_{xy}^{II}} > 0 \text{ always} \tag{9}$$

It is observed from Eq.(9),that the ratio of the weights obtained is always positive as the product $\Delta x \Delta y$ is always negative in the quadrant II while it is always positive in the quadrant III. This ratio goes to zero or infinity when either one of the quadrants is empty which goes against our requirement that the region contributing to the derivative must have atleast a point. A similar procedure can then be applied to find positive weights for other connectivity stencils.

By doing the SVD (Singular Value Decomposition) analysis of the least squares matrix of a node with good and a bad connectivity [6], it is observed that the rank deficiency of the LS matrix of the node with bad connectivity is overcome to a good extent, thus enabling the method to work effectively on a bad point distribution thereby enhancing the robustness of the meshfree method.

4 Higher Order Accuracy in meshfree methods

The issue of the accuracy of the numerical method is equally important. For a numerical scheme to be of useful value for the industrial and design purposes, it should be able to provide accurate results to the user. There are two main methods of obtaining higher order accurate formulae by least squares meshfree method : the Direct Method and the Defect Correction Method [11, 12]. The Direct method has been developed by Liszka and Orkisz [13] and it is called by them as Finite Difference Method on arbitrary irregular grids. In the Direct method, the higher order accurate formulae of the derivatives are obtained by minimizing the sum of the squares of the deviation of the appropriately truncated Taylor series. It is observed that by using this method, we have to solve a larger system of equations even to get first order derivatives in one dimension [5]. The LS matrix obtained using the direct approach is in many cases rank deficient. Liszka and Orkisz [13] have also noted problems about deriving well conditioned FD formulae. The matrix therefore has a greater sensitivity to the connectivity thereby affecting the robustness of the least squares meshfree method. The second approach which is different from the direct method is called the defect correction approach [11, 12]. In this approach higher accuracy is achieved by employing a PADE type of method for approximating derivatives. The LS formulae derived using this approach are implicit in nature because values of derivatives at a node P_o depends on their values at other nodes.

In 2-D, the Taylors Series, for higher order accuracy, is truncated up to $O\left(\Delta x^3, \Delta x^2 \Delta y, \Delta x \Delta y^2, \Delta y^3\right)$. Hence sum of the squares of the deviation in 2-D is given by :

$$E = \sum_{i=1}^{p} \left(\Delta F_i - \Delta x_i F_{xo} - \Delta y_i F_{y_o} - \frac{\Delta x_i^{\,2}}{2} F_{xxo} - \Delta x_i \Delta y_i F_{yy_o} - \frac{\Delta y_i^{\,2}}{2} F_{yy_o} \right)^2 \tag{10}$$

The Taylors series in 2-D truncated up to $O\left(\Delta x^2, \Delta x \Delta y, \Delta y^2\right)$ is

$$F_i = F_o + \Delta x_i F_{xo} + \Delta y_i F_{y_o} + O\left(\Delta x^2, \Delta x \Delta y, \Delta y^2\right) \tag{11}$$

Taking the derivative of this series with respect to x, we get

$$F_{xi} = F_{xo} + \Delta x_i F_{xxo} + \Delta y_i F_{xy_o} + HOT \tag{12}$$

Similarly taking the derivative of the series with respect to y, we get

$$F_{y_i} = F_{y_o} + \Delta x_i F_{xy_o} + \Delta y_i F_{yy_o} + HOT \tag{13}$$

Multiplying Eq.(12) by Δx_i and Eq.(13) by Δy_i, adding them and neglecting higher order terms, we get

$$\Delta x_i^{\,2} F_{xxo} + 2\Delta x_i \Delta y_i F_{xy_o} + \Delta y_i^{\,2} F_{yy_o} = \Delta x_i \Delta F_{xi} + \Delta y_i \Delta F_{y_i}$$

which can be simplified to get

$$\frac{\Delta x_i^{\,2}}{2} F_{xxo} + \Delta x_i \Delta y_i F_{xy_o} + \frac{\Delta y_i^{\,2}}{2} F_{yy_o} = \frac{\Delta x_i}{2} \Delta F_{xi} + \frac{\Delta y_i}{2} \Delta F_{y_i} \tag{14}$$

Substituting Eq.(14) in Eq.(10), the sum of the squares of the deviation becomes

$$E = \sum_{i=1}^{p} \left(\Delta F_i - \Delta x_i F_{xo} - \Delta y_i F_{y_o} - \frac{\Delta x_i}{2} \Delta F_{xi} - \frac{\Delta y_i}{2} \Delta F_{y_i} \right)^2 \tag{15}$$

Define the modified difference as :

$$\Delta \widetilde{F}_i = \Delta F_i - \frac{\Delta x_i}{2} \Delta F_{xi} - \frac{\Delta y_i}{2} \Delta F_{y_i}$$

In terms of this modified difference, Eq.(15) becomes

$$E = \sum_{i=1}^{p} \left(\Delta \widetilde{F}_i - \Delta x_i F_{xo} - \Delta y_i F_{y_o} \right)^2 \tag{16}$$

Now proceeding as before, on minimizing the deviation, we get second order accurate least squares formulae for the derivatives.

$$F_{xo}^{\;(2)} = \frac{\sum \Delta y_i^{\,2} \sum \Delta x_i \Delta \widetilde{F}_i - \sum \Delta x_i \Delta y_i \sum \Delta y_i \Delta \widetilde{F}_i}{\sum \Delta x_i^{\,2} \sum \Delta y_i^{\,2} - \left(\sum \Delta x_i \Delta y_i\right)^2} \tag{17}$$

$$F_{y_o}^{\;(2)} = \frac{\sum \Delta x_i^{\,2} \sum \Delta y_i \Delta \widetilde{F}_i - \sum \Delta x_i \Delta y_i \sum \Delta x_i \Delta \widetilde{F}_i}{\sum \Delta x_i^{\,2} \sum \Delta y_i^{\,2} - \left(\sum \Delta x_i \Delta y_i\right)^2} \tag{18}$$

It is observed that the value of the modified difference $\Delta\widetilde{F}_i$ in Eqs. 17 and 18 depends upon the value of the second order accurate derivatives at the node and at its neighbours, thus requiring some inner iterations to get formally second order accurate derivatives at the node under consideration. However, this approach is simple as compared to the direct method described above and the formulae used to get the higher order accurate value of the derivatives has the same form as first order formula.

5 Kinetic Meshfree Method for Euler Equations

The least squares meshfree method has been applied to the Euler Equations via the kinetic theory approach [14] using the moment method strategy. The starting point of the kinetic method for the Euler equations is the Boltzmann equation governing the motion of the molecules of the gas at the microscopic level. The meshfree upwind scheme is first developed for the Boltzmann equation in the Eulerian limit at the microscopic level. Applying the moment method strategy to the discretized Boltzmann equation, we get the corresponding meshfree scheme for the Euler equations [11] at the macroscopic level. The robustness of the meshfree method for the Euler equations can be obtained as described above by using the weighted least squares method. The positive weights in the weighted least squares method are suitably chosen that the x-y direction in which upwinding is done becomes one of the eigendirections of the LS matrix A along which the multidimensional LS formulae reduce to the corresponding one dimensional formulae. The defect correction method too can be easily applied to obtain higher order accurate meshfree method for the Euler equations. However, it is to be noted that by using the defect correction method (even with inner iterations), we will not be able to get uniform higher order accuracy in the entire domain. The reason for this is the fact that the defect correction method requires the computation of the values of the corresponding derivatives at the neighbouring nodes in the domain. While this can be easily done for the interior nodes in the domain, it is not possible for the boundary nodes as the connectivity stencil for calculation of the corresponding derivatives may be outside the computational domain and hence will be totally empty. This problem in case of the Euler equations has been solved by making use of the q-variables (also called the Entropy variables). [7,9,10]

6 Higher order accuracy by combining Defect Correction with Entropy Variables (q-LSKUM)

The entropy variables are obtained while transforming the Euler equations to the symmetric hyperbolic form [8]. The entropy variables (also called the q-variables) are so called because of their relation to the Boltzmann H Function

: $q = \dfrac{\partial H}{\partial U}$. The Maxwellian is uniquely determined by the q-variables as well as the primitive variables. The linear interpolation of the q-variables is also another q-variable from which a new Maxwellian can be determined. This property has been used to construct a fully second order accurate q-LSKUM (Least Squares Kinetic Upwind Meshfree method using q-variables) solver. In q-LSKUM, the defect correction step of second order LSKUM is modified so as to obtain positive velocity distributions. The perturbed entropy variables are defined in a manner similar to the modified variables in case of the defect correction step described earlier. The perturbed entopy variable is defined as

$$\widetilde{q_i} = q_i - \frac{1}{2}\left(\Delta x_i q_{x_i}^{(2)} + \Delta y_i q_{y_i}^{(2)}\right) \tag{19}$$

The second order accurate derivatives $q_{x_i}^{(2)}$ and $q_{y_i}^{(2)}$ in the Eq.19 are obtained by using the full stencil in the least squares formulae. The perturbed q-variables (ie. $\widetilde{q_i}$) are then used to construct the modified Maxwellians which are then used to obtain formally second order accurate derivatives in the split equations. The q-LSKUM thus inherits all the good properties of the first order LSKUM such as robusness, positivity and convergence characteristics.

7 Results and Discussion

The enhanced robustness of the meshfree method using eigendirections over the conventional meshfree method has been demonstrated by running both the codes on a good and tampered point distribution for the subsonic flow test case past NACA0012 aerofoil. Figures 3 and 4 show the tampered point distribution over the NACA0012 aerofoil, used to demonstrate the enhanced robustness of the meshfree method with eigendirections. Figures 5 and 6 show that while the conventional meshfree method fails for a tampered point distribution, the meshfree method with eigendirections successfully runs for the same tampered point distribution. The results clearly show the enhanced robustness of the meshfree method with eigendirections. But a question arises regarding the reason for the enhanced robustness of the meshfree method with eigendirections where the difference between the two approaches is the diagonalization of the LS matrix by using appropriate weights. The answer becomes clear when we perform the Singular Value Decomposition (SVD) of the LS matrix for the bad as well as the good connectivity of the node in the computational domain. The least squares matrix A and the corresponding singular value matrix S for the node with *good connectivity* in the computational domain is :

$$A = \begin{bmatrix} 0.002404 & 0.000938 \\ 0.000938 & 0.003815 \end{bmatrix} \qquad S = \begin{bmatrix} 0.0042832 & 0.0000000 \\ 0.0000000 & 0.0019350 \end{bmatrix}$$

The least squares matrix A and the corresponding singular value matrix S for the node with *tampered connectivity* in the computational domain is :

Point Distribution and Residue Drop comparison for good and bad connectivity using conventional and new meshfree method with eigendirections

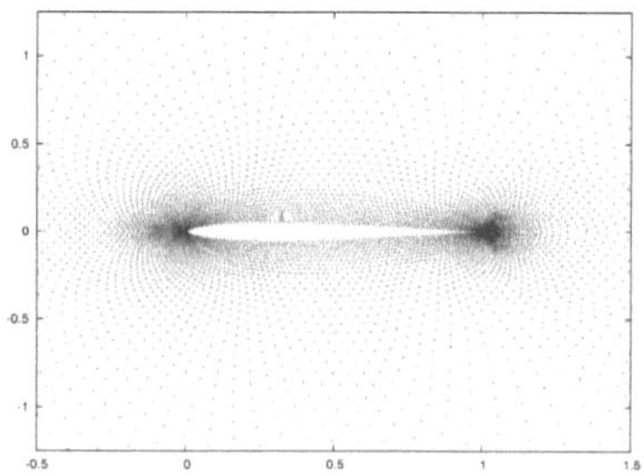

Fig. 3. Tampered Point Distribution

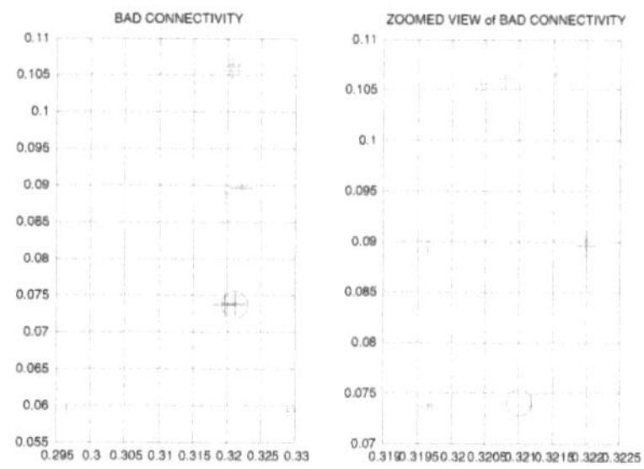

Fig. 4. Zoomed View

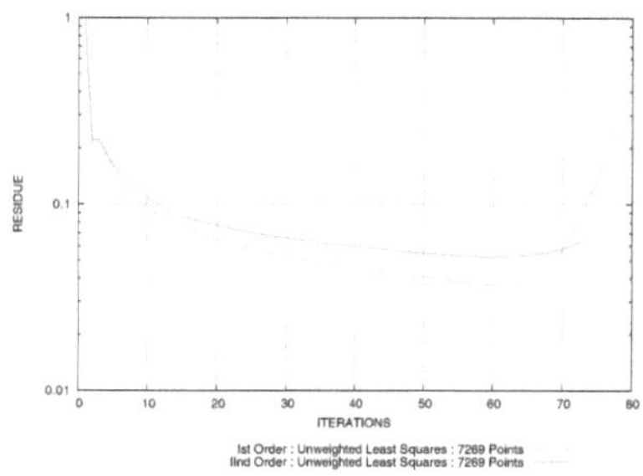

Fig. 5. Residue drop for Conventional Meshfree method

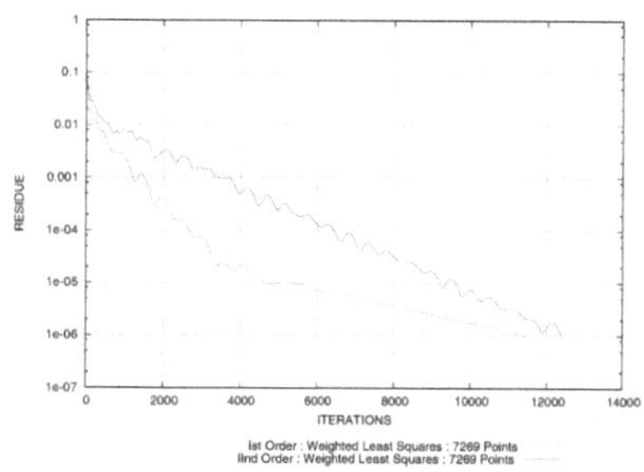

Fig. 6. Residue drop for Meshfree method with eigendirections

$$A = \begin{bmatrix} 0.000066 & -0.000068 \\ -0.000068 & 0.003815 \end{bmatrix} \qquad S = \begin{bmatrix} 0.000065 & 0.0000000 \\ 0.000000 & 0.003816 \end{bmatrix}$$

It is observed that one of the singular values of the LS matrix for the node with a bad connectivity becomes nearly zero resulting in the LS matrix to be a rank deficient matrix. The LS matrix for the node with a good connectivity on the other hand is found to be a full rank matrix. How does the meshfree method with eigendirections tackle the rank deficiency problem? The answer lies in the fact that the diagonalization of LS matrix by suitable weights reduces the LS formulae for the derivatives to the 1D like formulae yielding LED state update scheme. It is observed that the accumulation of error in fluxes with iterations in case of conventional meshfree method using bad connectivity eventually results in blowing off of the solution. But the positive weights in meshfree method with eigendirections preserve its LED property [2] providing stability and keeping the error in the fluxes bounded, thus enhancing its robustness by preventing divergence.

The impact of the use of inner iterations coupled with defect correction method on the accuracy of the results of the meshfree method has been demonstarted by the test case of Linear convective equation (LCE) [18]. The enhancement of the accuracy of the solution of the Eulers' equations of gas dynamics by using entropy variables in place of the primitive variables too has been demonstrated by the standard test cases of subsonic, transonic and supersonic flow past NACA 0012 aerofoil using an unstructured point distribution. It is observed that the use of defect correction method coupled with inner iterations and entropy variables results in an accurate prediction of drag coefficient (close to the theoretical value) and capture of crisper discontinuities.

- The first test case consists of a rectangular domain $[-1,1] \times [0,1]$ on which we consider the linear convective equation :

$$\frac{\partial u}{\partial t} + y\frac{\partial u}{\partial x} - x\frac{\partial u}{\partial y} = 0$$

The boundary conditions for this test case are :

$$\begin{aligned} u(x,0) &= 0, && x < -0.65 \\ u(x,0) &= 1, && -0.65 \le x \le -0.35 \\ u(x,0) &= 0, && -0.35 < x < 0 \\ u(-1,y) &= 0, && 0 < y < 1 \\ u(x,1) &= 1, && 0 < x < 1 \end{aligned}$$

The exact solution of this steady state problem is :

$$\begin{aligned} u_{ex}(x,y) &= 1, && 0.35 < \sqrt{(x^2+y^2)} < 0.65 \\ u_{ex}(x,y) &= 0, && \text{otherwise} \end{aligned}$$

showing that it contains a **contact discontinuity**. Figs. 7 to 8 show the contours of the numerical solution of this test case obtained on a 64^2 structured point distribution using the method described above. Fig. 7 shows the solution contours for this test case where in the second order accuracy has been obtained by the defect correction method without any inner iterations. Fig. 8 shows the solution contours obtained by using two inner iterations coupled with defect correction method. It is clearly observed that the contact discontinuity is crisply captured by using the inner iterations coupled with the defect correction approach on the same point distribution as compared to the solution obtained without the use of inner iterations in defect correction method.
- The second test case is of subsonic flow past NACA 0012 aerofoil. The flow conditions for this test case are a subsonic Mach number of 0.63 at angle of attack(α) = 2.0°. The number of nodes in the computational domain of this test case is 10058.Fig. 9 shows the comparison of the Cp plot over the NACA 0012 aerofoil for the subsonic flow test case, when the defect

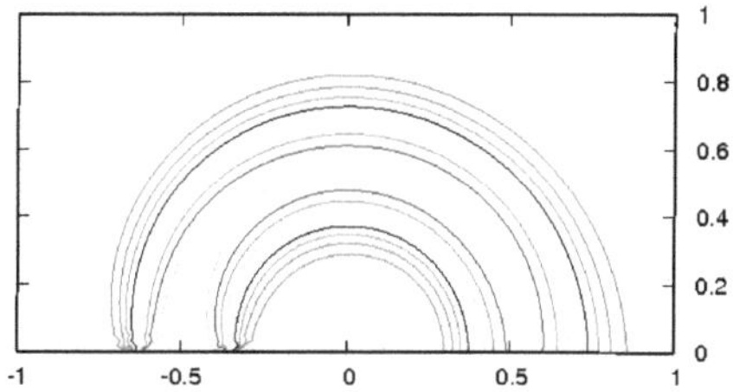

Fig. 7. Structured point distribution with 64^2 nodes in the computational domain. Solution Contours for the LCE with defect correction method without inner iterations

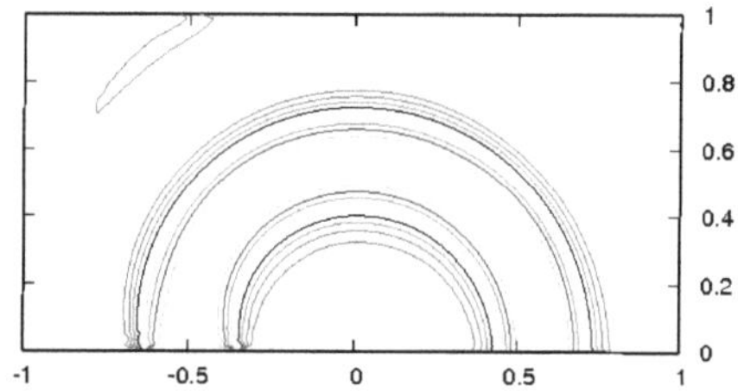

Fig. 8. Structured point distribution with 64^2 nodes in the computational domain. Solution Contours for the LCE with defect correction method with 2 inner iterations

correction method has been applied using q-variables coupled with and without inner iterations. It is observed that use of inner iterations coupled with defect correction method enables the better capture of the suction peak in this case. The pressure contours for the corresponding test case are shown in Figures 10 and 11. Table 1 shows that the drag value is accurately predicted when inner iterations coupled with defect correction method using q-variables are used to obtain higher order accuracy.

- The third test case is of weak transonic flow past NACA 0012 aerofoil. The flow conditions for this test case are a Mach number of 0.80 at angle of attack(α) = 1.25°. The number of nodes in the computational domain of this test case is 10058. Fig. 12 shows the comparison of the Cp plot over the NACA 0012 aerofoil for this test case, when the defect correction method has been applied using q-variables with and without use of inner iterations. The pressure contours for this test case are shown in Figures 13 and 14. It is observed that a crisp strong shock on the upper surface and a weak shock on the bottom surface of the aerofoil are captured when defect correction is applied using q-variables coupled with inner iterations.
- The fourth test case is of supersonic flow past NACA 0012 aerofoil. The flow conditions for this test case are a Mach number of 1.20 at angle of attack(α) = 0.00°. The number of nodes in the computational domain of this test case is 10058. Fig. 15 shows the comparison of the Cp plot over the NACA 0012 aerofoil for this test case, when the defect correction method has been applied using q-variables coupled with and without inner iterations. The pressure contours for the corresponding test case are shown in Figures 16 and 17. Table 1 shows that the lift coefficient value is accurately predicted when inner iterations coupled with defect correction method using q-variables are used to obtain higher order accuracy.

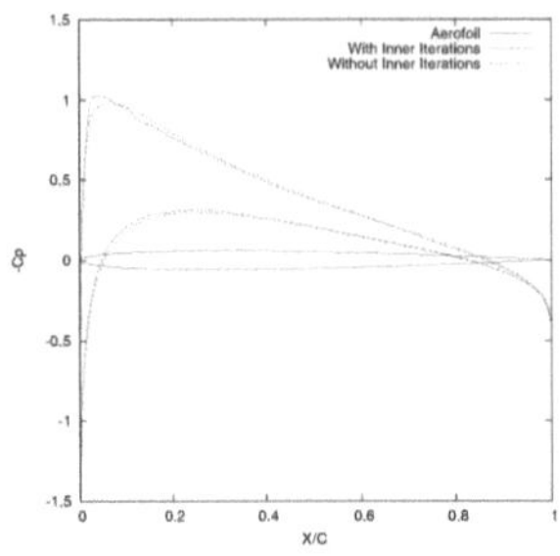

Fig. 9. Cp distribution for Subsonic flow test case over NACA 0012 Aerofoil. Comparison of Cp using q-LSKUM with and without inner iterations

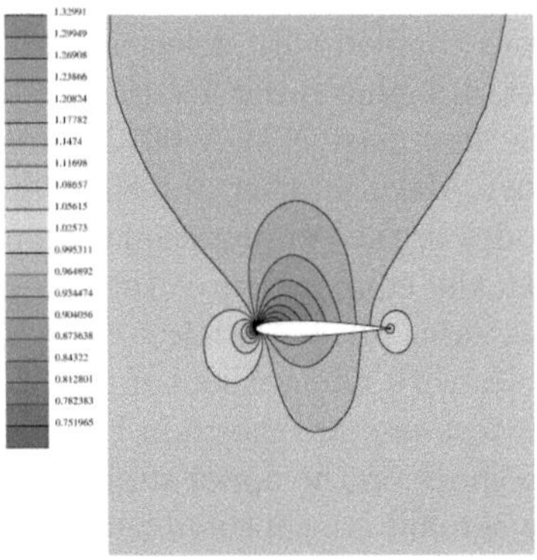

Fig. 10. Pressure Contours for Subsonic flow test case over NACA 0012 Aerofoil using qLSKUM without inner iterations

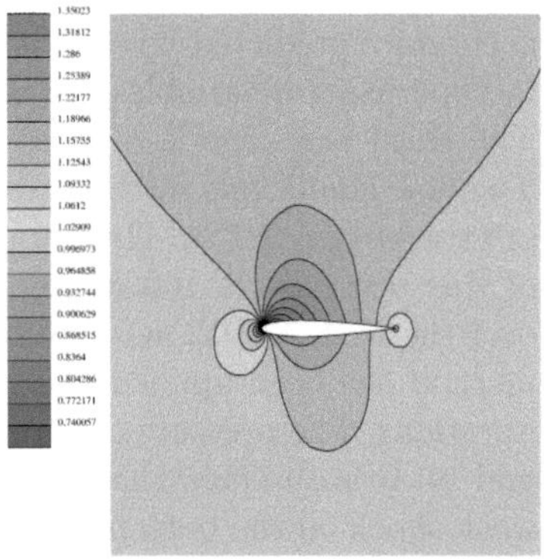

Fig. 11. Pressure Contours for Subsonic flow test case over NACA 0012 Aerofoil using qLSKUM with inner iterations

8 Conclusion

The usual least squares Kinetic Upwind Meshfree method can be made robust by using weighted LSKUM(WLSKUM). The weights are determined by enforcing x y directions as eigendirections of Least Squares (LS) matrix. The resultant WLSKUM method is LED at Boltzmann level. Further the WLSKUM can be be made more accurate by using entropy (q) variables and inner iterations. The robustness and accuracy of the resultant q-LSKUM along eigendirections has been tested on some standard test cases. It is observed

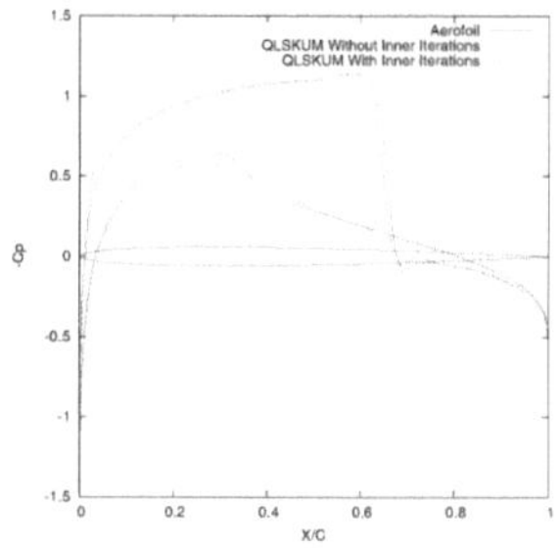

Fig. 12. Cp distribution for Weak Transonic flow test case over NACA 0012 Aerofoil. Comparison of Cp using q-LSKUM with and without inner iterations

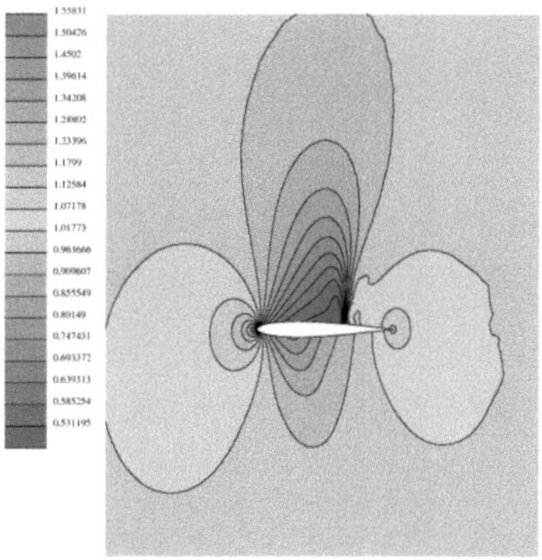

Fig. 13. Pressure Contours for Weak Transonic flow test case over NACA 0012 Aerofoil using qLSKUM without inner iterations

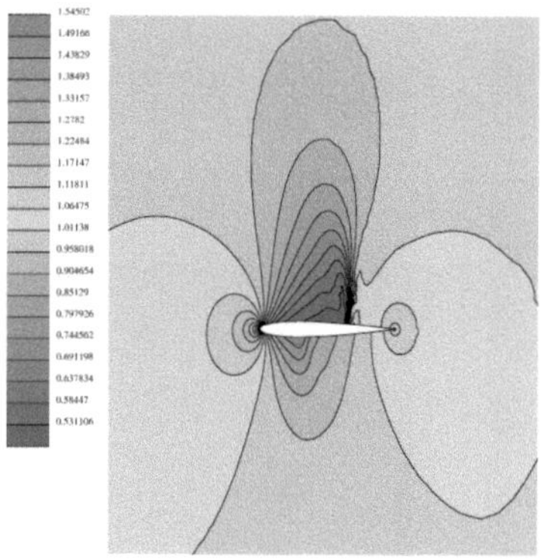

Fig. 14. Pressure Contours for Weak Transonic flow test case over NACA 0012 Aerofoil using qLSKUM with inner iterations

that the inner iterations coupled with defect correction method using entropy variables not only results in accurate prediction of lift and drag coefficients but also enables the capture of crisper discontinuities.

References

1. Anandhanarayanan K., *Development and Applications of a Gridfree Kinetic Upwind Solver to Multibody Configurations* ,PhD. Thesis, Department of

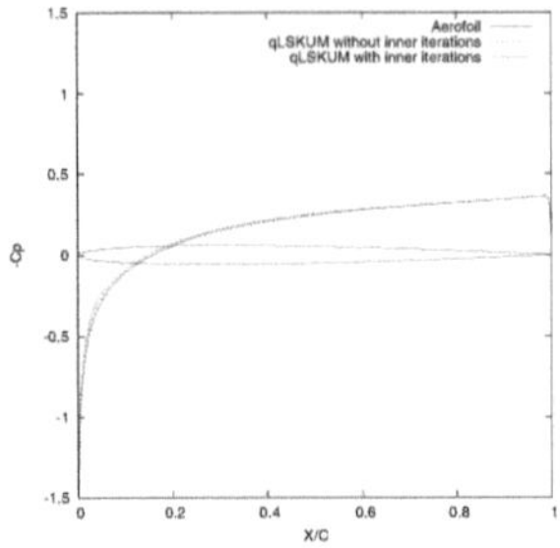

Fig. 15. Cp distribution for Supersonic flow test case over NACA 0012 Aerofoil. Comparison of Cp using q-LSKUM with and without inner iterations

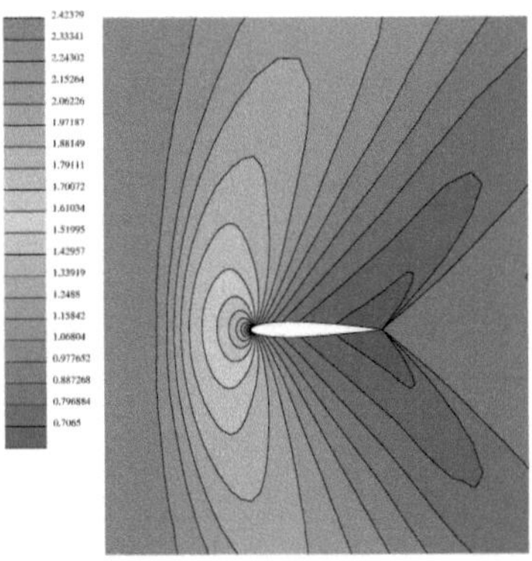

Fig. 16. Pressure Contours for Supersonic flow test case over NACA 0012 Aerofoil using qLSKUM without inner iterations

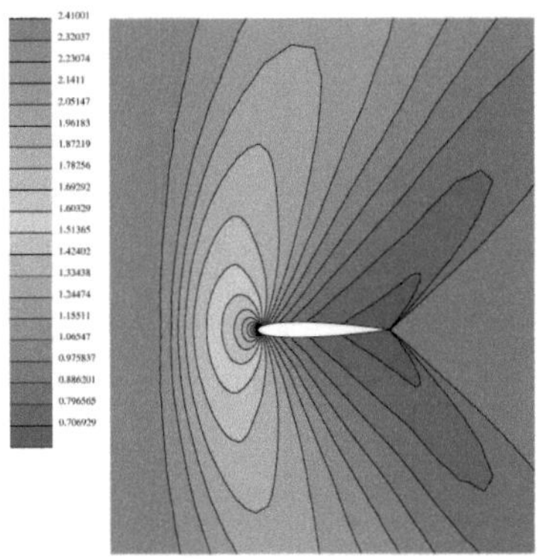

Fig. 17. Pressure Contours for Supersonic flow test case over NACA 0012 Aerofoil using qLSKUM with inner iterations

Aerospace Engineering, Indian Institute of Science, Bangalore, India.

2. Arora K. and Deshpande S. M., *Weighted Least Squares Kinetic Upwind Method using Eigenvector Basis*, FM Report No. 2004 FM 17, Department of Aerospace Engineering,Indian Institute of Science, Bangalore, India.
3. Arora K., Rajan N. K. S. and Deshpande S. M., *Weighted Least Squares Kinetic Upwind Method (WLSKUM) using Eigenvector Basis*, 8th Annual Aesi CFD Symposium, 11th-13th August,2005,Bangalore.
4. Arora K., *Weighted Least Squares Kinetic Upwind Method using Eigendirections (WLSKUM-ED)*, PhD. Thesis, Department of Aerospace Engineering, Indian

Table 1. Comparison of computed C_L and C_D with those of AGARD/GAMM workshop values

Explicit Test Case : 10058 Nodes in the computational domain				
Flow Regime	C_L/C_D *Values*	qLSKUM without inner iterations	qLSKUM with inner iterations	AGARD/GAMM Values
SUBSONIC	C_L	0.301666	0.281740	0.329-0.336
	C_D	0.003745	0.000897	0.003-0.0007
WEAK	C_L	0.355055	0.371242	0.3632
TRANSONIC	C_D	0.023945	0.022119	0.0187-0.02698
SUPERSONIC	C_L	-0.001278	0.000676	0.0000
	C_D	0.095882	0.094695	0.0946-0.0960

Institute of Science, Bangalore, India.

5. Arora K., Rajan N. K. S. and Deshpande S. M., *On the Order of Accuracy of Gridfree Methods using Defect Correction with Inner Iterations*, 7th ACFD Conference, November 26-29, 2007, Bangalore, India.
6. Arora K., Rajan N. K. S. and Deshpande S. M., *On the Robustness and Accuracy of Least Squares Kinetic Upwind Method (LSKUM)*,12th Asian Congress of Fluid Mechanics (ACFM), August 18-21, 2008, Daejeon, Korea.
7. Dauhoo M. Z., Ghosh A. K., Ramesh V. and Deshpande S. M., *q-LSKUM - A new Higher Order Kinetic Upwind Method for Euler Equations using Entropy Variables*, Computational Fluid Dynamics Journal **9** (2000).
8. Deshpande S. M., *On the Maxwellian distribution, symmetric form and entropy conservation for the Euler equations*, NASA TP-2583, 1986.
9. Deshpande S. M., *Some Recent Developments in Kinetic Schemes based on Least Squares and Entropy Variables*, Conference on Solutions of PDE, held in honour of Prof. Roe on the occasion of his 60th birthday, July 1998, Arcacon, France.
10. Deshpande S. M., Anandhanarayanan K., Praveen C. and Ramesh V., *Theory and Applications of 3-D LSKUM based on Entropy Variables*, First ICFD, Oxford, March 2001.
11. Ghosh A. K., *Robust Least Squares Kinetic Upwind Method for Inviscid Compressible Flows*, PhD. Thesis, Department of Aerospace Engineering, Indian Institute of Science, Bangalore, India.
12. Ghosh A. K., and Deshpande S. M., *Least Squares Kinetic Upwind Method for Inviscid Compressible Flows*, AIAA Paper No. 95-1735.
13. Liszka T. and Orkisz J., *The Finite Difference Method At Arbitrary Irregular Grids And Its Application In Applied Mechanics*, Computers & Structures, **Vol 11**, pp. 83-95 (1979).
14. Mandal J. C. and Deshpande S. M., *Kinetic Flux Vector Splitting for Euler Equations*, Computers and Fluids **Vol 23, No. 2** (1994), 447–478.
15. Praveen C., *Development and Applications of Kinetic Meshless Methods for Euler Equations*, PhD. Thesis, Department of Aerospace Engineering, Indian Institute of Science, Bangalore, India.
16. Praveen C., Ghosh A. K. and Deshpande S. M., *Positivity preservation, stencil selection and applications of LSKUM to 3-D inviscid flows*, Computers and Fluids **doi:10.1016/j.compfluid.2008.04.017** (2009).

17. Ramesh V., *Least Squares Grid-Free Kinetic Upwind Method*, PhD. Thesis, Department of Aerospace Engineering, Indian Institute of Science, Bangalore, India.
18. Sperkreijse S., *Multigrid Solution of Monotone Second-Order Discretization of Hyperbolic Conservation Laws*, Mathematics of Computation, **Vol 49, No. 179** (1987), 135–155.

Kinetic meshless methods for unsteady moving boundaries

V. Ramesh[1], S. Vivek[2], and S. M. Deshpande[3]

[1] Computational & Theoretical Fluid Dynamics Division, National Aerospace Laboratories, CSIR, Bangalore-560017, India vr@ctfd.cmmacs.ernet.in
[2] Department of Mechanical Engineering, Indian Institute of Technology, Madras, Chennai- 600036, India svivek2006@gmail.com
[3] Engineering Mechanics Unit, Jawaharlal Nehru Centre for Advanced Scientific Research, Jakkur, Bangalore-560 064, India smd@jncasr.ac.in

Summary. Least squares Kinetic Upwind method for Moving Nodes (LSKUM-MN) is a kinetic theory based gridfree method capable of handling the unsteady flow past multiple moving boundaries. In the present work this capability of LSKUM-MN has been demonstrated by computing flow past multiple oscillating blades encountered in turbomachinary flows. Flutter prediction in such a flow scenario has also been carried out on standard cascade configurations and compared with available results. Energy method has been used to predict flutter. The method has been further applied to 2D store separation problem. We have considered the NACA0012 airfoil section for both wing and the store. Chimera cloud approach has been used to generate the point distribution around each airfoil. As the store moves dynamic blanking and de-blanking of points entering into and out of the solid bodies has been carried out in efficient way by using bounding boxes. Effect of relative time scales of flow and store movement have been studied as well.

Key words: Meshfree, unsteady, flutter prediction, turbomachinery, store separation

1 Introduction

In the present work, we present the use of a grid free method in unsteady flows with moving boundaries like flutter prediction in turbomachinery blades by computing unsteady flow past multiple oscillating blades and store separation. A code based on Least Squares Kinetic Upwind Method(LSKUM) [1, 2] has been developed to tackle multiple oscillating bodies with moving nodes. This is an upwind code which uses kinetic Flux vector splitting(KFVS). LSKUM is a kinetic theory [3] based grid free scheme for solving the inviscid compressible Euler equations of gas dynamics. This method has also been applied to compute viscous flows [4]. LSKUM has been extended to applications with

M. Griebel, M.A. Schweitzer (eds.), *Meshfree Methods for Partial Differential Equations V*, Lecture Notes in Computational Science and Engineering 79,
DOI 10.1007/978-3-642-16229-9_12,

moving nodes(LSKUM_MN) [5–7]. In the present work we use the Modified CIR splitting(MCIR) [8, 9] to obtain spatially higher order accuracy in LSKUM_MN. Essentially the dissipation term present in the first order scheme is suitably chosen to get an equivalent higher order scheme where the dissipation terms are comparable to the usual second order schemes. This leads to a single step(without defect correction step [1]) higher order scheme, thereby reducing the computational costs. Apart from the implementation of MCIR in LSKUM_MN, we have also adopted the weighted least squares approach based on Eigenvector basis [10]. In this approach the least squares approximations for all the derivatives reduce to an equivalent 1-D form. This again helps in further reducing the computational time. For the unsteady calculations we have used the well known dual time stepping procedure [11].

In order to validate the method for aeroelastic applications as well as demonstrate the power of the method to deal with multiple oscillating bodies we have computed the unsteady flow for the 4th standard aeroelastic test case [12]. Results from the present computations are also compared with other Euler computations of Gruber and Cartsens [13], Ji and Liu [14] and Mani Sadeghi [15]. All of them predict the flutter using energy method [16]. This is based on the net energy transfer from the airstream to the blade. In general for turbo machinery blades the aerodynamic forces are much smaller compared to the inertial and stiffness forces. Thus this allows one to do flutter predictions without the need for an active coupling between the CFD computations and a structural module. As such, the prediction of the unsteady flow field for a given blade vibration mode is of essential importance. We have also used the same approach to predict the flutter. The 2D stores separation problem has been solved by using NACA0012 airfoil sections for the wing and the store. Our strategy uses a chimera cloud of points around the bodies without any need for insertion of new points into the changing domain. Dynamic blanking and de-Blanking of points which fall inside the solid boundaries is done using a very simple method and connectivity is generated dynamically using a quad tree preprocessor. We consider the need for this kind of unsteady solver over and above the commonly used quasi steady approaches quasi steady approaches. We explore the effect of relative time scales based on the timescale of the store movement and the flow and how much this dictates the use of true time accurate solvers.

2 Least Squares Kinetic Upwind Method on Moving Nodes

Least squares kinetic upwind method (LSKUM) [1] is a kinetic theory based meshless solver. Meshless solvers are a class of numerical algorithms which do not require generation of mesh or grid. It only needs point distribution over the domain of interest and connectivity information. Connectivity is the set of nearest neighbors of a given node which are used for computation at the node.

The meshless methods do not need any topological information. LSKUM is based on the fact that suitable moments of Boltzmann equation lead to Euler equations. Numerical schemes are derived at Boltzmann level [17]. An upwind scheme is achieved at Boltzmann level by splitting molecular velocity into positive and negative parts using the Courant-Isaacson-Reeves (CIR) splitting. Spatial derivatives are approximated by using least squares approximation. Then update scheme at Boltzmann level is mapped to Euler level through moment method. The second order accuracy can be achieved through defect correction using q-variables. Alternatively a modified CIR (MCIR) [8] splitting can be used to achieve higher order accuracy. The advantage of MCIR is that it involves only single step. The time derivative can be suitably discretized to arrive at update scheme. The LSKUM has been very successful in computing flow past bodies of complex configuration. Much advancement has taken place since its inception. Recently, Konark [10] developed weighted LSKUM (WLSKUM) which increases the robustness of the LSKUM. The MCIR of Ramesh [8] achieves higher order accuracy in single step. These have been implemented in the code being used for the work. A multistage R-K method is used for time stepping. The modification of LSKUM for inclusion of moving nodes(LSKUM_MN) [7] is used.

The details of the various modified split moving fluxes which arise in this boundary treatment are given in reference [7]. Time step for pseudo time marching in the dual time stepping procedure is chosen as given by the stability criterion of Ghosh [2]. The treatment of other boundary conditions using this gridfree method can also be found in [2].

3 Formulation of LSKUM_MN

Here we briefly describe the formulation of 2-D LSKUM_MN. Consider the 2-D Boltzmann equation

$$\frac{\partial f}{\partial t} + v_1 \frac{\partial f}{\partial x} + v_2 \frac{\partial f}{\partial y} = J \tag{1}$$

where f is the velocity distribution function, v_1 and v_2 are the Cartesian components of the molecular velocity. J represents a collision term which vanishes in the Euler limit, when f is a Maxwellian distribution, F, which in two dimensions is given by

$$F = \frac{\rho}{I_0}\frac{\beta}{\pi} \exp\left[-\beta(v_1 - u_1)^2 - \beta(v_2 - u_2)^2 - I/I_0\right] \tag{2}$$

where $\beta = 1/(2RT)$, ρ is the fluid density, I is internal energy variable, I_0 is the internal energy due to non-translational degrees of freedom, $I_0 = \frac{2-\gamma}{\gamma-1}RT$ and u_1 and u_2 are the Cartesian components of the fluid velocity, R is the gas constant and T is the absolute temperature of the fluid. Therefore in the Euler limit it is enough to consider

$$\frac{\partial F}{\partial t} + v_1 \frac{\partial F}{\partial x} + v_2 \frac{\partial F}{\partial y} = 0 \tag{3}$$

Now let w_1 and w_2 represent the Cartesian components of the velocity of any moving node.

In order to deal with problems involving moving nodes, we define the derivative of F along the path of the node as

$$\left(\frac{dF}{dt}\right)_{mov} = \left(\frac{\partial F}{\partial t} + w_1 \frac{\partial F}{\partial x} + w_2 \frac{\partial F}{\partial y}\right)$$

Substituting for $\left(\frac{\partial F}{\partial t}\right)$ in Eq.(3) we get

$$\left(\frac{dF}{dt}\right)_{mov} + (v_1 - w_1) \frac{\partial F}{\partial x} + (v_2 - w_2) \frac{\partial F}{\partial y} = 0 \tag{4}$$

Let $\bar{v_1} = v_1 - w_1$, $\bar{v_2} = v_2 - w_2$ be the components of the particle velocity relative to the moving node. Then Eq.(4) can be compactly written as

$$\left(\frac{dF}{dt}\right)_{mov} + \bar{v_1} \frac{\partial F}{\partial x} + \bar{v_2} \frac{\partial F}{\partial y} = 0 \tag{5}$$

Using MCIR [8] splitting, $\bar{v_1}$ and $\bar{v_2}$ are written as

$$\bar{v_1} = \frac{\bar{v_1} + |\bar{v_1}|\, \phi_1}{2} + \frac{\bar{v_1} - |\bar{v_1}|\, \phi_1}{2}, \qquad v_2 = \frac{\bar{v_2} + |\bar{v_2}|\, \phi_2}{2} + \frac{\bar{v_2} - |\bar{v_2}|\, \phi_2}{2} \tag{6}$$

where ϕ_1, ϕ_2 are dissipation control parameters corresponding to two components of molecular velocity. The dissipation control parameters are conveniently chosen as $\phi_1 = \phi_2 = \Delta r^p$, where $0 < p < 1$ and Δr is the distance between a node and any point in its neighbourhood. We usually choose the closest point.

Using MCIR splitting for both the components of molecular velocity, the Boltzmann Eq.(5) can be written as

$$\left(\frac{dF}{dt}\right)_{mov} + \frac{\bar{v_1} + |\bar{v_1}|\phi_1}{2}\frac{\partial F}{\partial x} + \frac{\bar{v_1} - |\bar{v_1}|\phi_1}{2}\frac{\partial F}{\partial x} + \frac{\bar{v_2} + |\bar{v_2}|\phi_2}{2}\frac{\partial F}{\partial y} + \frac{\bar{v_2} - |\bar{v_2}|\phi_2}{2}\frac{\partial F}{\partial y} = 0 \tag{7}$$

We define moment vector function Ψ by

$$\Psi = \left[1, v_1, v_2, I + \frac{v_1^2 + v_2^2}{2}\right]^T, \tag{8}$$

and define the Ψ moment as

$$\langle \Psi, F \rangle \equiv \int_0^\infty dI \int_{-\infty}^{\infty} dv_1 \int_{-\infty}^{\infty} dv_2\, \Psi F \tag{9}$$

Now the Ψ moment of the Eq.(7) will lead to the Modified Moving Kinetic Flux vector split Euler equations

$$\left(\frac{dU}{dt}\right)_{mov} + \frac{\partial}{\partial x}(GX_M^+) + \frac{\partial}{\partial x}(GX_M^-) + \frac{\partial}{\partial y}(GY_M^+) + \frac{\partial}{\partial y}(GY_M^-) = 0 \quad (10)$$

where, $U \equiv \langle \Psi, F \rangle$, $GX_M^{\pm} \equiv \left\langle \Psi, \frac{\bar{v}_1 \pm \phi_1 |\bar{v}_1|}{2} F \right\rangle$ and $GY_M^{\pm} \equiv \left\langle \Psi, \frac{\bar{v}_2 \pm \phi_2 |\bar{v}_2|}{2} F \right\rangle$

U is the state vector given by $U = (\rho, \ \rho u_1, \ \rho u_2, \ \rho e)^T$, e is the internal energy per unit mass given by $e = \frac{p}{\rho(\gamma-1)} + \frac{1}{2}(u_1^2 + u_2^2)$, $GX_M^{\pm}$ and $GY_M^{\pm}$ are the modified split fluxes for the moving nodes. These modified moving split fluxes are expressed in terms of moving fluxes [5,6] as

$$GX_M^{\pm} = \tfrac{1}{2} \ \{(1+\phi_1) \, GX_m^{\pm} + (1-\phi_1) GX_m^{\mp}\} \quad (11)$$
$$GY_M^{\pm} = \tfrac{1}{2} \ \{(1+\phi_2) \, GY_m^{\pm} + (1-\phi_2) GY_m^{\mp}\} \quad (12)$$

In order to develop an update scheme we need to evaluate the space derivatives of various moving split flxues. In LSKUM_MN the space derivatives are evaluated using least squares approximation. Assume that values of F are available at a node P_0 and its immediate surrounding nodes, refered to as connectivity of point P_0. A first order approximation to the derivatives $\frac{\partial F}{\partial x}$ and $\frac{\partial F}{\partial y}$ using weighted least squares approach [2] is then given by the following formulae

$$F_{x_o}^{(1)} = \frac{\sum w_i \Delta y_i^2 \sum w_i \Delta x_i \Delta F_i - \sum w_i \Delta x_i \Delta y_i \sum w_i \Delta y_i \Delta F_i}{\sum w_i \Delta x_i^2 \, \sum w_i \Delta y_i^2 - (\sum w_i \Delta x_i \Delta y_i)^2}$$
$$F_{y_o}^{(1)} = \frac{\sum w_i \Delta x_i^2 \sum w_i \Delta y_i \Delta F_i - \sum w_i \Delta x_i \Delta y_i \sum w_i \Delta x_i \Delta F_i}{\sum w_i \Delta x_i^2 \, \sum w_i \Delta y_i^2 - (\sum w_i \Delta x_i \Delta y_i)^2}, \quad (13)$$

where, $\Delta x_i = x_i - x_o$, $\Delta y_i = y_i - y_o$, $\Delta F_i = F_i - F_o$, w_i is a weight function. The weights can be chosen such that they lead to the diagonalisation of the least squares matrix [10]. The immediate consequence of this is, all the 2-D least squares formulae reduce to equivalent 1-D form for all the derivatives. $\sum$ represents the summation over all the points in the neighbourhood(i.e. connectivity) of P_o.

Discretising the time derivative to first order and using least squares formulae [2] to approximate the derivatives of the various split fluxes, an update scheme for 2-D Modified moving KFVS split Euler equations can be written as,

$$U_0^{n+1} = U_0^n - \Delta t \left[\left\{ \frac{\partial}{\partial x}(GX_M^+) \right\}_{\Delta x_i < 0} + \left\{ \frac{\partial}{\partial x}(GX_M^-) \right\}_{\Delta x_i > 0} + \left\{ \frac{\partial}{\partial y}(GY_M^+) \right\}_{\Delta y_i < 0} + \left\{ \frac{\partial}{\partial y}(GY_M^-) \right\}_{\Delta y_i > 0} \right]^n \quad (14)$$

The spatial derivatives in the above equation are evaluated using least squares formulae Eq.(13) over a suitable subset of points in the connectivity (refered to as sub-stencil) to ensure that the signal [5] propagation property is not violated. The subscripts to the various flux derivative approximations indicates the sub-stencil chosen from the full connectivity set.

4 Advantages of LSKUM_MN

It is appropriate to note some adavantages of LSKUM_MN over other popular approaches to handle moving boundaries. Arbitrary Lagrangian Eulerian (ALE) method, which is quite popular mesh based algorithm to handle moving body problems, suffers from complexities of computing new cell volume, addition of mass, momentum and energy convective fluxes to Lagrangian fluxes [18]. The other grid based approach which has enjoyed considerable success is dynamic mesh method due to Batina [19]. It uses spring analogy. The mesh is updated for every iteration by solving equilibrium equations. One has to be careful about the quality of mesh that gets generated. These methods involve edge, face and volume data and these keep changing every iteration due to the motion of the body. The LSKUM_MN is a meshless algorithm and requires only the data at neighbouring nodes. Hence the data would consist of only node data. Hence data structure is very easy to handle in case of LSKUM_MN compared to mesh based solvers. Further the motion of nodes is very easy to implement. The nodes can be assigned velocities which decay with distance from moving boundary. This is possible because each node can move with its own velocity independent of any other node. This is very characteristic of LSKUM_MN. In LSKUM_MN one would face problems concerning the quality of connectivity as the nodes move around. This problem is less complicated than the problems mentioned above with other methods. Further the present method naturally blends to both moving as well as static nodes without the need for any sort of complex interpolation [6] as needed by all other conventional grid based solvers. This approach we feel has a tremendous potential to do numerical solution of the more challenging applications involving multiple moving boundaries.

5 Results and Discussion

5.1 Turbomachinery cascades

A set of cascade standard configurations, with emphasis on the unsteady aerodynamics, was established by Bolcs and Fransson in 1986 [12]. The original report was further updated in 1991 by Fransson and Verdon [20]. These reports contains comprehensive set of experimental data for various test flow conditions related to the aeroelastic problem. All the results are categorized into 11 standard configurations. A detailed presentation for each test case can be found in these references. The test case we have chosen is the 4th Standard Aeroelastic test case. This test case refers to an annular turbine cascade in which the blades oscillate sinusoidally in the first bending mode. The same oscillation amplitude and frequency are imposed for all the blades whereas the IBPA(Inter Blade Phase Angle) is a variable. A multipassage computational domain is employed, using a number of blade passages, $n_p (n_p > 1)$ given by

$$n_p = \frac{360^\circ z}{|\mathrm{IBPA}|} + 1, \quad \mathrm{IBPA} \neq 0$$

z is the minimum integer which leads to an integer value for n_p. This approach allows us to use periodic boundary conditions even for cases with IBPA $\neq 0$. This also avoids the direct store [21] method used with single passage. One of the main difficulties for turbomachinery flutter calculations is the grid generation as well as grid movement strategy required in an environment where multiple blades are in oscillation. Usual approach is to use multi-block overlapped meshes with inter block boundary conditions imposed by bilinear interpolation [22]. However with our present TKFMG code we do not face any such difficulties because the grid free method has the capability to just operate on any arbitrary distribution of points with random node velocities. In our present work, the nodes lying exactly on the blades are moved according to the motion prescribed for the blades. The nodes on the inflow, exit and the periodic boundaries are held fixed. For the rest of the nodes the motion exponentially decreases from the blades towards the fixed boundaries. As the blades oscillate, the points in the domain also move, hence the connectivity continuously changes. In our code we have an inbuilt connectivity generation preprocessor which updates the connectivity continuously as the points move.

The inlet flow Mach number M_1 is 0.28, with exit isentropic Mach number $M_2 = 0.9$ Fig.1 shows the details of the various inflow/exit and periodic boundary conditions. The inlet flow angle is $\beta_1 = 45^\circ$ inclined downward with respect to the axial direction and the corresponding exit flow angle $\beta_2 = 72^\circ$. The chordal stagger angle $\gamma = 56.6^\circ$ and the direction of vibration with respect to the chord line is given by $\delta = 60.4^\circ$. This corresponds to the harmonic motion of the blades in first bending mode. The amplitude of vibration is 0.33% of the chord length and reduced frequency k, based on the half-chord length and the outlet conditions, is equal to 0.107.

The total number of nodes used for computation is 16976 with 306 nodes on each blade. As already mentioned in the previous section, we use eigen direction based weights for the least squares approximation of derivatives. For every point in the connectivity, the weight chosen is of the form $w_i = \frac{C_i}{d_i^{2n}}$, where i represents any point in the connectivity and d represents the distance between i and the node at which update of solution is carried out. In this particular choice of weight function, the constant C_i is suitably chosen to diagonalise the least squares matrix [7]. The factor $\frac{1}{d_i^{2n}}$ ensures the local property of the derivative by giving more weight to closer points in the

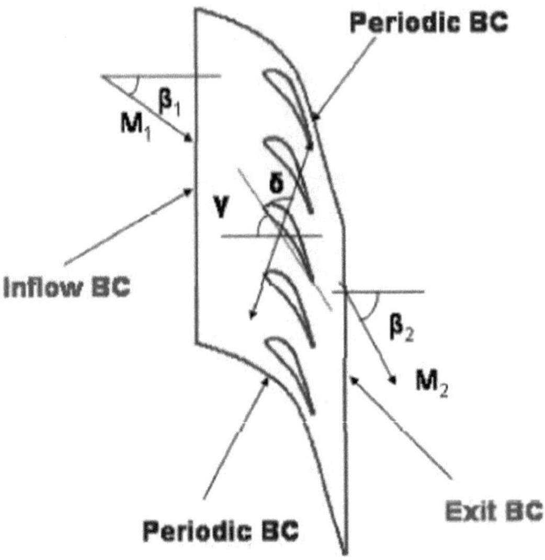

Fig. 1. Problem definition

connectivity. Apart from these weights, at every node a dissipation control parameter ϕ defined as $\phi = (\Delta r)^p$ is used, where Δr is the distance between the node and its closest point in the connectivity. This parameter is quite different from the weights mentioned previously. The dissipation control parameter is used to modify the formal order of accuracy of the scheme while the Eigen direction based weights are used for derivative approximation. In the present computations we use $n = 3$ and $p = 0.3$. The steady state solution is used as the initial condition for unsteady computations. For steady flows, the solution is marched in time using a four-stage Runge-Kutta scheme. For unsteady flow computations, the time derivative is discretized using a two level second-order accurate Crank-Nicholson method and the dual time stepping procedure is used. We have used sixty physical time-steps to compute each cycle. For each physical time step, in the inner loop we use 1000 iterations to converge in the pseudo local time steps. All the unsteady computations are typically carried out for about five cycles in which the periodic behavior of the flow is captured. Fig.2 shows the points distribution used for the computations. Fig.3 shows the steady state pressure contours for single blade computation while Fig.4 shows similar contours for the cascade with five blades. It can be observed that we have captured essentially all the features exactly same in both the cases. This can be further verified in Fig.5 which shows the isentropic Mach number distribution for single blade as well as cascade computations. Exactly similar solutions for all the blades can be obtained only when the periodic boundary condition is accurately implemented. It is important that

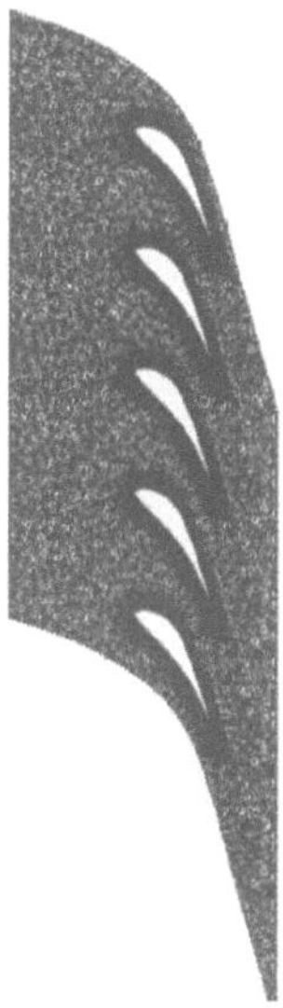

Fig. 2. Points distribution for the STCF4 cascade configuration

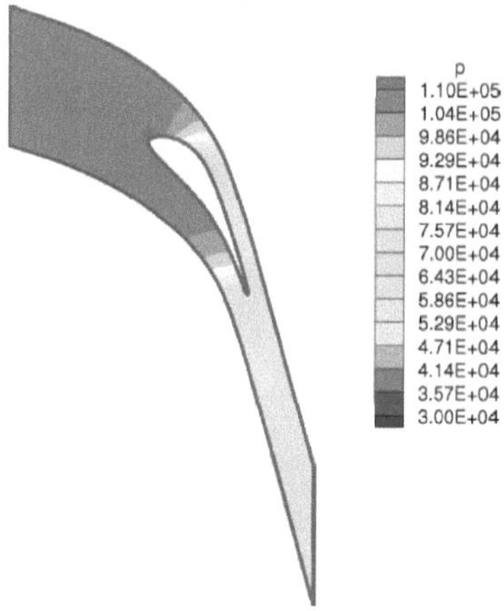

Fig. 3. Steady state pressure contours - Single blade

we ensure this symmetric steady solution, because this solution in turn is used as an initial condition for the unsteady flow computations.

Unsteady computations have been done for various IBPA's. Unsteady $C_p(t)$ is defined as [20]

$$C_P(t) = \frac{p(t)}{\bar{h}_{max}(p_o - p_\infty)_{inlet}}$$

where p_o is the stagnation pressure at the inlet, p_∞ is the static pressure at inlet, $\bar{h}_{max} = \frac{h_{max}}{C}$, C is the chord length of the blade and h_{max} is the amplitude of vibration. Unsteady $C_p(t)$ is expressed in a fourier series, as

$$C_P(t) = C_{P_0} + C_{P_1}\sin(wt + \epsilon_1) + C_{P_2}\sin(2wt + \epsilon_2) + \cdots$$

where $w = 2\pi f$, f is the frequency of the vibration in the first bending mode, C_{P_i} is the amplitude of the ith component and ϵ_i is the phase shift. The experimental results are available in terms of the amplitude and phase of the unsteady C_P for the first harmonic. In our present work we have compared our predictions with the experiments in terms of these two quantities. However amplitude and phase of the higher harmonics are available from the computations. But these are not plotted since in general the amplitudes of the higher harmonics are very small compared to the fundamental. Comparison of the phase assumes significance in the view that it is crucial in the correct prediction of the flutter instability. In the present work we have chosen IBPA's

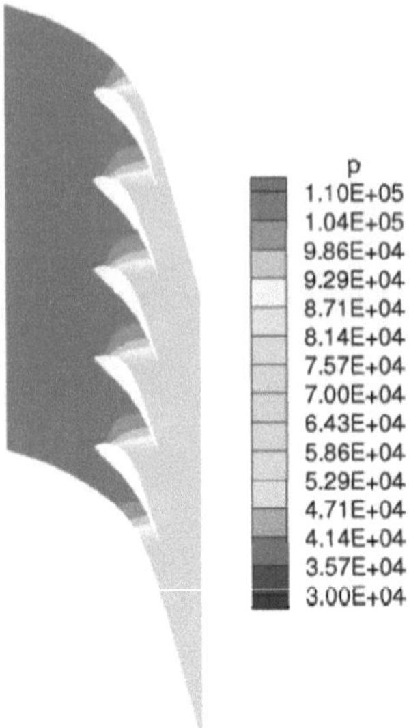

Fig. 4. Steady state pressure contours - cascade

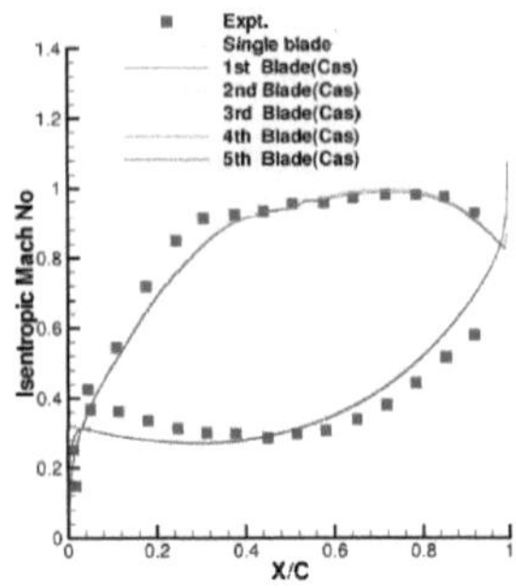

Fig. 5. Isentropic Mach number plot

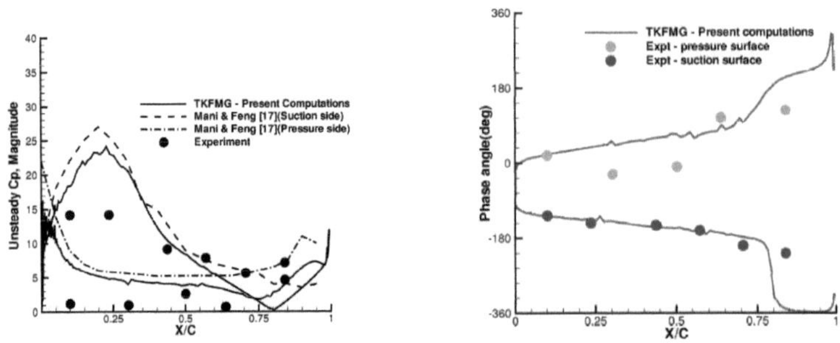

(a) Amplitude (b) Phase

Fig. 6. First harmonic of unsteady pressure for IBPA = 90°

corresponding to $-90°, 90°, 180°$ and $0°$. These cases corresponds to the test case no.'s 3,6,7 and 8 respectively for the fourth standard configuration. It is for these test cases extensive computations have been reported in the literature.

Fig.6a shows a comparison of the computations with the experiment for the first harmonic of the unsteady C_p for IBPA $= 90°$. It can be noticed that the calculation shows higher amplitudes over the front half chord. However the trend is predicted well. This behavior has been reported by many previous computations. In the figure we also show a comparison with the results

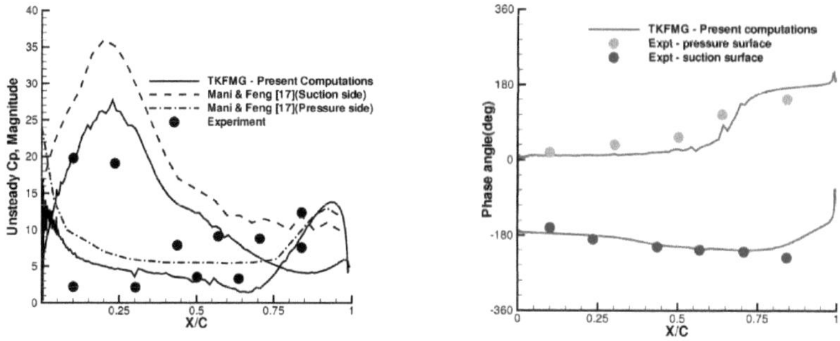

(a) Amplitude (b) Phase

Fig. 7. First harmonic of unsteady pressure for IBPA $= 180°$

of [15] obtained by finite volume Euler computations. Very similar behavior is also reported by Ji and Liu [14] using Euler and N-S computations. As already pointed out by Mani [15]for this test case higher prediction of amplitudes have been mentioned in the original reports [12, 20]. The exact reason is not known, however report by fransson [20] suggests the discrepancy could be due to uncertainty of the inflow angle, leakage flow and boundary layer growth in the test facility. The mismatch is in general more pronounced for the transonic flow conditions. In fact the computations by [14, 15] uses a stream-tube contraction model in the calculations to compensate for leakage flow and boundary layer growth in the test facility. In our present work we do not use any such corrections but still we are able to predict the behavior comparable to other methods as mentioned earlier. Euler and N-S computations by Napalitano [22] is another example where flutter computations for this test case have been carried out without any such corrections for the stream-tube contractions. Even in this work the amplitudes are not consistently predicted for all the IBPA's. Fig .6b shows the phase plot for IBPA $= 90°$. The discrepancies in phase is much less, and this ensures the correct stability range prediction. Similar amplitude and phase plots can be seen in figures 7 and 8 for IBPA $= 180°$and $-90°$ respectively.

Finally the most important parameter needed to predict flutter is the computation of the aerodynamic damping from the unsteady data. This represents the work done per cycle on the blade by the aerodynamic forces. A negative value of this coefficient represents the energy transfer from the flow to the blades indicating flutter. Net energy transfer to the blade in one period is calculated as

$$W_A = \oint \int_0^T -p(x,t) \ \hat{n} \ ds \ \cdot (i\frac{dX_B}{dt} + j\frac{dY_B}{dt})dt$$

where $p(x,t)$, is the pressure at any point on the blade, X_B, Y_B, co-ordinates of the point on the blade non dimensionalised with chord length, $\hat{n}$, normal

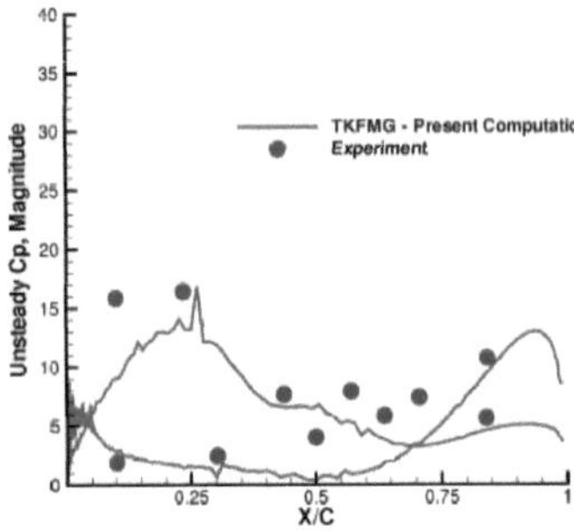

(a) Amplitude

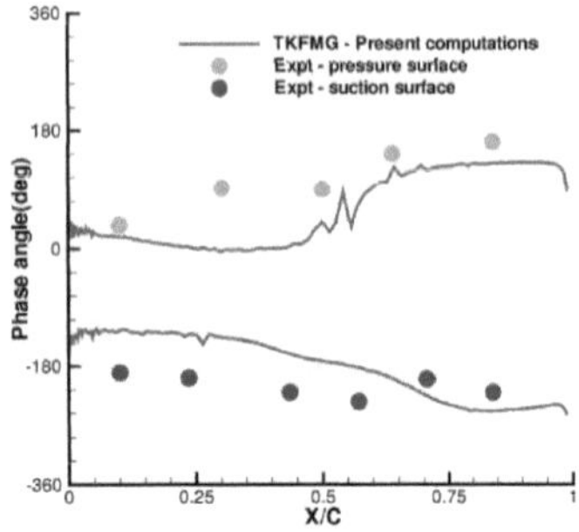

(b) Phase

Fig. 8. First harmonic of unsteady pressure for IBPA = $-90°$

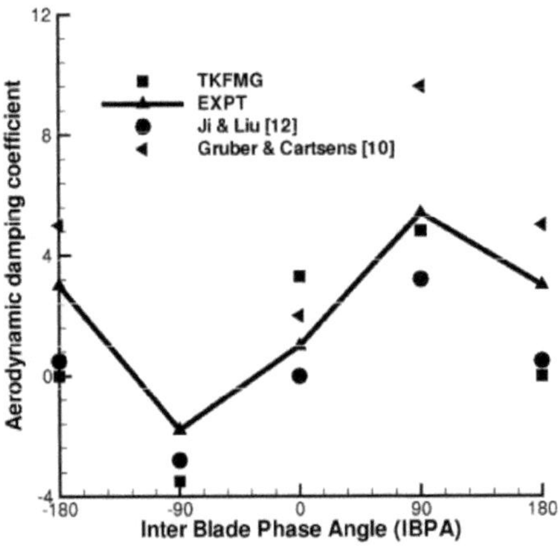

Fig. 9. Aerodynamic damping

at the point on the blade, T is the time period of each oscillation, and s arc length non dimensionalised with chord length

Aerodynamic damping coefficient is defined as

$$\Xi = \frac{-W_A}{(p_o - p)_{inlet}\bar{h}^2_{max}},$$

where $(p_o - p)$, is the compressible dynamic pressure at the upstream.

Fig.9 shows the plot for aerodynamic damping for various IBPA's. It can be seen that our TKFMG code clearly shows that flutter is present for IBPA=$-90°$. Also in the figure we have shown the results of Euler computations by Ji and Liu [14], Gruber and Carstens [13]. It can be observed that the trend is captured correctly, that is, regions of positive and negative aerody-

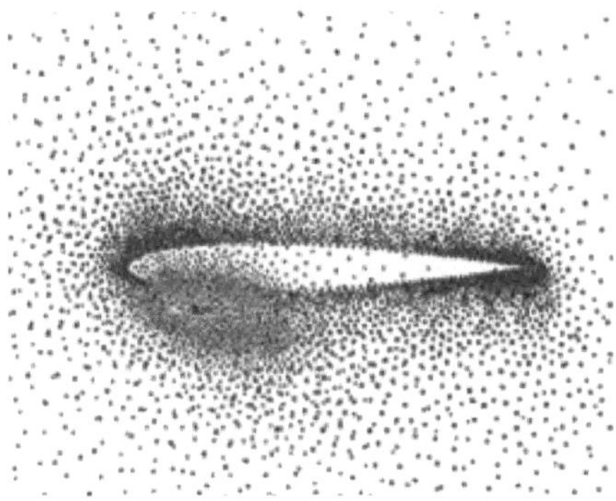

Fig. 10. The Initial Unblanked Chimera cloud

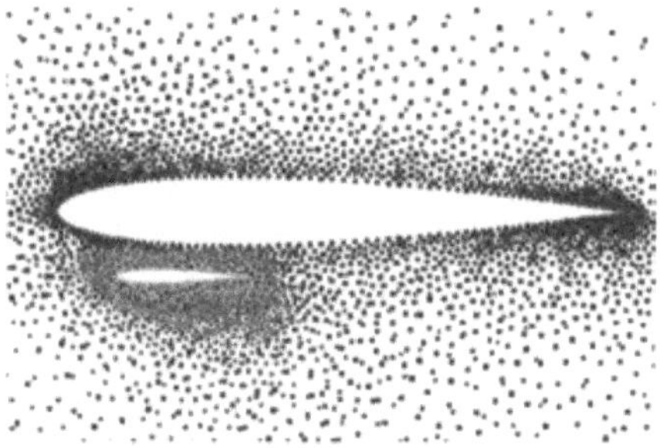

Fig. 11. The Initial Chimera cloud after Blanking

namic damping given by computations are in agreement with those observed by experiments.

5.2 Store separation

We consider the initial point distribution which was generated around each body to be attached rigidly to that body (Fig.10). The green points which are the points from the cloud around the store are attached rigidly to the store as the store moves; while the red one is rigidly attached to the main body which is anyway fixed. So as the store moves the total distribution (colored red in Fig.11) at any time is the superposition of the main body's cloud and the store's cloud at that time.

An issue to be considered is the blanking of solid points, points inside the boundaries of the bodies. For this people have used various techniques like the surface normal test [23] to check whether a point is inside or outside a given closed boundary. Such techniques are very involved. But we have used a rather simplistic yet efficient method to tackle that.

We use the concept of a "Bounding box". The bounding box of a body (here a 2D curve) is the smallest rectangle (in 2D) which completely encloses the curve. Just blank out the points belonging to the store's generated cloud(green) inside the bounding box of the main body and the ones of the main body inside the store's bounding box (Fig.11). Here we have used a box slightly larger than the bounding box in order to prevent the cluttering of points from the store's cloud near the main body's boundary and vice versa. We still have a large cluster of points of each body's own cloud around it. So this method works well. Of course, this will be problematic if the bounding box of a body contains the other body. This might happen for unsymmetrical airfoils(main body) having stores too close to them initially. In those cases we return to the use of the surface normal test. But such a procedure is not required for the most of the time when the store is sufficiently outside the bounding box of the main body.

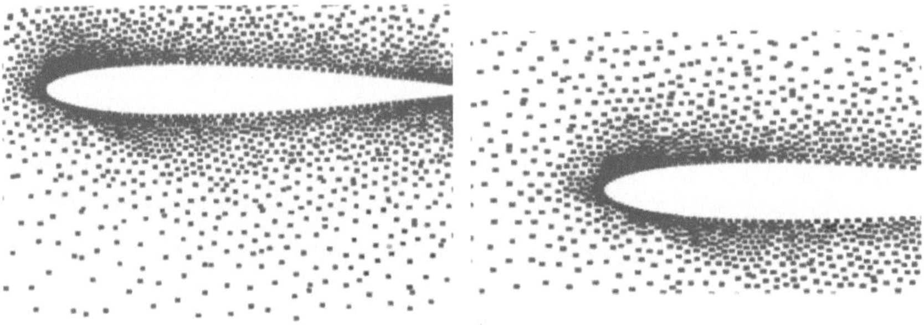

(a) point away from bodies (b) same point near store

Fig. 12. Dynamic update of the connectivity(blue) of a typical fixed point(green) as the store moves down across the fixed window shown (2 different instants)

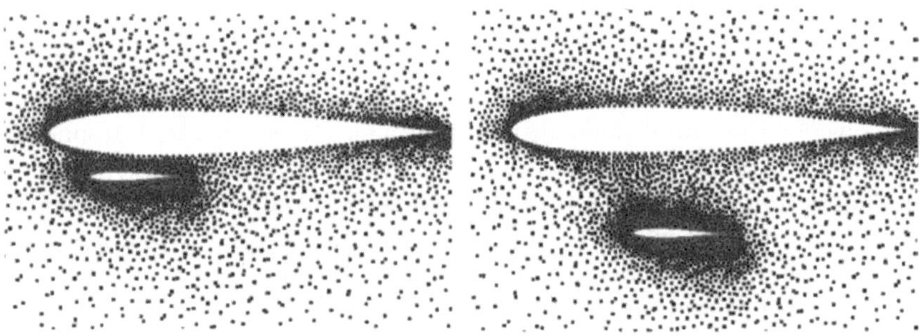

(a) Initial position (b) Final position

Fig. 13. Point movement strategy

We need to do this process of Blanking dynamically. Furthermore points which were blanked in the previous instant may have to be de-blanked in this instant. This is a process of Dynamic Blanking and De-Blanking. The Bounding box method saves time and effort for this process. It is much more efficient than regenerating new points at newer regions as they come out. For the solver we dynamically update the connectivity of each point, during the separation, using the quad tree preprocessor (Fig.12).

As a preliminary attempt to study the capability of our grid free solver past multiple moving boundaries with large displacements as opposed to our earlier work involving just small angular displacements [7], we have used a predefined relative motion of the store (Fig.13). These equations are similar to the ones chosen in [24].

We solve the store separation problem for angle of attack $\alpha = 1.0°$. For the initial configuration of the main body and the store we have obtained a steady state solution. With that as a starting point, we obtain the unsteady

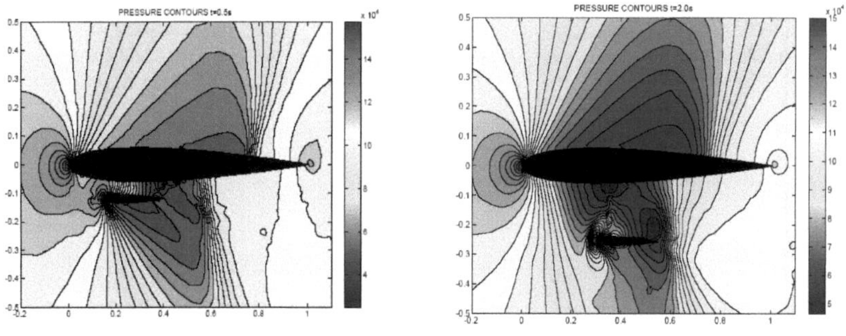

(a) t=0.5s (b) t=2.0s

Fig. 14. Pressure contours at 2 instants for the slow separation of store

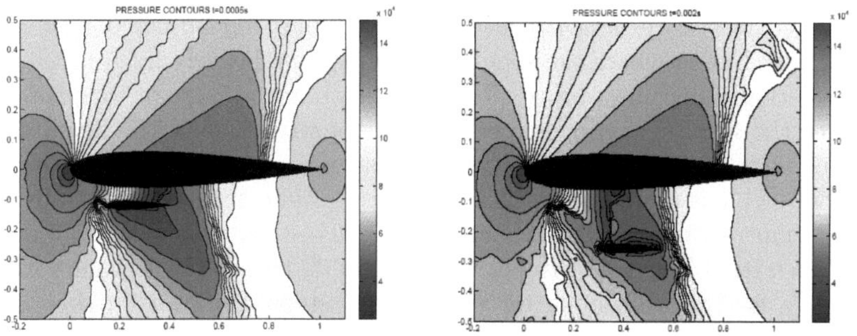

(a) t=0.0005s (b) t=0.002s

Fig. 15. Pressure contours at 2 instants for the fast separation of store

solution. For the store, the total X displacement after 20 timesteps is 0.2m to the right, total Y displacement is 0.13m downwards, the total angular displacement being 8.6° clockwise.

The relative timescale ratio of flow to store movement is given by the ratio k = U_{store}/U_{flow}. When $k \ll 1$, the timescale of the flow is much faster than that of the store, which gives the flow enough time to adjust to the new position of the store. In these cases a quasi steady solution like in [23] would suffice. However, otherwise (for other values of k) ie., if the timescale of movement of the store is comparable or much faster than that of the flow, then an unsteady solver is more appropriate. In the case we considered $k \sim 10^{-4} \ll 1$. Hence to illustrate the real advantage of our unsteady solver, we have considered the motion of the store along the same path but 1000 times faster. Here it covers the same distance, but in a total time of .002s (as opposed to 2s earlier), ie., $t_m \sim 10^{-2}s$.

Clearly we see the amount by which flow around the store is lagging behind it. The shock waves near the leading and trailing edges of the store move

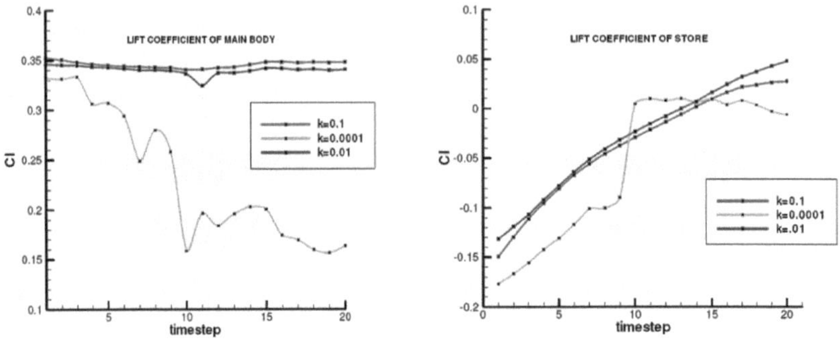

(a) main body (b) store

Fig. 16. Plots of Lift coefficient vs Timestep

relative to the store (Fig.15) as opposed to the well adjusted flow in the earlier case Fig.14 where the shock wave remains relatively fixed to the store and eventually becomes diffuse. The flow has not had enough time to adjust itself to the fast separating store. This demonstrates, that the flow features can be entirely different for the cases of fast movement of the stores. In such cases we feel it is more appropriate to compute time accurate unsteady flows, rather than rely on quasi steady analysis.

We see from Fig. 15 for the case of $k = 10^{-4}$, there is significant variation in the C_L for the main body for various time steps. However for the cases of k=.01 and k=0.1, there is very little variation of C_L. This demonstrates that for the fast moving store, it has less effect on the main body as compared to the slow moving case. The lift coefficient of the store is more negative in the initial stages for the slow moving case. This again suggests that the slow moving stores separates safely in the initial stages from the body rather than a fast moving store. These results clearly demonstrates the difference between aerodynamics of a slow and fast moving stores.

6 Conclusions

In this work we have demonstrated the power of the grid free method to compute unsteady flow in oscillating turbine cascade. The present method naturally blends to both moving as well as static nodes without the need for any sort of complex interpolation as needed by all other conventional solvers. This approach we feel has a tremendous potential to handle challenging applications in turbomachines such as multi-stage computations with rotor/stator interactions, tip flow analysis with casing etc. For the 4th standard aeroelastic test case we have obtained results comparing well with the experiments for the unsteady pressure coefficients. Finally we have been able to predict the flutter with our TKFMG code for the 4th standard aeroelastic test case by computing the aerodynamic damping coefficient. We have also illustrated the importance of our true time accurate approach in the context of the store separation problem.

7 Acknowledgements

The authors would like to thank the funding support received from the GTRE for carrying out this work.

References

1. Ghosh, A. K., and Deshpande, S. M., *Least squares kinetic upwind method for inviscid compressible flows*, AIAA Paper 95-1735, 1995.
2. Ghosh, A. K., *Robust least squares kinetic upwind method for inviscid compressible flows*, Ph.D. Thesis, Indian Institute of Science, Bangalore, 1996.
3. Deshpande, S. M., *Kinetic theory based new upwind methods for inviscid compressible flows*, AIAA Paper 86-0275, 1986.
4. Mahendra, A.K., *Application of least squares kinetic upwind method to strongly rotating viscous flows*, MSc(Engg) thesis , Indian Institute of Science, Bangalore, 2003.
5. Ramesh, V., *Least Squares Grid Free Kinetic Upwind Method*, PhD thesis , Indian Institute of Science, Bangalore, July 2001.
6. Ramesh, V. and Deshpande, S.M., *Least Squares Kinetic Upwind Method on moving grids for unsteady Euler computations*, Computers and Fluids, Vol. 30/5, pp. 621-641, May 2001.
7. Ramesh, V. and Deshpande, S.M., *Unsteady flow computations for flow past multiple moving boundaries using LSKUM*, Computers and Fluids, Vol. 36/10, pp. 1592-1608, Dec 2007.
8. Ramesh, V. and Deshpande, S. M., *Low dissipation grid free upwind kinetic scheme with modified CIR splitting* , Fluid Mechanics Report 2004 FM 20, Centre of Excellence in Aerospace CFD, Dept. of Aero. Engg., Indian Institute of Science, Bangalore.
9. Ramesh,V. and Deshpande,S.M., *Least squares kinetic upwind method with modified CIR splitting*, proceedings of 7th Annual CFD symposium, August 11-12 2004, National Aerospace Laboratories, Bangalore, India.
10. Konark Arora, *Weighted Least Squares Kinetic Upwind method using Eigenvector basis*,PhD Thesis, Indian Institute of Science, Bangalore, 2006.
11. Hong Luo, Joseph D. Baum and Rainald Löhner, *An accurate, fast, matrix-free implicit method for computing unsteady flows on unstructured grids*, Computers & Fluids 30(2001) 137-159.
12. Bolcs, A., and Fransson, T.H., *Aeroelasticity in Turbomachines, Comparison of Theoretical and Experimental Cascade Results*, communication du Laboratoire de Thermique Appliquee et de Turbomachines, No. 13, EPFL, Lausanne, Switzerland.
13. Gruber, B., and Cartsens, V., *Computation of unsteady transonic flow in harmonically oscillating turbine cascades taking into account viscous effects*, ASME J. Turbomach., 120, pp. 104-120, 1998.

14. Ji, S., and Liu, F., *Flutter Computation of Turbomachinery Cascades Using a Parallel Unsteady Navier-Stokes Code*, AIAA J. 37(3), pp.320-327, 1999.
15. Mani Sadeghi, and Feng Liu, *Computation of Mistuning Effects on Cascade Flutter*, AIAA Journal, Vol. 39, No. 1, January 2001.
16. Carta, F.O., *Coupled Blade-Disc-Shroud flutter Instabilities in Turbo-jet engine Rotors, Journal of Engineering for Power*, Vol. 89, No. 3, 1967, pp. 419-426.
17. Mandal, J.C. and Deshpande, S.M., *Kinetic flux vector splitting for Euler equations*, Computers and Fluids, Vol. 23, pp. 447-478, 1994.
18. Serge Piperno, *Numerical methods used in Aeroelasticity simulations*, Report No. 92-5, 1992, CERMICS, INRIA, France.
19. Batina JT, *Unsteady airfoil solutions using unstructured dynamic meshes*, AIAA Journal, Vol. 28, No. 8, pp 1381-1388, 1990.
20. Fransson, T.H., and Verdon, J.M., *Updated report on standard configurations for unsteady flow through vibrating axial-flow turbomachine-cascades*, http://www.egi.kth.se .
21. Erdos, J.I., and Alzner,E., *Numerical solution of periodic transonic flow through a fan stage*, AIAA J., 15, pp. 1559-1568, 1977.
22. Cinnella, P., De Palma, P., Pascazio, G., and Napolitano, M., *A numerical method for turbomachinery aeroelaticity*, Transactions of the ASME, Vol. 126, pp. 310-316, April 2004.
23. Anandhanarayanan, *Development and Applications of a Gridfree Kinetic Upwind Solver to Multibody Configurations*, Ph.D Thesis, IISc, Bangalore, 2003.
24. Arif Masud et al, *An adaptive mesh rezoning scheme for moving boundary flows and fluid-structure interaction*, Computers and Fluids, Vol. 36/1, pp.77-91, Jan 2007.

Efficient cloud refinement for kinetic meshless methods

M. Somasekhar[1], S. Vivek[2], K. S. Malagi[2], V. Ramesh[2], and S. M. Deshpande[3]

[1] Computational & Theoretical Fluid Dynamics Division, National Aerospace Laboratories, CSIR, Bangalore-560017, India somumachani@gmail.com, keshav@ctfd.cmmacs.ernet.in, vr@ctfd.cmmacs.ernet.in

[2] Department of Mechanical Engineering, Indian Institute of Technology, Madras, Chennai-600036, India svivek2006@gmail.com

[3] Engineering Mechanics Unit, Jawaharlal Nehru Centre for Advanced Scientific Research, Jakkur, Bangalore-560 064, India smd@jncasr.ac.in

Summary. In the present work efficient cloud refinement to carry out adaptation in meshless framework is proposed. Locally refining point density in meshless adaptation modifies the connectivity of points in this region. Previously [1] the initial connectivity was ignored and quadtree based algorithms [2] generated new connectivity for every node from scratch. This is particularly inefficient and an efficient connectivity update must exploit the fact that the node distribution would be largely unaffected except in the region of adaptation. Hence connectivity modification and generation needs to be done locally, only in these regions. We present a fast, efficient algorithm which uses the existing connectivity from the initial cloud to modify or generate the connectivity over the refined cloud.

Key words: adaptive cloud refinement, automatic connectivity update, meshfree, LSKUM

1 Introduction

Mesh adaptation is a technique to reduce the error in approximations used to solve the PDE's. This can be done by redistributing the grid (r-refinement), refining the grid (h-refinement) or increasing the order of approximation (p-refinement). In the conventional solvers h-adaptation is done by subdividing the cells or elements into finer ones. These techniques, called Automatic Mesh Refinement (AMR), are fairly well developed. The idea is to have a finer discretization. In the case of meshless methods there are no cells or elements. The adaptation is carried out by locally refining the point density in the regions demanding higher resolution. This results in an adapted and enriched cloud

M. Griebel, M.A. Schweitzer (eds.), *Meshfree Methods for Partial Differential Equations V*, Lecture Notes in Computational Science and Engineering 79,
DOI 10.1007/978-3-642-16229-9_13,

of points. We call this method as Adaptive cloud refinement (ACR) [1]. Adaptation involves obtaining the flow solution on an initially discretized domain. Then suitable sensors are used to identify regions requiring the adaptation. Refinement is carried out in these regions. This may lead to a remeshing or regeneration of grid in the case of conventional grid-based schemes. The Least Squares Kinetic Upwind Method (LSKUM) [3] which is a meshless solver offers a considerable advantage in such a situation. In the present work we use LSKUM to obtain flow solutions. Connectivity is the crucial component of meshless solvers. Here connectivity means a set of neighboring nodes or points. Generally the neighboring nodes or points fall in a ball of certain specified radius but for highly stretched clouds the ball can be highly anisotropic. It is not necessary to take all points in the ball as neighbors for connectivity. Also the radius of the ball can vary from point to point. There are several guiding principles to choose good connectivity. To name a couple of them one can consider minimum number of nodes in each quadrant and test for quality of approximation. A detailed discussion on connectivity issues can be found in [4]. When we refine the point distribution by increasing the local point density the connectivity of the nodes in that region gets altered. The new nodes which are added do not have any connectivity. Hence the connectivity has to be modified for the nodes from initial cloud and has to be generated for new nodes. One approach proposed earlier is to look at this refined cloud as one new cloud and generate connectivity using quadtree based algorithms [1]. This approach is obviously inefficient. The point distribution will have changed only in some regions of higher gradients while the rest of domain is largely unaffected. In the quadtree based connectivity algorithm the connectivity updating is done even in unaffected regions, thus leading to computational inefficiency. Hence the connectivity regeneration issue on the refined cloud should be addressed locally in regions where refinement is carried out. In the present work we have developed an algorithm which addresses this connectivity issue for new nodes and old nodes in the area of refinement locally. The algorithm uses the information of parent nodes (nodes marked for the enrichment) and connectivity data from initial cloud. This reduces the time significantly. We call this algorithm as Automatic Connectivity Update (ACU). In contrast to using a quadtree based method for connectivity generation the ACU is extremely efficient owing to its local nature.

2 LSKUM: a meshfree solver

Least squares kinetic upwind method (LSKUM) [3] is a kinetic theory based meshless solver. Meshless solvers are a class of numerical algorithms which do not require generation of mesh or grid. It only needs point distribution over the domain of interest and connectivity information. Connectivity is the set of nearest neighbors of a given node which are used for computation at the node. The meshless methods do not need any topological information. LSKUM is based on the fact that suitable moments of Boltzmann equation lead to Euler equations. Numerical schemes are derived at Boltzmann level [5]. An upwind scheme is achieved at Boltzmann level by splitting molecular velocity into positive and negative parts using the Courant-Isaacson-Reeves (CIR) splitting. Spatial derivatives are approximated by using least squares approximation. Then update scheme at Boltzmann level is mapped to Euler level through moment method. The second order accuracy can be achieved through defect correction using q-variables [6]. Alternatively a modified CIR (MCIR) [7] splitting can be used to achieve higher order accuracy. The advantage of MCIR is that it involves only single step. The time derivative can be suitably discretized to arrive at update scheme. The LSKUM has been very successful in computing flow past bodies of complex configuration. Much advancement has taken place since its inception. Recently, Konark [8] developed weighted LSKUM (WLSKUM) which increases the robustness of the LSKUM. The MCIR of Ramesh [7] achieves higher order accuracy in single step. These have been implemented in the code being used for the work. A multistage R-K method is used for time stepping. In the present work, we use this code to obtain solutions that will be used along with sensors to identify regions requiring adaptation.

3 Adaptive Cloud Refinement (ACR)

Adaptation through h-refinement essentially involves obtaining solutions on an initial discretized domain. In the present work we are using LSKUM, which is a meshless method, to obtain flow solutions. Then a suitable sensor is used to identify regions where refinement is necessary. Refinement is carried out by increasing point density in such areas. Refinement in these regions obviously alters the domain discretization locally. The solver is run on the adapted domain to obtain a better solution. Several cycles of adaptation can be carried out to achieve good results. In the conventional solvers i.e., FVM, FDM or FEM, h-refinement is achieved by subdividing the cells or elements into finer ones. The idea is to have finer discretization of domain. This is fairly well developed in conventional solvers. In this work we focus on using meshless methods. The meshless methods do not contain any cells or elements, they contain only node data. Hence the concepts used in conventional methods of adaptation are not relevant in this work. In meshless methods we achieve the finer discretization by increasing the local point density. i.e., adaptation using h-refinement is done by adding more points locally in the regions of rapid flow variations. Suitable sensor is used to mark nodes for refinement. The nodes so marked are called parent nodes. Refinement is carried out by increasing point density around parent nodes. The new nodes added around such parent nodes are called child nodes. Several options are open to increase the point density around parent node. In this work we are using (X) stencil to add four points at some percentage of average radial distance in the connectivity of parent node as shown in the Fig.1. The refinement process might lead to some child nodes crossing the domain of interest. Such nodes are deleted by using blanking algorithm [2]. In AMR one needs to take care of hanging nodes and edges [9]. This would lead to refinement even in cells or elements not marked for refinement causing excessive refinement. Such a problem does not arise in ACR. The points can be added arbitrarily if one decides to do so. Data structure requirements for ACR present a good case for it. The meshless methods require only connectivity information. Hence the data structure of meshless methods would consist of nodes in the connectivity set and flow variables at these nodes. The ACR also uses similar data structure and only needs node data. This data structure is highly local. The data structure of AMR generally needs node, edge, element, data structure [9]. The data structure becomes further complicated in case of hybrid grids and hierarchical grids [10]. Obviously the overheads due to data structure of adaptation in ACR are expected to be far less when compared to AMR. In the present work we have not made any attempt to study error estimator and impact of cloud refinement on it. We have used entropy based sensor for refinement.

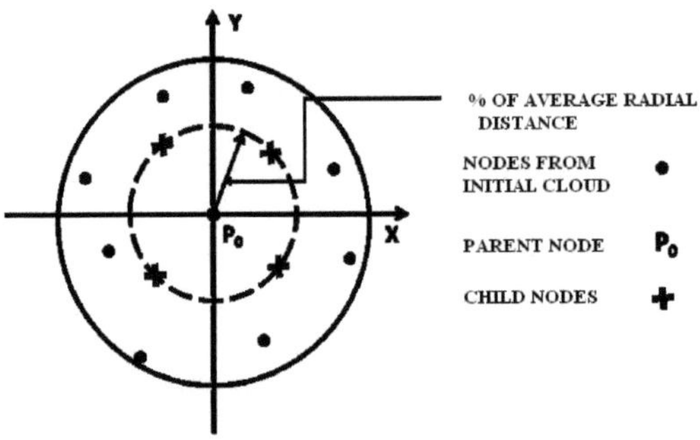

Fig. 1. Point enrichment with (X) stencil.

4 Automatic Connectivity Update(ACU)

Central to the meshless solver is the connectivity. Connectivity is defined as the set of nearest neighboring nodes to a given node which are used for computations at that node. When we refine the local point density, the neighborhood of the nodes in that region gets modified. Hence the connectivity of nodes in such regions has to be modified accordingly to reflect the refinement. Also there are new nodes added to the domain. The connectivity of these new nodes has to be generated. The node distribution would be largely unaffected excepting the region of adaptation. Hence connectivity modification and generation needs to be done only locally in the region by of adaptation. Such an approach will be more efficient. In this work we propose an algorithm which locally modifies or generates connectivity only in the region of adaptation. The algorithm makes use of existing information of connectivity from initial cloud to modify or generate the connectivity over refined cloud. This is achieved by considering the connectivity of a given node and connectivities of nodes in the connectivity of given node. This forms a superset from which we can deduce the connectivity set required. We call this method of generating or modifying the connectivity as Automatic Connectivity Update (ACU). It is explained further here. Consider any node P_o which is marked for refinement, shown in Fig.2 by Green Square. Let $C(P_o)$ be its connectivity from initial cloud, shown by black circles in Fig.2 and $N_C(P_o)$ be the set of newly added nodes, shown in Fig.2 by red dots. These are child nodes of P_o. $C(P_o) = \{P_i, i = 1, .m\}$ where m is the number of nodes in connectivity of P_o. Similarly, $C(P_i)$ is the connectivity of any node P_i in the $C(P_o)$. Modify these connectivities from initial cloud by adding child nodes. The modified connectivities will then be

$$C'(P_o) = C(P_o) \cup N_C(P_o) \tag{1}$$

$$C'(P_i) = C(P_i) \cup N_C(P_i) \tag{2}$$

Fig.3 shows the modified connectivity for P_o. Let $C_S(P_o)$ be the super set of connectivities which includes the above mentioned connectivities.

$$C_S(P_o) = C'(P_o) \cup \{C'(P_i), i = 1, .., m\} \tag{3}$$

This set is shown in Fig.4 by blue squares. The connectivity of any child node of P_o will be a subset of $C_S(P_o)$. Hence the connectivities of all children of P_o can be deduced from the $C_S(P_o)$. Fig.5 shows the connectivity generated for a child node using connectivity superset $C_S(P_o)$. The connectivity of P_o can also be modified using this superset of connectivity to reflect the finer distribution of nodes due to adaptive refinement of nodes in the domain. Fig.6 shows the modified final connectivity of parent node. Thus the connectivity can be generated for new points as well as connectivity can be modified for parent nodes using this algorithm.

5 Results and Discussions

The ACR along with new connectivity generation algorithm, ACU has been tested on the standard 2D test cases to assess the performance of refinement and connectivity generation. The test cases considered are

i. Transonic test case NACA0012 $M_\infty = 0.85, \alpha = 10^\circ$

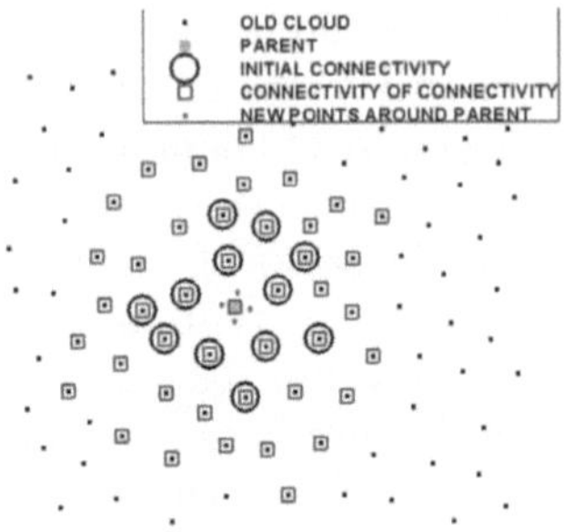

Fig. 2. Set of old points to build connectivity

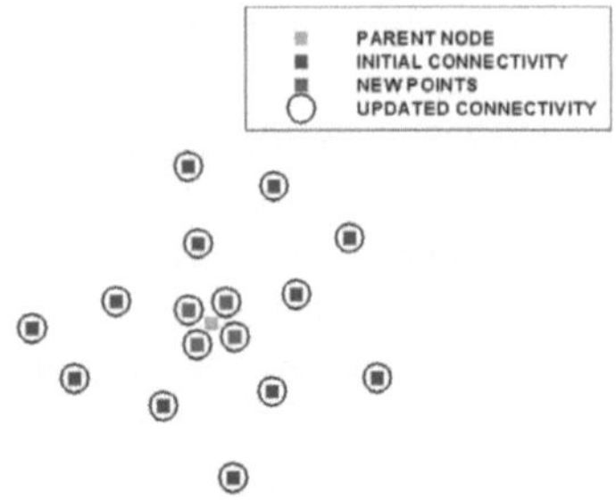

Fig. 3. Modified Connectivity including children

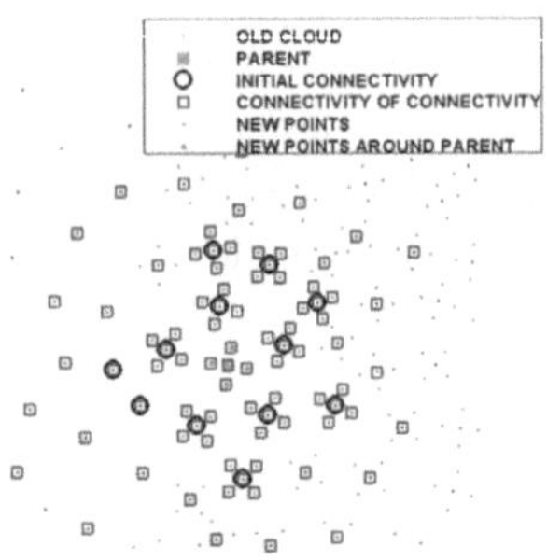

Fig. 4. Connectivity super set

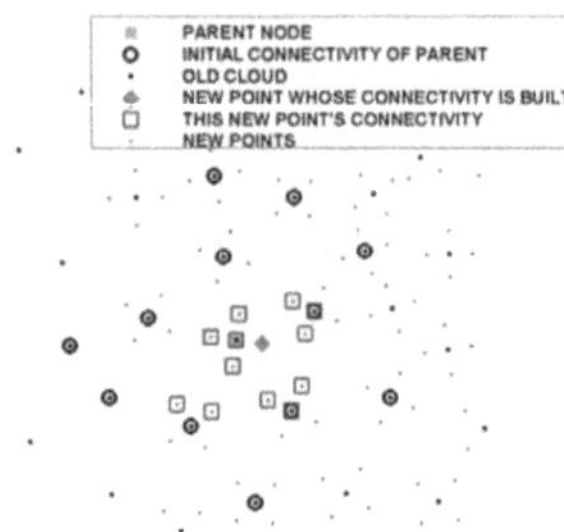

Fig. 5. Generated connectivity of a child

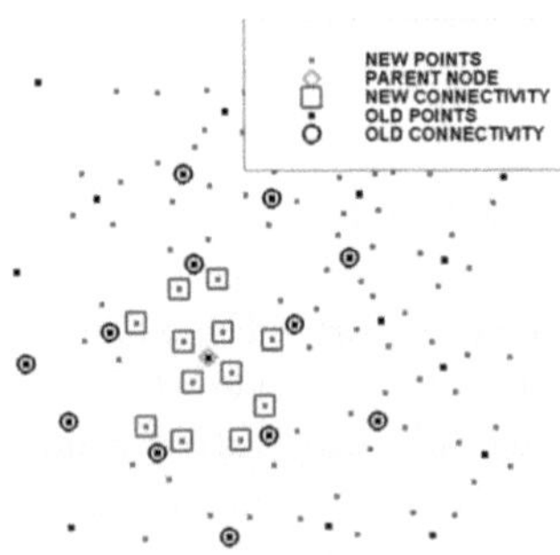

Fig. 6. Final modified connectivity of parent node

ii. Supersonic test case NACA0012 $M_{\infty} = 1.20, \alpha = 0^{\circ}$
iii. Subsonic test case NACA0012 $M_{\infty} = 0.63, \alpha = 20^{\circ}$

The LSKUM solver has been used to obtain the solution. The initial cloud of points for the NACA0012 cases is obtained using unstructured grid with 49,046 points. The number of points on the body are 240 and number of points on the outer boundary are 120. In this work standard entropy based sensor is used. The solution on the initial cloud is obtained using LSKUM solver. The sensor then identifies the nodes for refinement. Refinement and connectivity update are carried out to generate the adapted cloud of points with connectivity information.

5.1 Transonic test case NACA0012

The transonic test case has a shock on the upper surface and a weak shock on the lower surface. For this test case AGARD results are available for comparison. Thus both qualitatively and quantitatively this is ideal for assessing the performance of our adaptation by point refinement strategy. Fig.7 & Fig.8 show the initial and adapted point distribution after one cycle of adaptation respectively. It clearly shows that refinement has taken place in the regions having dominant flow features. The total number of points after adaptation is 83226. Fig.9 & Fig.10 show the pressure contours on initial cloud and adapted cloud. Fig.11 shows comparison of C_P plots for adapted cloud and initial cloud. We can easily see that the flow features have been captured more accurately on the adapted cloud. The Table1 shows the C_L and C_D comparison with AGARD results. The C_D has been predicted well on the refined cloud.

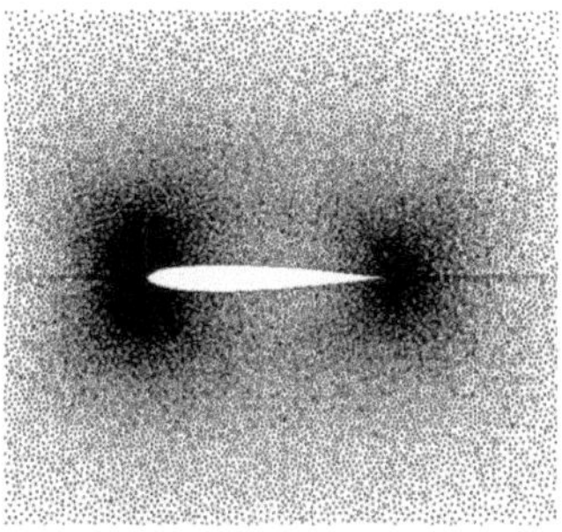

Fig. 7. Initial Cloud

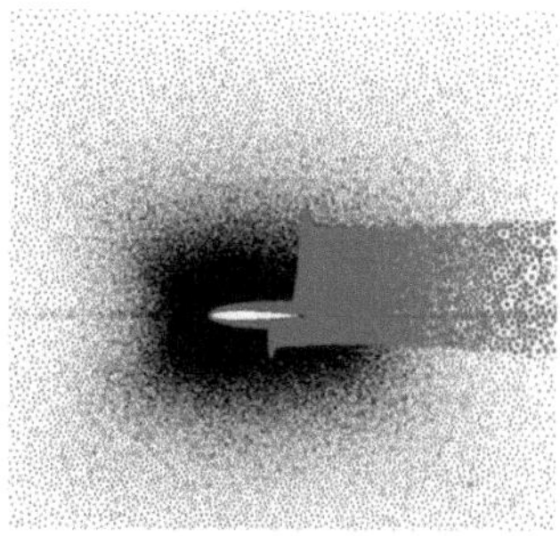

Fig. 8. Adapted Cloud

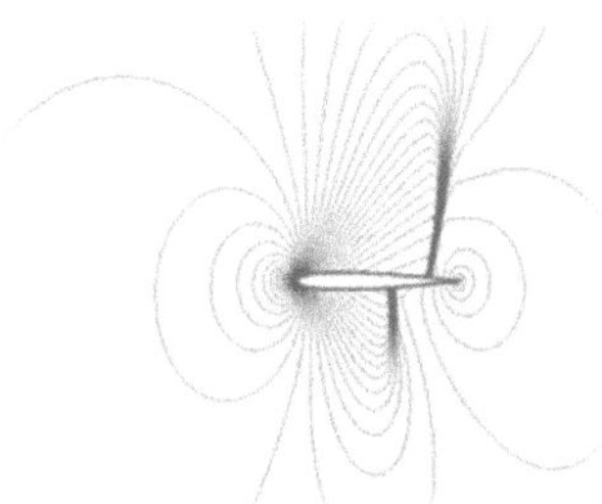

Fig. 9. Pressure contours on initial cloud

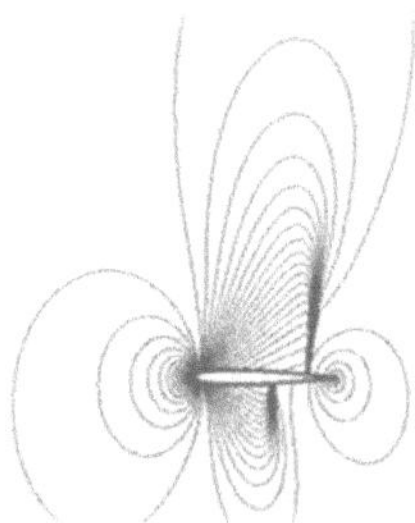

Fig. 10. Pressure contours on adapted cloud

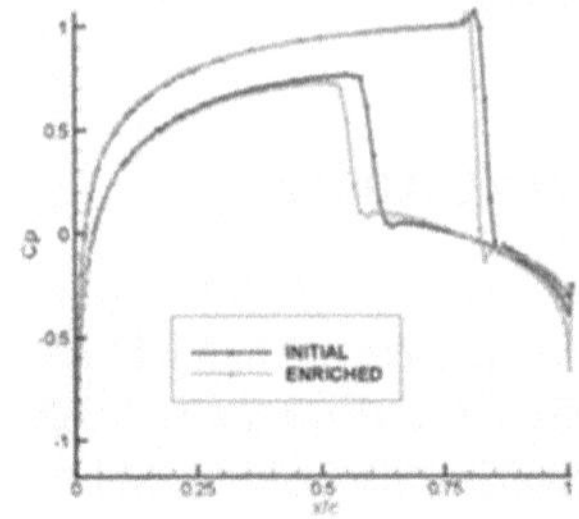

Fig. 11. C_P plots for Transonic case

Table 1. C_L, C_D for $M_\infty = 0.85$, Transonic Case

Cloud from Unstructured Grid	C_L	C_D
Initial	0.379	0.062
Enriched	0.343	0.055
AGARD	0.330-0.389	0.0464-0.0590

5.2 Supersonic test case NACA0012

The supersonic test case has a bow shock ahead of airfoil and fish tail shock at trailing edge. Fig.12 shows the adapted cloud of points. The total number of points after refinement is 136654. It is clearly seen that refinement has taken place in the regions of bow shock and fish tail shocks. Pressure contours (Fig.13 & Fig.14) indicate improved performance on the adaptive cloud. Table2 shows C_L and C_D comparison with AGARD results. C_L should be zero for this case which is predicted well by the adapted cloud.

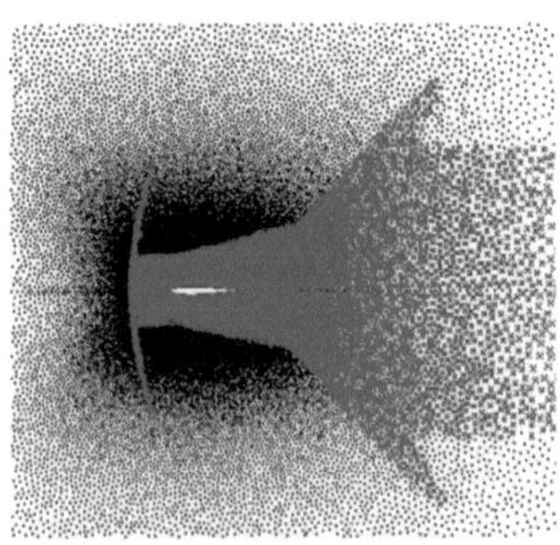

Fig. 12. Adapted point Distribution for supersonic case

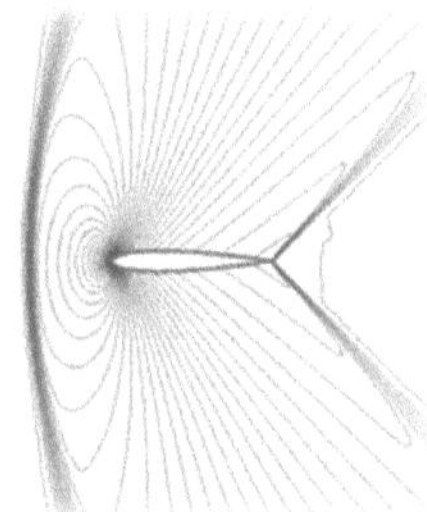

Fig. 13. Pressure contours on initial cloud

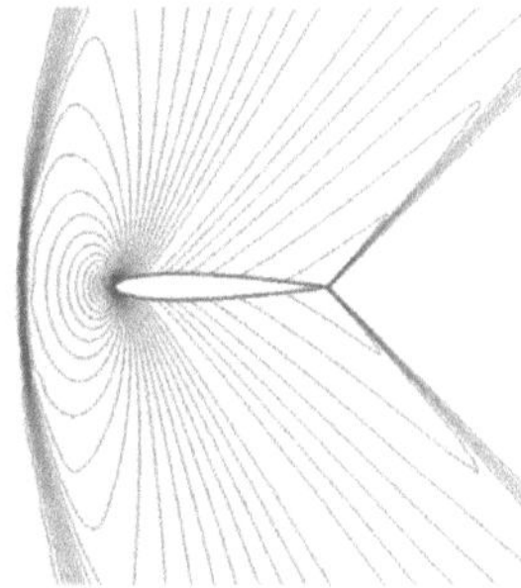

Fig. 14. Pressure contours on adapted cloud

Table 2. C_L, C_D for $M_\infty = 1.20$, Supersonic Case

Cloud from Unstructured Grid	C_L	C_D
Initial	0.000975	0.0956
Enriched	-0.000399	0.0953
AGARD	0	0.0946-0.096

5.3 Subsonic test case NACA0012

In this case there are rapid flow variations near the leading edge. The suction peak appearing near leading edge is of interest. Fig.15 shows refinement in this region. The C_P plot Fig.16 shows that suction peak is captured well by adapted cloud. The total number of points after enrichment in this case is 74222. C_D should be zero for this case which is predicted well by the adapted cloud.

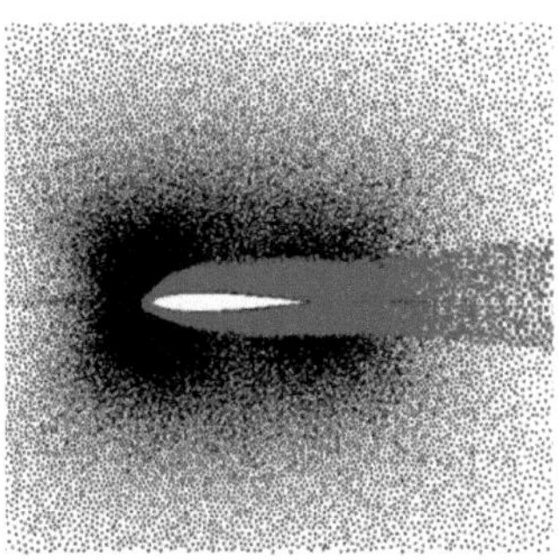

Fig. 15. Adapted point Distribution for subsonic case

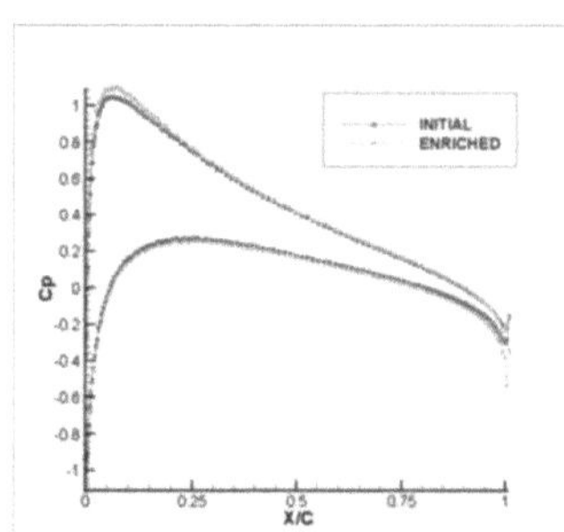

Fig. 16. C_P plots for subsonic case

Table 3. C_L, C_D for $M_\infty = 0.63$, Subsonic Case

Cloud from Unstructured Grid	C_L	C_D
Initial	0.346	0.005
Enriched	0.38	0.00036
AGARD	0.329-0.336	0.0003-0.0007

6 Conclusions

We have successfully demonstrated the new connectivity generation algorithm along with adaptive cloud refinement. The new connectivity generation algorithm is extremely efficient. It reduces the time required to generate or modify connectivity of the adapted cloud of points dramatically -This efficiency is due to the fact that we are able to exploit locality of refinement and prior connectivity information from initial cloud while generating connectivity. The data structure overheads due to adaptation are minimal as it requires only node data and is highly local. The ACR also avoids problems of hanging nodes and edges which cause unintended refinement. The capture of sharp gradients and better values of coefficients over adapted cloud in the standard test case of NACA 0012 airfoil have demonstrated the success and efficiency of ACR with ACU. The algorithm is directly applicable to adaptation in 3D.

7 Acknowledgements

The authors would like to thank the funding support received from the GTRE, Bangalore, India for carrying out this work.

References

1. M.Somasekhar, S.Vivek, V.Ramesh, S.M. Deshpande, *Automatic cloud refinement for gridfree Euler solver*, Proc. of the 10th Annual CFD symposium of the Aeronautical Society of India, Aug 11-12, National Aerospace Laboratories, Bangalore, 2008.
2. Ramesh, V., *Least Squares Grid Free Kinetic Upwind Method*, PhD thesis, Indian Institute of Science, Bangalore, July 2001.
3. Ghosh, A. K., and Deshpande, S. M., *Least squares kinetic upwind method for inviscid compressible flows*, AIAA Paper 95-1735, 1995.
4. C.Praveen, A. K. Ghosh, S. M. Deshpande, *Positivity preservation, stencil selection and applications of LSKUM to 3-D inviscid flows*, Computers & Fluids, Vol. 38, No. 8, pp.1481-1494, 2009.
5. Mandal, J.C. and Deshpande, S.M., *Kinetic flux vector splitting for Euler equations*, Computers & Fluids, Vol. 23, pp.447-478, 1994.
6. Deshpande S.M., Anandanarayanan, Praveen C., V. Ramesh, *Theory and Applications of 3-D LSKUM based on entropy variables*, International Journal for Numerical methods in Fluids Volume 40, Issue 1-2, pp.47 - 62, Sep 2002.

7. Ramesh, V. and Deshpande, S. M., *Low dissipation grid free upwind kinetic scheme with modified CIR splitting* , Fluid Mechanics Report, FM 20, Centre of Excellence in Aerospace CFD, Dept. of Aero. Engg., Indian Institute of Science, Bangalore, 2004.
8. Konark Arora, *Weighted Least Squares Kinetic Upwind method using Eigen-vector basis*, PhD Thesis, Indian Institute of Science, Bangalore, 2006.
9. Yannis Kallinderis, Christos Kavouklis, *A dynamic adaptation for general 3-D hybrid meshes*, Computer methods in applied mechanics and engineering, 194, pp.5019-5050, 2005.
10. D.J. Mavriplis, *Adaptive meshing techniques for viscous flow calculations on mixed element unstructured meshes*, International journal for numerical methods in fluids, 34, pp.93-111, 2000.

Fast exact evaluation of particle interaction vectors in the finite volume particle method

Nathan J. Quinlan[1] and Ruairi M. Nestor[1,2]

[1] Mechanical and Biomedical Engineering, College of Engineering and Informatics, National University of Ireland, Galway, Ireland nathan.quinlan@nuigalway.ie
[2] Irish Centre for High-End Computing, Trinity Technology & Enterprise Campus, Grand Canal Quay, Dublin 2, Ireland ruairi.nestor@ichec.ie

Summary. The Finite Volume Particle Method (FVPM) is a mesh-free method which inherits many of the desirable properties of mesh-based finite volume methods. It relies on particle interaction vectors which are closely analogous to the intercell area vectors in the mesh-based finite volume method. To date, these vectors have been computed by numerical integration, which is not only a source of error but is also by far the most computationally expensive part of the algorithm. We show that by choosing an appropriate particle weight or kernel function, it is possible to evaluate the particle interaction vectors exactly and relatively quickly. The new formulation is validated for 2D viscous flow, and shown to enable modelling of free-surface flow.

Key words: Finite volume particle method, Meshfree method. Free surface.

1 Introduction

The Finite Volume Particle Method (FVPM) is a mesh-free arbitrary Lagrangian-Eulerian (ALE) method which inherits many of the desirable properties of both particle methods such as smoothed particle hydrodynamics (SPH) [12] and conventional mesh-based finite volume methods (FVM) [10]. In this article, we present an enhancement to FVPM which improves both the speed and the accuracy of the method.

In meshfree methods, the computational nodes have no fixed connectivity and are usually Lagrangian. These methods are attractive in principle not only because they do not require mesh generation (a costly stage in the computational simulation of complex problems). In Lagrangian or near-Lagrangian mode, they enable the handling of moving interfaces without the problems of mesh deformation or tangling. For example, SPH has proven highly successful in the modelling of engineering free-surface flows [3]. FVPM [5] has some features of SPH, but can also be interpreted as a generalisation of the

M. Griebel, M.A. Schweitzer (eds.), *Meshfree Methods for Partial Differential Equations V*, Lecture Notes in Computational Science and Engineering 79,
DOI 10.1007/978-3-642-16229-9_14,

conventional mesh-based FVM [7], and can accommodate any finite volume flux function in its mesh-free framework. Unlike SPH, it is locally conservative regardless of any variation in particle size and interparticle distance, and incorporates boundary conditions without the use of fictitious particles.

In FVM, exchanges between mesh cells are weighted by a vector directed normal to the cell-cell interface, of magnitude equal to the interface area. In FVPM, the discrete finite volume cells of FVM are replaced with overlapping particles. FVPM depends crucially on a particle interaction vector $\boldsymbol{\beta}_{ij}$ which is closely analogous to the area vector in mesh-based FVM. To date, this vector has been computed by numerical integration. Inexact evaluation of this vector results in numerical error and, in some cases, a loss of exact conservation. When acceptable accuracy is achieved, evaluation of the interaction vector is by far the most costly part of the FVPM algorithm and makes FVPM more expensive than mesh-based methods and SPH.

In this work we will present a formulation based on a simple choice of particle weight function which allows particle interaction vectors to be determined both exactly and quickly. In section 2 the finite volume particle method is briefly introduced. The new formulation is presented in section 3, and other details of the FVPM implementation used in this work are given in section 4. In section 5, the original and new formulations of FVPM are compared for accuracy and computation time on a validation problem. In section 6, it is demonstrated that the new method enables application of FVPM to free surface problems.

2 The Finite Volume Particle Method

2.1 Derivation and properties

The FVPM was introduced by Hietel *et al.* [5], and a similar method was introduced under a different name by Ismagilov [6]. The derivation of the FVPM is outlined here very briefly. More detailed discsussions are given by Keck and Hietel [9] and Teleaga and Struckmeier [16].

FVPM is a numerical scheme for PDEs written in conservation form as

$$\frac{\partial \mathbf{U}}{\partial t} + \nabla \cdot \mathbf{F}(\mathbf{U}) = 0 \tag{1}$$

where $\mathbf{U}$ is the vector of conserved variables, $\mathbf{F}$ is the flux function, and $\nabla \cdot \mathbf{F}$ is the divergence of the flux. In two-dimensional weakly compressible flow (as in numerical examples presented later in this paper) $\mathbf{U} = (\rho, \rho u, \rho v)^T$, where ρ is density and u and v are Cartesian velocity components. The classical finite volume method can be derived by partioning the domain Ω into non-overlapping cells, integrating the governing equation over each cell, and applying the divergence theorem to convert the volume integral of $\mathbf{F}$ into a surface integral. To derive the FVPM, the governing PDE is multiplied by a test function $\psi_i(\mathbf{x})$ and then integrated over the domain. The following weak formulation results:

$$\int_\Omega \psi_i(\mathbf{x}) \frac{\partial \mathbf{U}}{\partial t} \mathrm{d}\mathbf{x} - \int_\Omega \nabla \psi_i(\mathbf{x}) \cdot \mathbf{F}(\mathbf{U}) \mathrm{d}\mathbf{x} = 0 \,. \tag{2}$$

In the finite volume particle method, the scalar-valued test function $\psi_i(\mathbf{x})$ (called a particle) is compactly supported on Ω_i. The particle supports Ω_i cover the whole domain, and may overlap. Each is associated with a reference particle location $\mathbf{x}_i$ and a particle volume $V_i = \int_\Omega \psi_i(\mathbf{x})\mathrm{d}\mathbf{x}$. The particles may move according to an arbitrary velocity $\dot{\mathbf{x}}_i$, and therefore do not necessarily represent fixed masses of material.

If $\psi_i(\mathbf{x})$ is defined such that $\psi_i(\mathbf{x}) = 1$ for $\mathbf{x} \in \Omega_i$ and $\psi_i(\mathbf{x}) = 0$ otherwise, then Eq. (2) yields the classical finite volume method with Ω_i as cells [7].

The test functions are defined such that they form a partition of unity at every point, i.e. $\sum_i \psi_i(\mathbf{x}) = 1$. (The value of ψ_i may then be interpreted as the fraction of local volume allocated to particle i in a region of overlap.) This property is ensured by first defining a compactly supported particle weight (or kernel) function $W_i(\mathbf{x})$, and then constructing $\psi_i(\mathbf{x})$ with a Shepard normalisation as

$$\psi_i(\mathbf{x}) = \frac{W_i(\mathbf{x})}{\sum\limits_k W_k(\mathbf{x})} \,. \tag{3}$$

Examples of particle weight and test functions are illustrated in Figure 1.

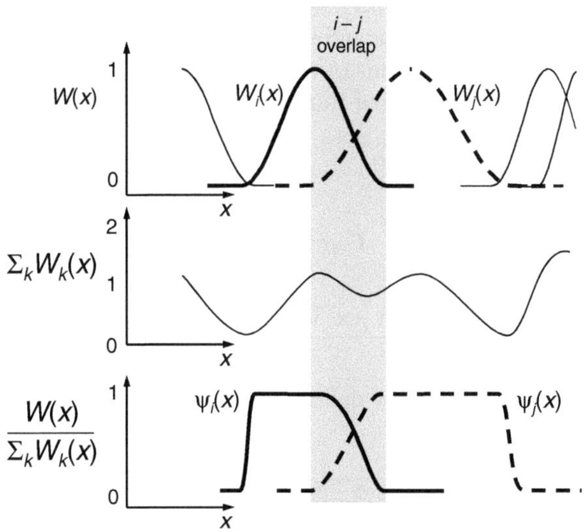

Fig. 1. Schematic diagram of particle weight functions, normalisation function and test functions for overlapping 1D particles i and j and neighbours.

Since $\psi_i(\mathbf{x}) = 0$ outside Ω_i, Eq. (2) can be rewritten after application of the divergence theorem as

$$\frac{\mathrm{d}}{\mathrm{d}t}\int_{\Omega_i} \psi_i(\mathbf{x})\mathbf{U}\mathrm{d}\mathbf{x} - \int_{\Omega_i}\left(\nabla\psi_i(\mathbf{x})\cdot\mathbf{F}(\mathbf{U}) + \mathbf{U}\frac{\partial\psi_i}{\partial t}\right)\mathrm{d}\mathbf{x} = 0\,. \tag{4}$$

Expansion of $\nabla\psi_i(\mathbf{x})$ yields

$$\begin{aligned}\frac{\mathrm{d}}{\mathrm{d}t}\int_{\Omega_i} \psi_i(\mathbf{x})\mathbf{U}\mathrm{d}\mathbf{x} + \sum_j \int_{\Omega_i} (\mathbf{\Gamma}_{ij} - \mathbf{\Gamma}_{ji})\cdot\mathbf{F}(\mathbf{U})\mathrm{d}\mathbf{x} \\ + \sum_j \int_{\Omega_i} (\mathbf{\Gamma}_{ji}\cdot\dot{\mathbf{x}}_i - \mathbf{\Gamma}_{ij}\cdot\dot{\mathbf{x}}_j)\,\mathbf{U}\mathrm{d}\mathbf{x} = 0\,,\end{aligned} \tag{5}$$

where $\mathbf{\Gamma}_{ij}$ is defined as

$$\mathbf{\Gamma}_{ij}(\mathbf{x}) = \frac{W_i(\mathbf{x})\nabla W_j(\mathbf{x})}{\sigma^2(\mathbf{x})}\,, \tag{6}$$

and

$$\sigma(\mathbf{x}) = \sum_{k=1}^{n} W_k(\mathbf{x})\,. \tag{7}$$

The vector field $\mathbf{\Gamma}_{ij}(\mathbf{x})$ is non-zero only in the region where both $W_i(\mathbf{x})$ and $W_j(\mathbf{x})$ are non-zero, that is, in the overlap of particles i and j.

The third integral in Eq. (5) above represents interparticle exchanges due to relative motion. Particle velocity $\dot{\mathbf{x}}_i(t)$ may be chosen independently of the fluid velocity $\mathbf{u}(\mathbf{x}, t)$. Two possible choices are $\dot{\mathbf{x}} = 0$ for an Eulerian scheme and $\dot{\mathbf{x}} = \mathbf{u}$ for a Lagrangian scheme. The interparticle flux and particle motion terms are combined and written in terms of a modified flux $\mathbf{G} = \mathbf{F}(\mathbf{U}) - \dot{\mathbf{x}}\mathbf{U}$, which is approximated with a numerical flux $\boldsymbol{\mathcal{G}}(\mathbf{U}_i, \mathbf{U}_j)$, abbreviated in the following as $\boldsymbol{\mathcal{G}}_{ij}$.

The vectors $\boldsymbol{\gamma}_{ij}$ and $\boldsymbol{\beta}_{ij}$ are now defined as follows.

$$\boldsymbol{\gamma}_{ij} = \int_{\Omega} \mathbf{\Gamma}_{ij}\mathrm{d}\mathbf{x} \tag{8}$$

$$\boldsymbol{\beta}_{ij} = \int_{\Omega} \frac{W_i(\mathbf{x})\nabla W_j(\mathbf{x}) - W_j(\mathbf{x})\nabla W_i(\mathbf{x})}{\sigma^2(\mathbf{x})}\mathrm{d}\mathbf{x} = \boldsymbol{\gamma}_{ij} - \boldsymbol{\gamma}_{ji} \tag{9}$$

The following resulting expression is the defining equation of the semi-discrete FVPM far from boundaries:

$$\frac{\mathrm{d}}{\mathrm{d}t}\int_{\Omega} \psi_i(\mathbf{x})\mathbf{U}\mathrm{d}\mathbf{x} + \sum_j \boldsymbol{\beta}_{ij}\cdot\boldsymbol{\mathcal{G}}_{ij} = 0\,. \tag{10}$$

This is closely analogous to the classical finite volume method. The term $\boldsymbol{\beta}_{ij}$ is a particle interaction vector which has the same role as the cell interface area in the mesh-based finite volume method, weighting the flux between neighbouring particles i and j.

Monotonicity [15], stability [15] and Lax-Wendroff consistency [8] of the scheme have been shown for the special case of a scalar conservation law in one dimension. Keck and Hietel [9] have proven that $\mathbf{U}_i = \int_\Omega \psi_i(\mathbf{x})\mathbf{U}(\mathbf{x})\mathrm{d}\mathbf{x}$ is a second-order accurate approximation to $\mathbf{U}(\mathbf{b}_i)$, where b_i is the particle barycentre, defined as

$$\mathbf{b}_i V_i = \int_\Omega \mathbf{x}\psi_i(\mathbf{x})\mathrm{d}\mathbf{x}\,. \tag{11}$$

This suggests that the barycentre should be used when it is necessary to associate field variables with a point location. In practice, the particle location $\mathbf{x}_i$ is used only as an origin for computation of the particle weight function.

2.2 The particle interaction vectors

It follows from definition (9) that the particle interaction vectors satisfy the conditions

$$\sum_j \boldsymbol{\beta}_{ij} = 0 \tag{12}$$

$$\boldsymbol{\beta}_{ij} = -\boldsymbol{\beta}_{ji} \tag{13}$$

(implying $\boldsymbol{\beta}_{ii} = 0$). The summation condition (12) ensures that a uniform field remains exactly constant, and condition (13) ensures exact conservation in any exchange between i and j.

Junk [7] has shown that the classical finite volume method may be considered a limiting special case of FVPM in which the interparticle overlap region collapses to a thin surface. The particle supports then become non-overlapping cells. In this case, $\boldsymbol{\beta}_{ij}$ is equal to the cell interface area $\mathbf{A}_{ij}$. Conditions (12) and (13) are satisfied by the $\mathbf{A}_{ij}$ of the finite volume method as a result of the geometric meaning of the cell face areas. Thus, the vector $\boldsymbol{\beta}_{ij}$ may be understood as a FVPM analogue of $\mathbf{A}_{ij}$. Because of its definition in terms of $\nabla W_i(\mathbf{x})$ and $\nabla W_j(\mathbf{x})$, $\boldsymbol{\beta}_{ij}$ points approximately from $\mathbf{x}_i$ to particle $\mathbf{x}_j$.

Many choices are possible for the weight $W_i(\mathbf{x})$ used to generate $\boldsymbol{\beta}_{ij}$, but to date, no systematic analysis has been published to guide the choice. Cubic [5], quadratic [13], piecewise quadratic [16] and piecewise linear [7] functions have been used successfully.

The integration of $\boldsymbol{\Gamma}_{ij}$ to determine $\boldsymbol{\beta}_{ij}$ is usually performed by numerical quadrature. This is a major component of the method's overall computational cost; we have found that it accounts for 85% of CPU time. It is also an additional source of error. It has been demonstrated [4, 7] that inaccurate integration can result in significant errors, such as non-physical oscillations near a shock wave. Hietel and Keck [4] developed a correction procedure by which the non-zero error $\sum_j \boldsymbol{\beta}_{ij}$ is successively transferred from particle i to particle $i+1$ to enforce conditions (12) and (13). Teleaga and Struckmeier [16] showed that addition of a corrective flux $\mathbf{F}(\mathbf{U}_i)\sum_j \boldsymbol{\beta}_{ij}$ for each particle ensures preservation of a uniform state. However, the latter method violates global conservation.

2.3 Boundary conditions

For a particle i near a boundary, Eq. (10) must be augmented to include a boundary flux $\mathcal{G}_i^b$.

$$\frac{\mathrm{d}}{\mathrm{d}t}\int_{\Omega}\psi_i(\mathbf{x})\mathbf{U}\mathrm{d}\mathbf{x}+\sum_j \boldsymbol{\beta}_{ij}\cdot\mathcal{G}_{ij}+\boldsymbol{\beta}_i^b\cdot\mathcal{G}_i^b=0\,. \tag{14}$$

The boundary interaction vector $\boldsymbol{\beta}_i^b$ is given by [9]

$$\boldsymbol{\beta}_i^b=\int_{\partial\Omega}\psi_i(\mathbf{x})\mathbf{n}\mathrm{d}s, \tag{15}$$

where $\mathbf{n}$ is an outward normal on the boundary $\partial\Omega$ and s is the length along the boundary. Often, $\boldsymbol{\beta}_i^b$ is determined more conveniently from $\sum_j \boldsymbol{\beta}_{ij}+\boldsymbol{\beta}_i^b=\mathbf{0}$, a modified form of condition (12). The flux $\mathcal{G}_i^b$ must be determined from the appropriate boundary condition. At a stationary no-slip wall, for example, the mass flux is zero and the momentum flux is equal to pressure.

3 A new choice for the particle weight function

The weight function of particle i, $W_i(\mathbf{x})$, is now defined as the characteristic function of the particle's support volume:

$$W_i(\mathbf{x})=\begin{cases}1 & (\mathbf{x}\in\Omega_i)\\ 0 & \text{otherwise}\end{cases}\,. \tag{16}$$

This top-hat function is illustrated for a 1D case in Figure 2 along with the resulting normalisation and test functions. The normalisation function $\sigma(\mathbf{x})=\sum_k W_k(\mathbf{x})$ is simply equal to the number of particle supports which cover the point $\mathbf{x}$. The test functions are piecewise constant with discontinuities at each particle support boundary.

The classical finite volume method (FVM) can be considered a version of FVPM, using this definition of $W_i(\mathbf{x})$, with the further condition that particles do not overlap. This suggests that there is no special requirement for continuity of the weight function, and that the FVPM with the new top-hat weight function can be regarded as intermediate between FVPM with smooth weight functions and the mesh-based FVM.

This choice of $W(\mathbf{x})$ allows $\boldsymbol{\beta}_{ij}$ to be evaluated analytically, exactly and inexpensively. For any choice of $W(\mathbf{x})$, $\boldsymbol{\Gamma}_{ij}$ in Eq. (9) is non-zero only in the

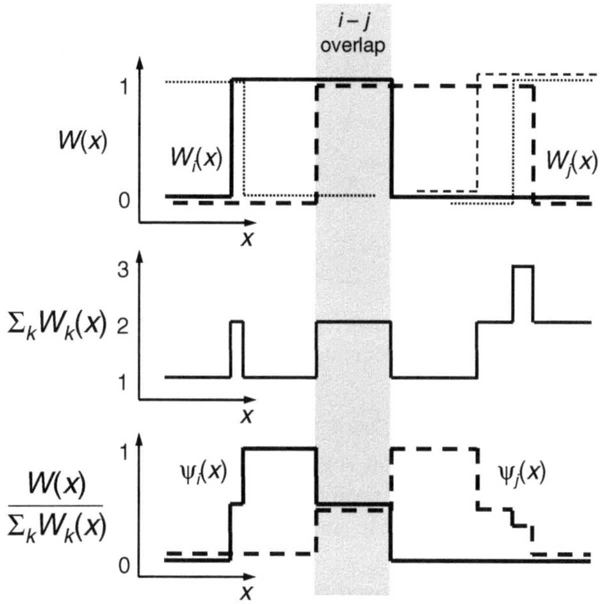

Fig. 2. Schematic diagram of top-hat particle weight functions $W(x)$, normalisation function $\sigma(x) = \sum_k W_k(x)$, and test function $\psi(x) = W(x)/\sigma(x)$ for particles in 1D. (For clarity, some curves are offset in the vertical direction.)

overlap of particles i and j, as depicted in Figure 3. With top-hat weight functions, furthermore, the term $W_i(\mathbf{x})\nabla W_j(\mathbf{x})$ is non-zero only at points that lie on the curve $\partial\Omega_{ji}$, defined as the part of the boundary of Ω_j that lies inside Ω_i, the support of particle i. $\mathbf{\Gamma}_{ij}$ is undefined on $\partial\Omega_{ji}$, but only its integral is required to determine γ_{ij}. In one dimension, its integral along the x direction can be shown to be

$$\int_{-\infty}^{\infty} \mathbf{\Gamma}_{ij} \mathrm{d}x = \int_{-\infty}^{\infty} \frac{W_i(\mathbf{x})}{\sigma^2(\mathbf{x})} \frac{\partial W_j(\mathbf{x})}{\partial x} \mathrm{d}x = \frac{1}{\sigma_-} - \frac{1}{\sigma_- + 1} . \tag{17}$$

With $W(\mathbf{x})$ defined as in Eq. (16), the normalisation function $\sigma(\mathbf{x})$ is simply the number of particle supports covering the point $\mathbf{x}$, and σ_- is defined as the value of σ immediately adjacent to the boundary outside particle j.

The vector γ_{ij} is computed by integration of Eq. (17) in the y direction. The curve $\partial\Omega_{ji}$ is partitioned into m segments delimited by intersections with the support boundaries of other particles at $\mathbf{x} = \mathbf{x}^*_{k-1}, \mathbf{x}^*_k$ for $k = 1...m$. The 1D integral defined by Eq. (17) is constant within each segment, and thus γ_{ij} can be represented as a summation:

$$\gamma_{ij} = \int\int_{\Omega} \frac{W_i(\mathbf{x})}{\sigma^2(\mathbf{x})} \nabla W_j(\mathbf{x}) \mathrm{d}x\mathrm{d}y = \sum_{k=1}^{m} (\mathbf{x}_k - \mathbf{x}_{k-1}) \left(\frac{1}{\sigma_{-,k}} - \frac{1}{\sigma_{-,k} + 1} \right) \tag{18}$$

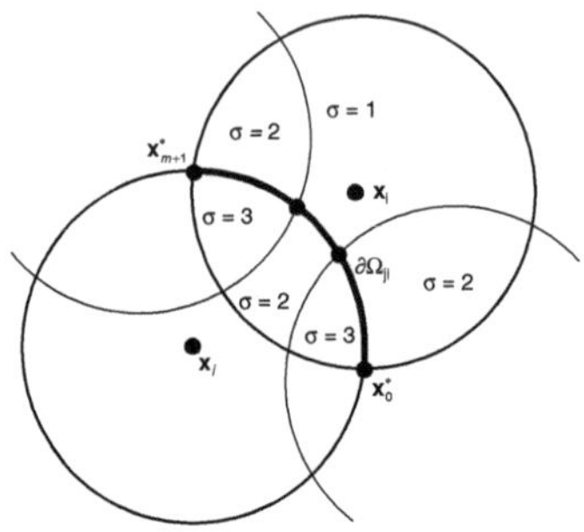

Fig. 3. Schematic diagram of the overlap of particles i and j and two neighbours in 2D, illustrating the notation for calculation of γ_{ij}. The highlighted arc is the region on which $W_j \nabla W_i \neq 0$.

where $\sigma_{-,k}$ is the value of σ_- on segment k. This approach allows γ_{ij} to be determined exactly and inexpensively. The most computationally taxing part of the procedure is the geometric analysis to determine the intersections of the particle support edges.

A similar approach may be used to evaluate Eq. (15) for boundary conditions, since the integrand is piecewise constant along the boundary. Corners can be handled straightforwardly with this method.

4 Implementation

The FVPM framework used in the present work includes an adaptation of van Leer's MUSCL approach [17]. For every particle pair, field variables are reconstructed from each particle barycentre to the particle-particle interface (located at the midpoint of a line segment between the particle centres). This reconstruction is based on gradients estimated at each barycentre using the corrected SPH formulation of Bonet and Lok [1]. The AUSM$^+$ scheme of Liou *et al.* [11] is then used to compute the inviscid fluxes from the reconstructed values on either side of the interface. Viscous stresses are calculated using the same SPH gradient approximations. Full details are given by Nestor *et al.* [13].

FVPM is an arbitrary Lagrange-Euler (ALE) method. For some applications, a fully Lagrangian scheme is desirable, but may lead to the evolution of strongly non-uniform or anisotropic particle distributions, which in turn can lead to degraded accuracy. Schick [14] introduced a near-Lagrangian ALE scheme for 1D FVPM, in which the particle motion is corrected slightly to prevent the formation of particle voids or clumps. Nestor *et al.* [13] developed a similar concept for multi-dimensional flows. After each time step, the convection velocity $\dot{\mathbf{x}}_i$ of particle i is corrected according to

$$\dot{\mathbf{x}}_i = \mathbf{u}_i + C\frac{\bar{r}_i}{\Delta t}\sum_j \left(\frac{\bar{r}_i}{r_{ij}}\right)^2 \mathbf{n}_{ij} \tag{19}$$

where $\mathbf{u}_i$ is the field velocity, r_{ij} is the distance between particles i and j, $\bar{r}_i$ is the average distance from i to its neighbours, $\mathbf{n}_{ij}$ is a unit vector pointing from $\mathbf{x}_i$ to $\mathbf{x}_j$, and the sum includes only the neighbours of i. C is a constant chosen in the range 10^{-4} to 10^{-2}, depending on the application. This particle motion correction acts like a weak repulsive force that influences the position of particles, rather than a spurious physical force. It is correctly accounted for by the $\dot{\mathbf{x}}$ term in Eq. (5).

Pressure is updated using the following equation of state for a weakly compressible fluid:

$$p = \frac{\rho_0 c_0^2}{\gamma}\left[\left(\frac{\rho}{\rho_0}\right)^{\gamma} - 1\right], \tag{20}$$

where ρ_0 is the reference density, c_0 is the speed of sound, and $\gamma = 7$. A second-order Runge-Kutta time stepping scheme is used, with a CFL criterion used to determine the timestep size.

For exact evaluation of the particle interaction vectors, the top-hat weight function is used, as described in section 3. For evaluation of $\boldsymbol{\beta}_{ij}$ by numerical integration (for comparison), the following quadratic weight function is used:

$$W\left(\mathbf{x} - \mathbf{x}_i(t), h\right) = \begin{cases} 4 - \left|\dfrac{\mathbf{x} - \mathbf{x}_i}{h}\right|^2, & \text{if } \left|\dfrac{\mathbf{x} - \mathbf{x}_i}{h}\right| < 2 \\ 0, & \text{otherwise} \end{cases}, \tag{21}$$

where h is the smoothing length, defined (following the SPH convention) as half the radius of the particle support. The numerical integration is performed by Gaussian quadrature with an array of 6×6 integration points in the particle overlap region. When using the top-hat weight function of Eq. (16) to evaluate $\boldsymbol{\beta}_{ij}$ in the present work, the quadratic weight function is used with numerical integration to determine particle barycentres and volume. For all versions of the method, the smoothing length is chosen as 0.8 times the initial particle spacing, so that in two dimensions, each particle overlaps with 36 neighbours.

Boundary conditions used in the present work are no-slip walls and periodic boundaries. Walls are treated by the method outlined in section 2.3. Periodic boundaries are implemented simply by populating a halo region outside the boundary with copies of particles from the opposite side of the domain.

5 Validation

The new formulation has been applied to the 2D Taylor-Green flow to assess its accuracy and computational efficiency. This flow field consists of a decaying vortex described by the following exact solution of the Navier-Stokes equations:

$$u(x, y, t) = -u_0 \cos(2\pi x/L)\sin(2\pi y/L)e^{-\frac{8\nu\pi^2}{L^2}t} \tag{22}$$

$$v(x, y, t) = u_0 \sin(2\pi x/L)\cos(2\pi y/L)e^{-\frac{8\nu\pi^2}{L^2}t}, \tag{23}$$

where u and v are the x and y components of velocity, u_0 is the maximum initial speed, ν is kinematic viscosity, and the domain has periodic boundaries at $x = \pm L/2, y = \pm L/2$.

To simulate the Taylor-Green flow in FVPM, particles are distributed on a regular isotropic Cartesian grid at an initial spacing Δx and initialised according to Eqs. (22) and (23) at $t = 0$ with Reynolds number $u_0 L/\nu = 100$. The timestep is determined by maintaining a Courant number of 0.1 based on Δx. The maximum Mach number u_0/c_0 is 0.1, so that the flow is practically incompressibile. Results are presented for Eulerian particles and for corrected Lagrangian particle motion with $C = 0.01$ in Eq. (19). The accuracy of FVPM solutions is characterised by the L_2 norm error in the magnitude of velocity after the (theoretical) maximum velocity has decayed to $u_0/10$.

Results for the variation of L_2 norm error with particle spacing are shown in Figure 4 for Eulerian (stationary) particles and corrected Lagrangian particle motion, and for β_{ij} evaluated both by quadrature and by the new exact method. For all four variants, the solution displays convergence at better than second order over most of the range investigated. Exact evaluation of β_{ij} results in slighlty larger error than quadrature in many cases. This may be due to the fact that the top-hat weight function gives greater weight to distant neighbours than the domed quadratic function, resulting in a wider effective numerical stencil. However, quadrature results in slower convergence for the finest discretisations, whereas with exact β_{ij}, the method maintains second-order convergence or better.

In many cases, the simulations with near-Lagrangian corrected particle motion show smaller error than those with fixed particles. This may be a reflection of the lower dissipation inherent in a purely Lagrangian scheme, in which convective terms need not be approximated.

L_2 norm error is plotted as a function of CPU time in Figure 5. This demonstrates clearly that exact integration results in lower error for a given computational cost, or conversely, lower time for the same error, in all conditions tested. The speedup ratio achieved by exact analytical integration ranges between 2.7 and 3.5, as shown in Figure 6.

6 Application to free surface flow

The simulation of gravity-driven free-surface flow is a difficult but important problem in marine, coastal and hydraulic engineering. Lagrangian mesh-free methods have a natural adavantage over conventional mesh-based methods in such applications, since the computational points can track the free surface without dissipation. Although free-surface flows may be treated by modelling air and water phases, it is reasonable in many cases to model only the water phase and neglect the influence of air. The free surface is then an interface between the fluid and a void.

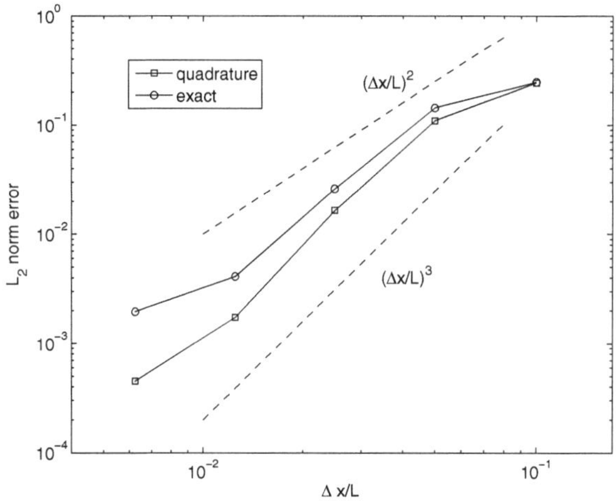

(a)

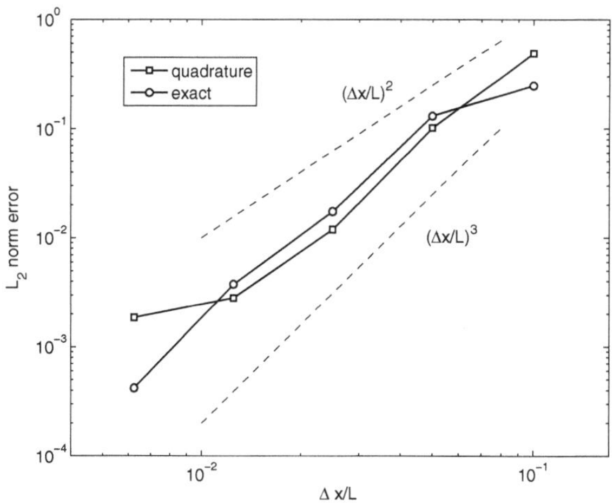

(b)

Fig. 4. Variation of L_2 norm error with particle spacing in FVPM simulations of Taylor-Green flow with (a) Eulerian particles and (b) corrected Lagrangian particle motion.

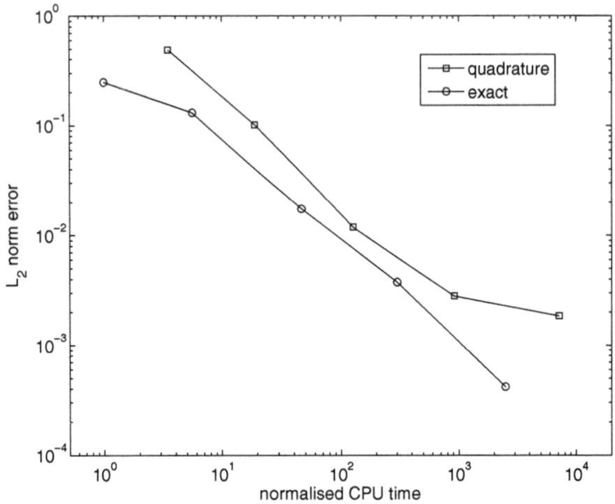

Fig. 5. Variation of L_2 norm error with CPU time in FVPM simulations of Taylor-Green flow.

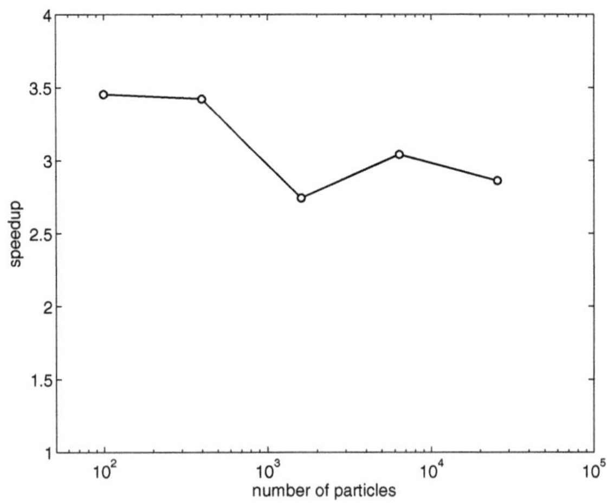

Fig. 6. Speedup ratio, defined as the ratio of CPU time with numerical quadrature to CPU time with exact β_{ij}, as a function of number of particles for Taylor-Green flow with corrected Lagrangian particle motion.

This presents a difficulty for quadrature-based FVPM. Each particle's interaction vectors $\boldsymbol{\beta}_{ij}$ are required to sum to zero for particles in the interior of the fluid, as expressed by Eq. (12). This condition follows from the definition of $\boldsymbol{\beta}_{ij}$, Eq. (9), since it can be shown that $\sum_j \boldsymbol{\beta}_{ij} = \int_\Omega \nabla\psi_i(\mathbf{x})\mathrm{d}\mathbf{x}$ and $\psi_i(\mathbf{x})$ is compactly supported. For particles at the free surface, however, $\sum_j \boldsymbol{\beta}_{ij} = 0$ is not required and not guaranteed, since $\sum_k W_k(\mathbf{x}) = 0$ and $\psi_i(\mathbf{x})$ is undefined on the interface between particle i and the void. As a result, correction methods based on enforcement of $\sum_j \boldsymbol{\beta}_{ij} = 0$ [4,16] will introduce large errors at a free surface. Without such a correction, errors due to numerical evaluation of $\boldsymbol{\beta}_{ij}$ contaminate the solution. However, with exact computation of $\boldsymbol{\beta}_{ij}$, these problems are avoided. In this section, the adaption of FVPM for free-surface flow is described, and a numerical example is presented.

Exact interaction vectors are computed by the method presented in section 3 above for all particles. Particles i for which $\sum_j \boldsymbol{\beta}_{ij} \neq \mathbf{0}$ (to within a small tolerance, allowing for round-off errors) are identified as free surface particles. Since $\boldsymbol{\beta}_{ij}$ typically points from $\mathbf{x}_i$ to $\mathbf{x}_j$, the non-zero $\sum_j \boldsymbol{\beta}_{ij}$ is expected to point away from the region where there are no neighbours – that is, it should be pointing into the fluid approximately normal to the free surface. Thus, $\sum_j \boldsymbol{\beta}_{ij}$ can be used to identify the location and orientation of the free surface.

Lagrangian particle motion is required for accurate resolution of the free surface. However, as discussed above, a small ALE component is often necessary to preserve a useful particle distribution in multi-dimensional flows. For free surface problems, the particle motion correction of Eq. (19) is computed for all particles with $C = 3 \times 10^{-4}$. For particles with $\sum_j \boldsymbol{\beta}_{ij} \neq \mathbf{0}$, the component of the correction velocity parallel to $\sum_j \boldsymbol{\beta}_{ij}$ (approximately normal to the free surface) is discarded. Thus, the kinematic boundary condition at the free surface is preserved while particles in the fluid interior are allowed a small ALE correction, and surface particles may redistribute along the surface. The dynamic boundary condition is enforced at particles on the free surface by prescribing zero pressure.

The free-surface forumulation is demonstrated for a wet-bottom dam-break experiment. A channel is initially divided by a gate into two regions of length L_0 and L_1, respectively, containing water at depth d_0 and d_1. When the gate is removed, a wave propagates from the deeper side into the shallow water. For this numerical test, $d_0 = 0.15$ m, $d_0/d_1 = 3.95$, $L_0/d_0 = 2.53$ and $L_1/d_0 = 6.67$. Acceleration due to gravity g is 9.81 m s^{-2} and the reference density ρ_0 is 1000 kg/m^3. The gate opening motion has been shown to have a significant effect on the subsequent flow [2], but here the gate is assumed to be removed instantaneously. The fluid is discretised with 2408 particles, intitially distributed on a Cartesian grid with spacing $\Delta x/d_0 = 0.0263$. The speed of sound c_0 is 10 m/s, giving a maximum Mach number less than 0.2 during the simulation, ensuring that compressibility effects are small. The Courant number, based on Δx, c_0 and a particle length scale $V_i/\sum_j \boldsymbol{\beta}_{ij}$, is 1.

Sample results are shown in Figure 7. The free-surface implementation of FVPM predicts the propagation and breaking of a primary wave, formation and breaking of a secondary wave, and run-up on the downstream wall. A close view of the breaking primary wave is shown in Figure 8 with $-\sum_j \beta_{ij}$ vectors, to illustrate the application of the free surface boundary condition. As expected, the vectors are non-zero only for particles which have no overlapping neighbours for some part of their support. The vectors point towards these void regions, in a direction approximately normal to the free surface. In the case of a void or concavity of size comparable to the particle size, nearby particles may have full overlap with neighbours and lose their free-surface identity. This occurs in the innermost part of the breaking wave shown.

7 Conclusions

By choosing a particle weight function which is uniform over the particle support, it is possible to evaluate the particle interaction vectors efficiently and exactly. In tests on a 2D flow problem, the method is faster by a factor of around 3 than FVPM with numerical quadrature, and displays improved convergence. The new formulation also facilitates the modelling of free surface flows.

Acknowledgement. The research leading to these results has received funding from the European Community's Seventh Framework Programme (FP7/2007-2013) under grant agreement 225967 NextMuSE. Ruairi Nestor was supported by the IRCSET Embark Initiative.

References

1. J. Bonet and T.-S. L. Lok, *Variational and momentum preservation aspects of smooth particle hydrodynamic formulations*, Computer Methods in Applied Mechanics and Engineering **180** (1999), 97–115.
2. A. J. C. Crespo, M. Gómez-Gesteira, and Robert A. Dalrymple, *Modeling dam break behavior over a wet bed by a SPH technique*, Journal of Waterway, Port, Coastal, and Ocean Engineering **134** (2008), no. 6, 313–320.
3. L. Delorme, A. Colagrossi, A. Souto-Iglesias, R. Zamora-Rodriguez, and E. Botia-Vera, *A set of canonical problems in sloshing. part i : Pressure field in forced roll. comparison between experimental results and SPH*, Ocean Engineering **36** (2009), no. 2, 168–178.
4. D. Hietel and R. Keck, *Consistency by coefficient correction in the finite volume particle method*, Meshfree Methods for Partial Differential Equations (M. Griebel, ed.), Lecture Notes in Computational Science and Engineering, Springer, 2003, pp. 211–221.
5. D. Hietel, K. Steiner, and J. Struckmeier, *A finite volume particle method for compressible flows*, Mathematical Models and Methods in Applied Science **10** (2000), no. 9, 1363–1382.

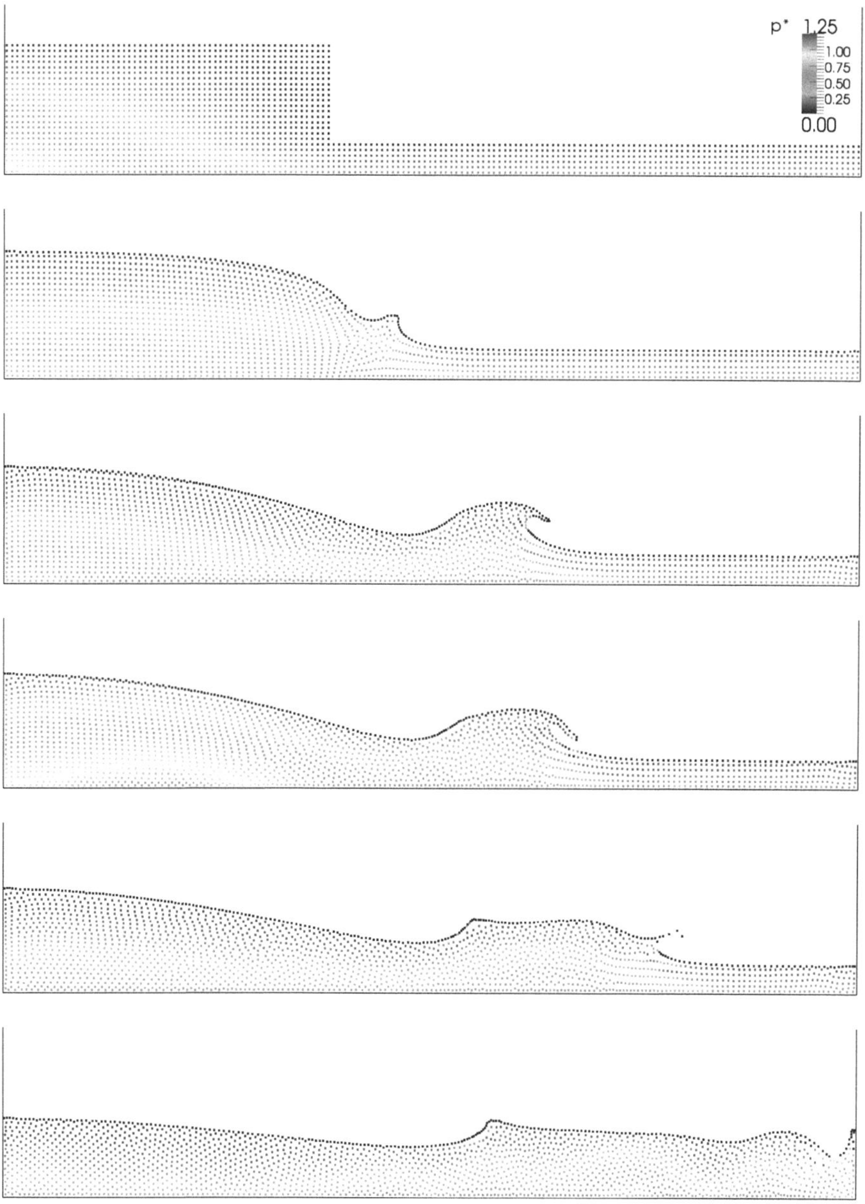

Fig. 7. Results at various times in FVPM simulation of a dam-break flow. Particles are coloured by dimensionless pressure $p^* = p/(\rho_0 g d_0)$.

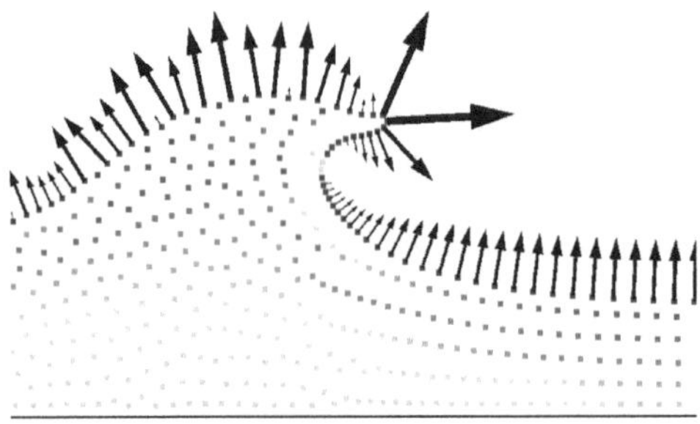

Fig. 8. A breaking wave in FVPM simulation of a dam-break flow. Particles are coloured by dimensionless pressure $p^* = p/(\rho_0 g d_0)$ and the vector for each particle i denotes $-\sum_j \beta_{ij}$. Particles without a visible vector have $\sum_j \beta_{ij} = 0$ within machine round-off error.

6. T. Ismagilov, *Smooth volume integral conservation law and method for problems in Lagrangian coordinates*, Computational Mathematics and Mathematical Physics **46** (2006), no. 3, 453–464.
7. M. Junk, *Do finite volume methods need a mesh?*, Lecture Notes in Computational Science and Engineering, Springer, 2003, pp. 223–238.
8. M. Junk and J. Struckmeier, *Consistency analysis of meshfree methods for conservation laws*, Mitteilungen der Gesellschaft für Angewandte Mathematik und Mechanik **24** (2002), no. 2, 99–126.
9. R. Keck and D. Hietel, *A projection technique for incompressible flow in the meshless finite volume particle method*, Advances in Computational Mathematics **23** (2005), no. 1, 143–169.
10. R. LeVeque, *Finite volume methods for hyperbolic problems*, Cambridge University Press, Cambridge, 1995.
11. Meng-Sing Liou, *A sequel to AUSM, part II: AUSM$^+$-up for all speeds*, Journal of Computational Physics **214** (2006), no. 1, 137–170.
12. J. J. Monaghan, *Smoothed particle hydrodynamics*, Reports on Progress in Physics **68** (2005), 1703–1759.
13. R. M. Nestor, M. Basa, M. Lastiwka, and N. Quinlan, *Extension of the finite volume particle method to viscous flow*, Journal of Computational Physics **228** (2009), 1733–1749.
14. C. Schick, *Adaptivity for particle methods in fluid dynamics*, Master's thesis, University of Kaiserslautern, 2000.
15. D. Teleaga, *A finite volume particle method for conservation laws*, Ph.D. thesis, University of Kaiserslautern, 2005.
16. D. Teleaga and J. Struckmeier, *A finite-volume particle method for conservation laws on moving domains*, International Journal for Numerical Methods in Fluids **58** (2008), no. 9, 945–967.
17. B. van Leer, *Towards the ultimate conservative difference scheme. V – A second-order sequel to Godunov's method*, Journal of Computational Physics **32** (1979), 101–136.

Parallel summation of symmetric inter-particle forces in smoothed particle hydrodynamics

Johannes Willkomm[1] and H. Martin Bücker[2]

[1] Institute for Scientific Computing, RWTH Aachen University, 52056 Aachen, Germany willkomm@sc.rwth-aachen.de
[2] Institute for Scientific Computing, RWTH Aachen University, 52056 Aachen, Germany bucker@sc.rwth-aachen.de

Summary. In the smoothed particle hydrodynamics (SPH) method, the forces between all particles are efficiently summed up in a serial environment as follows. Each pair of particles is considered once. The resulting inter-particle force is computed and then the contributions to both particles are updated, taking into account the symmetry of the problem. This algorithm is difficult to parallelise when concurrently accessing the same memory location in a multi-threaded process. We develop a parallel 1D summation algorithm consisting of two passes on a Cartesian grid of cells in which particles move freely. In a first pass, we consider all cells with an even index n. We compute the inter-particle forces for all pairs of particles located in cell n and those pairs where a particle is located in cell n and another particle is located in cell $n+1$. Each cell n is handled by a different thread since no data race can occur. In a second pass, we do the same for cells with odd n. This way, all inter-particle forces are computed. We generalise this algorithm to 2D, 3D, and arbitrarily high dimensions and report performance results on three different shared-memory platforms using the OpenMP programming paradigm.

Key words: Smoothed particle hydrodynamics, symmetry, computational fluid dynamics, parallel computation.

1 Introduction

Smoothed particle hydrodynamics (SPH) is a technique to approximate numerical solutions of the equations of fluid dynamics by replacing the fluid by a set of moving particles. It is extensively used in the areas of fluid flow and astrophysics. Since its introduction in the 1970s [4, 7], it was continuously developed further [5, 6, 9]. In SPH, the interactions between particles are local rather than global. Also, symmetric structure is available in the governing equations. While this structure is commonly exploited in serial SPH implementations [5, 6], it is more difficult to take this into account in a multi-threaded implementation as data races have to be avoided. In this note, we focus on a

M. Griebel, M.A. Schweitzer (eds.), *Meshfree Methods for Partial Differential Equations V*, Lecture Notes in Computational Science and Engineering 79, DOI 10.1007/978-3-642-16229-9_15,

new parallel algorithm that is used in a multi-threaded SPH implementation in the context of an ongoing collaboration with the Institute of Hydraulic Engineering and Water Resources Management of RWTH Aachen University. The overall SPH implementation described in [15] is based on a shared-memory parallelisation using OpenMP [1]. Rather than illustrating the overall SPH algorithm in more detail, the focus of the present paper is on the parallel summation of particle-particle interactions exploiting symmetry. These interactions are among the main computational building blocks of any SPH implementation. In the context of distributed-memory implementations based on the message passing paradigm (MPI) [14], the corresponding issues are more widely used and analysed in various parallel SPH programs [2,3,8,11–13].

This paper is organised as follows. In Section 2, the computational model involving SPH particles is described. A short description of the management to track the neigbourhood of particles using a Cartesian grid of cells is given in Section 3. The main contribution of this article is the parallel summation algorithm introduced in Section 4. In that section, we develop an algorithm for the computation of particle-particle interactions capable of handling an arbitrary number of spatial dimensions and show its correctness. Performance results on three different shared-memory systems are reported in Section 5 and conclusions are given in Section 6.

2 Implementation of smoothed particle hydrodynamics

In this section, we summarise a new implementation of the smoothed particle hydrodynamics method. A more detailed description of this implementation is given in [15]. A fluid particle in d spatial dimensions is described by a tuple

$$p = (\mathbf{x}, \mathbf{v}, \rho, u, m), \tag{1}$$

where $\mathbf{x}, \mathbf{v} \in \mathbb{R}^d$ denote its position and velocity, respectively. The symbol ρ represents the density. The internal heat energy is denoted by u. The mass m is constant whereas $\mathbf{x}$, $\mathbf{v}$, ρ and u are updated according to partial differential equations. We denote the number of particles by N and use subscripts to refer to an individual particle. Particles are represented by smoothing or kernel functions W, which are radial basis functions with local support. The kernel functions are parametrised by the so-called smoothing length h which determines the radius H of the local support.

Monaghan [9] introduced equations governing the movement of a particle and the change in density and heat. In addition to these particles, the new SPH code also implements an alternative particle type [3] which gives results of better quality, especially with regards to the density field. For the sake of brevity, we only sketch the particles by Monaghan. The differential change of the velocity $\mathbf{v}_i$ of the i-th particle is given by

$$\frac{\mathrm{d}\mathbf{v}_i}{\mathrm{d}t} = -\sum_{\substack{j=1\\ j\neq i}}^{N} m_j \left(\frac{P_i}{\rho_i^2} + \frac{P_j}{\rho_j^2} + \Pi_{ij} \right) \nabla_i W_{ij} + \mathbf{f}_i, \tag{2}$$

where $\mathbf{f}_i$ is the gravitational acceleration and P denotes the pressure. The symbol W_{ij} is a shorthand for $W(|\mathbf{x}_j - \mathbf{x}_i|)$, which is zero when the distance is larger than H. In practice, the sum will not have to iterate over all $N-1$ particles but only over those j which are actually within the radius of the support of W around $\mathbf{x}_i$. The symbol $\nabla_i W_{ij}$ denotes the spatial derivative of W_{ij} w.r.t. $\mathbf{x}_i$. The term Π_{ij} is the so-called artificial viscosity. The differential change of the density is governed by

$$\frac{\mathrm{d}\rho_i}{\mathrm{d}t} = \sum_{\substack{j=1\\ j\neq i}}^{N} m_j \left(\mathbf{v}_i - \mathbf{v}_j\right)^T \nabla_i W_{ij}. \tag{3}$$

The differential change of the heat energy is given by

$$\frac{\mathrm{d}u_i}{\mathrm{d}t} = \frac{1}{2}\sum_{\substack{j=1\\ j\neq i}}^{N} m_j \left(\frac{P_i}{\rho_i^2} + \frac{P_j}{\rho_j^2} + \Pi_{ij} \right) \left(\mathbf{v}_i - \mathbf{v}_j\right)^T \nabla_i W_{ij}. \tag{4}$$

The differential change of the position is described by

$$\frac{\mathrm{d}\mathbf{x}_i}{\mathrm{d}t} = \mathbf{v}_i. \tag{5}$$

The pressure is represented by the so-called Tait's or Gamma equation

$$P_i = B\left[\left(\frac{\rho_i}{\rho^{(0)}}\right)^{\gamma} - 1\right], \tag{6}$$

where $\rho^{(0)}$ is the reference density of the material, e.g., $\rho^{(0)} = 1000\,\frac{\mathrm{kg}}{\mathrm{m}^3}$ for water. The parameter $\gamma = 7$ is usually kept fixed. The symbol B can be interpreted as the spring constant of the system. The larger the value of B, the higher becomes the speed of sound and the lower the compressibility. So, a large B is desirable for realistic results, but will require small time steps in the time integration.

The new SPH code implements several explicit time integration methods such as different Runge-Kutta methods and predictor-corrector methods. From a conceptual point of view, time stepping is carried out as follows. Let $p^{(n)}$ denote the current state vector representing the agglomeration of all N particles. The state update function

$$y = F(\delta_{\mathrm{t}}, p^{(n)})$$

computes the update y to the state vector for a given time step size δ_{t}. For instance, if the explicit Euler method is used, then the new state is given by

$$p^{(n+1)} = p^{(n)} + y.$$

Here, the update of the i-th particle p_i from step n to step $n+1$ is represented by the i-th component of $y = (y_1, y_2, \ldots, y_N)$. That is, the update of p_i is given by

$$y_i = (\Delta \mathbf{x}, \Delta \mathbf{v}, \Delta \rho, \Delta u, \Delta m),$$

whose entries correspond to the notation used in (1).

3 Symmetric inter-particle forces on a grid of cells

The state update function F finds all pairs of particles (p_i, p_j) which are inside the support of each others' kernel function and computes the contribution to their respective updates y_i and y_j. A grid-based lookup data structure is used to find all neighbours of a certain particle, very similar to the approaches used in [3, 5, 9]. A grid cell in which a particle p_i at position $\mathbf{x}_i$ is located is represented by the index vector

$$\mathbf{I}_i = \text{floor}((\mathbf{x}_i - C_0)/H) \bmod M,$$

where $C_0 \in \mathbb{R}^d$ is the left bottom corner of a cuboid domain, H is the radius of the support of the kernel function, M is a user-given parameter specifying the number of cells in a single dimension, and the operations "floor" and "mod" are evaluated component-wise. Thus the total number of grid cells in d dimensions is given by M^d. In summary, the grid does not define the position of the particles. Rather it is used as a mechanism to track the neighbourhood of the particles that are allowed to move freely in space.

The update corresponding to the particle p_i is given as the sum of all contributions of neighbouring particles p_j, namely

$$y_i = \sum_{\substack{j=1 \\ j\neq i}}^{N} \delta t\, f\,(p_i, p_j) + (\delta t\, \mathbf{v}_i, \delta t\, \mathbf{f}_i, 0, 0, 0), \tag{7}$$

where the function f evaluates equations (2), (3) and (4). This formulation is geared toward symmetry in i and j. Symmetric terms are handled by the function f whereas non-symmetric terms are treated separately. For instance, the update of the position given by (5) is not symmetric in i and j and is handled separately from f in the first component, $\delta t\, \mathbf{v}_i$, of the last term in (7). Similarly, according to (2), the update of the velocity involves a non-symmetric contribution from gravity. This contribution, $\delta t\, \mathbf{f}_i$, is also treated separately from f. We stress that the artificial viscosity Π_{ij} is symmetric in i and j and so are the equations for density (3) and energy (4). Throughout the discussion, we assume the mass to be constant in time and identical for all particles. Since the velocity is antisymmetric, we can evaluate the contribution $f_{ij} := f\,(p_i, p_j)$ once for each pair (p_i, p_j) and update both, y_i and y_j, as follows:

$$y_i.\mathbf{v} \mathrel{+}= f_{ij}.\mathbf{v} \qquad y_j.\mathbf{v} \mathrel{-}= f_{ij}.\mathbf{v}$$
$$y_i.\rho \mathrel{+}= f_{ij}.\rho \qquad y_j.\rho \mathrel{+}= f_{ij}.\rho$$
$$y_i.u \mathrel{+}= f_{ij}.u \qquad y_j.u \mathrel{+}= f_{ij}.u$$

The notation $.\mathbf{v}$ denotes the velocity component. Similarly, $.\rho$ denotes the density component, and $.u$ the heat component of y_i, y_j, and f_{ij}.

Using the lookup data structure previously described, a naïve summation algorithm for the computation of the updates y would not exploit symmetry. In the naïve approach, each particle p_i is considered in turn. The neighbouring particles are found by looking in the particle's grid cell $\mathbf{I}_i$ and in the $3^d - 1$ surrounding cells and the contributions are summed up in y_i. The other cells do not have to be considered because particles contained in them are at a distance larger than H from p_i. This approach is trivially parallelisable, since the computations for each i are independent of each other. However, every pair of particles (p_i, p_j) is considered twice. Therefore it is advantageous to consider each pair of particles only once, updating both y_i and y_j. An algorithm described in [6] does exploit symmetry. However, it is not efficiently parallelisable. In the following section, we introduce a novel parallel algorithm exploiting symmetry in the computation of inter-particle forces.

4 Parallel symmetric summation algorithm

The new algorithm for d dimensions iterates over the grid cells, splitting them into 2^d disjoint subsets and handling each subset in a parallel pass. For the sake of simplicity, we first sketch the one-dimensional case as an illustrating example. For $d = 1$, the algorithm consists of 2 passes. In the first pass, we consider the even grid cells $k = 2k'$ with $k' \in \{0, 1, \ldots, M/2 - 1\}$. The first pass computes the following two steps:

i. the intra-cell relations of all particles located in cell k and
ii. the relations of all particle pairs where one particle is located in cell k and the other in cell $k + 1 \bmod M$.

For step 1, we define a helper function sumF1($\mathbf{I}$) which computes all intra-cell relations given the index $\mathbf{I}$ of a grid cell. For each pair of particles p_i and p_j with $\mathbf{I}_i = \mathbf{I}_j = k$, it updates both y_i and y_j. This helper function is represented graphically by a ball inside the cell, as shown in Fig. 1 (left).

For step 2 of the first pass, we define a helper function sumF2($\mathbf{I}, \mathbf{J}$) which computes all inter-cell relations given the indices $\mathbf{I}$ and $\mathbf{J}$ of two grid cells.

Fig. 1. Graphical representations of sumF1 (left) and sumF2 (right).

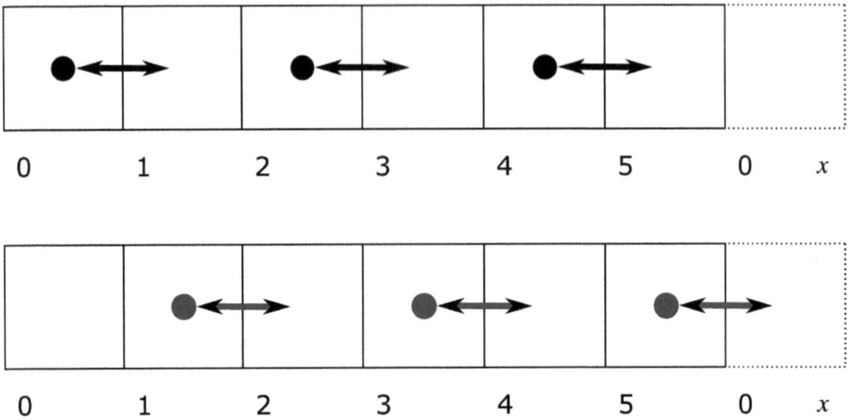

Fig. 2. First (top) and second (bottom) pass of the parallelisation scheme in 1D.

This helper function is represented graphically by an arrow shown in Fig. 1 (right). The modulo operation wraps around on the boundaries of the grid domain since there may be particles in cells $k = 0$ and $k = M - 1$ that are neighbours in the physical domain.

Within the first pass, the computational work for each k' is independent of each other. Thus, there are $M/2$ independent tasks each consisting of a call to the function sumF1 followed by a call to the function sumF2.

In the second pass, we consider the odd cells $k = 2k' + 1$ with $k' \in \{0, 1, \ldots, M/2 - 1\}$. The steps of the second pass are similar to those in the first pass. After the two passes are carried out, all particle relations are considered. The two passes of the algorithm are illustrated in Fig. 2 where the number of cells in one dimension is given by $M = 6$. The parallelisation of the second pass is identical to the one of the first pass. In summary, there are two passes depicted in different colours in Fig. 2 that are carried out one after another. Within each pass, $M/2$ tasks are carried out in parallel. Each task consists of intra-cell computations represented by the function sumF1 that are followed by inter-cell computations described by the function sumF2. The overall update scheme of a single pass on a subset of neighbouring cells is characterised by a "pattern" consisting of a ball (intra-cell computations) and a single arrow (inter-cell computations). This pattern is first applied to even cells and then moved by an "offset" to the odd cells.

To extend this idea to two spatial dimensions, we describe the problem graphically. The result is schematically depicted in Fig. 3. There are now $2^d = 4$ passes highlighted by colours in that figure. Each pass consists of a ball (intra-cell computations) and four arrows (inter-cell computations). The pattern describing the work of a task in a single pass on four neighbouring cells is shown to the right of the figure. In the first pass, this pattern is placed on all grid cells whose components of the index vectors are even, i.e., the offset is $(0,0)$. In the second pass, the patterns are moved by the offset $(0,1)$. The offsets of the third and fourth passes are given by $(1,0)$ and $(1,1)$, respectively. Again, the passes are carried out sequentially. Within each pass, the patterns are disjoint so that $M^d/2^d = M^2/4$ independent tasks can be processed in parallel.

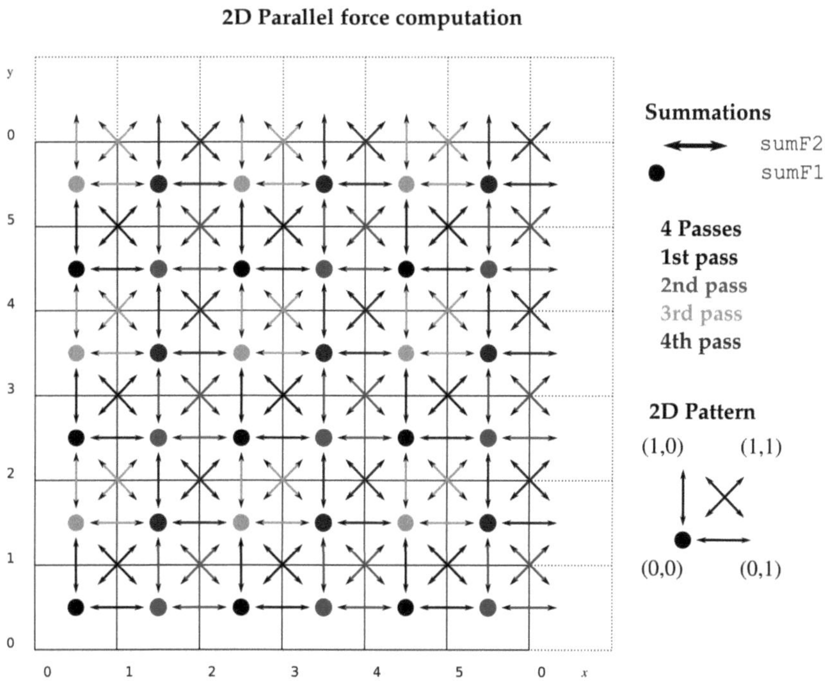

Fig. 3. Passes of the parallelisation scheme in 2D.

To extend the algorithm to any number of dimensions, we introduce a way to algorithmically describe the list of index pairs that make up the pattern. One property that they share and that describes them uniquely is the following. A pair of indices $\mathbf{I}$ and $\mathbf{J}$ is in the pattern if the relation $\mathbf{I}^T \cdot \mathbf{J} = 0$ holds, i.e., their scalar product is zero. The case where $\mathbf{I} = 0$ and $\mathbf{J} = 0$ is an exception needing special treatment by the function sumF1. Therefore, the new algorithm for an arbitrary number, d, of spatial dimensions is as follows:

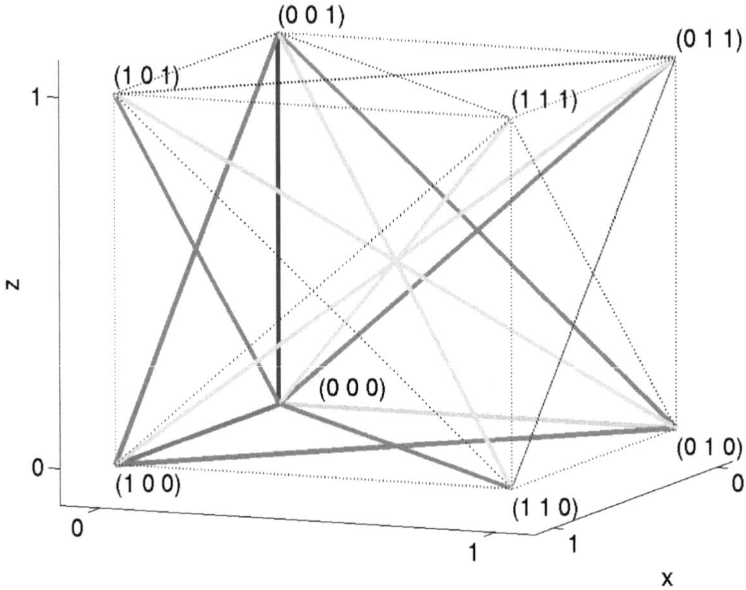

Fig. 4. The pattern of the parallelisation scheme in 3D.

$$\textbf{for } o \textbf{ in } \{0,1\}^d \tag{8}$$
$$\quad \textbf{parallel for } i \textbf{ in } \{0,\ldots,M/2-1\}^d \tag{9}$$
$$\quad\quad \mathrm{sumF1}(2*i+o) \tag{10}$$
$$\quad\quad \textbf{for } j \textbf{ in } \{0,1\}^d \tag{11}$$
$$\quad\quad\quad \textbf{for } k \textbf{ in } \{0,1\}^d \tag{12}$$
$$\quad\quad\quad\quad \textbf{if } k<j \textbf{ and } j^T\cdot k=0 \tag{13}$$
$$\quad\quad\quad\quad\quad \mathrm{sumF2}((2*i+o+j) \bmod M, (2*i+o+k) \bmod M) \tag{14}$$

In this algorithm, the loop indices o, i, j and k are d-dimensional integer vector variables. The test $k < j$ is a lexicographic comparison and is used in order to consider each pair of cells only once.

In Fig. 4, the pattern generated by the line (14) of the algorithm for three spatial dimensions is shown. The same pattern is also presented in [10] where it is used in the context of distributed memory to minimise the number of synchronisations across the boundaries of a domain decomposition approach.

The following property of the new algorithm ensures that all particle-particle relations are considered.

Lemma 1. *Each pair of grid cells is handled exactly once by the algorithm.*

Proof. Consider a pair of neighbouring grid cells $\mathbf{I}$ and $\mathbf{J}$, where $\mathbf{I} \leq \mathbf{J}$ lexicographically. For our purposes, $(\mathbf{I}, \mathbf{J})$ and $(\mathbf{J}, \mathbf{I})$ describe the same pair, because sumF2$(\mathbf{I}, \mathbf{J})$ performs the same computations and changes to the update vector y as sumF2$(\mathbf{J}, \mathbf{I})$. We also have sumF2$(\mathbf{I}, \mathbf{I})$ = sumF1$(\mathbf{I})$. The difference vector $\mathbf{d} = \mathbf{J} - \mathbf{I}$ contains only 1, 0 or -1 as entries since the two cells $\mathbf{I}$ and $\mathbf{J}$ are neighbours in the grid. Let $\mathbf{m} = \min(\mathbf{J}, \mathbf{I})$ taken component-wise. This vector can be decomposed bijectively into $\mathbf{m} = \mathbf{e} + \mathbf{o}$, where $\mathbf{e}$ is even and $\mathbf{o} \in \{0, 1\}^d$, as this can be done with any integer vector $\mathbf{m}$. The difference vectors $\mathbf{d}_\mathrm{I} = \mathbf{I} - \mathbf{m}$ and $\mathbf{d}_\mathrm{J} = \mathbf{J} - \mathbf{m}$ contain only 0 or 1 as entries. Moreover, we have $\mathbf{d}_\mathrm{I}^T \cdot \mathbf{d}_\mathrm{J} = 0$. In order to show that, we can distinguish the following three cases:

i. $\mathbf{I} = \mathbf{J} = \mathbf{m}$: Then $\mathbf{d}_\mathrm{I} = \mathbf{d}_\mathrm{J} = 0$ and $\mathbf{d}_\mathrm{I}^T \cdot \mathbf{d}_\mathrm{J} = 0$. This is the case that is handled by sumF1.
ii. $\mathbf{I} = \mathbf{m}$ (or $\mathbf{J} = \mathbf{m}$): Then $\mathbf{d}_\mathrm{I} = 0$ (or $\mathbf{d}_\mathrm{J} = 0$) and $\mathbf{d}_\mathrm{I}^T \cdot \mathbf{d}_\mathrm{J} = 0$.
iii. $\mathbf{I} \neq \mathbf{m}$ and $\mathbf{J} \neq \mathbf{m}$: Here we have to consider the vector components separately. Clearly, $\mathbf{d}_\mathrm{I}^T \cdot \mathbf{d}_\mathrm{J} = 0$ if and only if $[\mathbf{d}_\mathrm{I}]_i \cdot [\mathbf{d}_\mathrm{J}]_i = 0$ for all component indices $i \in \{0, \ldots, d-1\}$. Now for each i we have either
   a) $[\mathbf{I}]_i = [\mathbf{J}]_i = [\mathbf{m}]_i$: Then $[\mathbf{d}_\mathrm{I}]_i = [\mathbf{d}_\mathrm{J}]_i = 0$ so that $[\mathbf{d}_\mathrm{I}]_i \cdot [\mathbf{d}_\mathrm{J}]_i = 0$.
   b) $[\mathbf{I}]_i = [\mathbf{m}]_i$ (or $[\mathbf{J}]_i = [\mathbf{m}]_i$): Then $[\mathbf{d}_\mathrm{I}]_i = 0$ (or $[\mathbf{d}_\mathrm{J}]_i = 0$) so that $[\mathbf{d}_\mathrm{I}]_i \cdot [\mathbf{d}_\mathrm{J}]_i = 0$.

   It cannot happen that $[\mathbf{I}]_i \neq [\mathbf{m}]_i$ and $[\mathbf{J}]_i \neq [\mathbf{m}]_i$ as $[\mathbf{m}]_i = \min([\mathbf{I}]_i, [\mathbf{J}]_i)$.

Thus, any pair of neighbouring grid cells $(\mathbf{I}, \mathbf{J})$ can be decomposed bijectively as $(\mathbf{e}+\mathbf{o}+\mathbf{d}_\mathrm{I}, \mathbf{e}+\mathbf{o}+\mathbf{d}_\mathrm{J})$. The algorithm generates all grid cells $\mathbf{m} = \mathbf{e}+\mathbf{o}$ by iterating over all even grid cells $\mathbf{e}$ and all possible offsets $\mathbf{o} \in \{0, 1\}^d$. Then it calls sumF2$(\mathbf{m} + \mathbf{d}_\mathrm{I}, \mathbf{m} + \mathbf{d}_\mathrm{J})$ for all $(\mathbf{d}_\mathrm{I}, \mathbf{d}_\mathrm{J})$ where $\mathbf{d}_\mathrm{I} \in \{0, 1\}^d$, $\mathbf{d}_\mathrm{J} \in \{0, 1\}^d$, and $\mathbf{d}_\mathrm{I}^T \cdot \mathbf{d}_\mathrm{J} = 0$. Therefore, it will call sumF2 for every pair of neighbouring grid cells. This completes the proof.

The following lemma states that the algorithm is thread-safe.

Lemma 2. *The patterns in each parallel pass of the algorithm do not overlap.*

Proof. Recall from the algorithm that $\mathbf{o}$ is kept fixed during each of the parallel passes. Furthermore, the calls to sumF2 are decomposed into independent tasks according to the even component $\mathbf{e}$. As for both $\mathbf{d}_\mathrm{I} \in \{0, 1\}^d$ and $\mathbf{d}_\mathrm{J} \in \{0, 1\}^d$, it is not possible that $\mathbf{e} + \mathbf{d}_\mathrm{I}$ is equal to $\mathbf{e}' + \mathbf{d}_\mathrm{J}$ for any other even grid cell $\mathbf{e}'$. Thus, no two calls to sumF2 executed by different tasks will touch particles in the same grid cell.

The following theorem summarises the results of the previous discussion.

Theorem 1. *Given a Cartesian grid with M cells in each of d spatial dimensions, the algorithm correctly sums up all particle-particle computations needed in the first term of (7). The algorithms consists of 2^d passes carried out sequentially. In each pass, there are $(M/2)^d$ independent tasks that can be computed in parallel.*

5 Experimental results

We perform a simulation with different numbers of OpenMP threads. In particular, we compute five time steps on a 3D test case which has $1.3 \cdot 10^6$ particles. We run the test for the naïve summation algorithm with $M = 128$ and for the new algorithm exploiting symmetry with $M = 64$. A third algorithm is also tested with $M = 64$ for reasons of comparison. This multi-buffer algorithm uses a separate buffer for each thread and, thus, exploits symmetry similar to the new algorithm. Unlike the new algorithm, this algorithm requires a summation of these per-thread buffers after computing the particle-particle relations. The focus of this section is on the performance of the new summation algorithm in comparison with the naïve and the multi-buffer algorithm. The summation algorithm constitutes the major part of the overall SPH simulation. Experimental results with additional parallel performance measurements are reported in [15] for the overall SPH simulation.

We use fluid particles introduced by Ferrari et al. [3] and point symmetric boundary particles with the predictor-corrector time integration algorithm of order one. All tests were run on three different clusters with SUN Niagara 2 processors and Solaris operating system, Intel Xeon processors and Linux OS, and AMD Opteron processors and Linux OS, resp., at the Center for Computing and Communication of RWTH Aachen University, using the GCC compiler version 4.3.1.

In Fig. 5, the absolute times are shown for the naïve, the new, and the multi-buffer algorithm. Throughout the following figures, the new algorithm is denoted by "sym" to indicate that it exploits symmetry whereas the naïve algorithm does not. The multi-buffer algorithm is indicated by "sym, mb". The results reported in Fig. 5 indicate that all run times decrease when increasing the number of threads from 1 to 64 threads. The results also show that our new symmetric summation algorithm has a better performance in run time than the naïve algorithm for all numbers of threads and on all computing platforms. Furthermore, the multi-buffer algorithm needs more time than the new algorithm.

In Fig. 6, the parallel speedup is given. Here, the speedup is defined with respect to the run time of the parallel program with a single thread. This figure indicates that the new algorithm also achieves good speedups even for 64 threads. However, the speedups of the naïve algorithm are larger. The corresponding efficiency is given in Fig. 7.

In Fig. 8, we show the speedup of the new algorithm over the naïve version, i.e., the amount by which the new algorithm is faster. In these experiments, we also incorporate results for the naïve algorithm with $M = 64$. So, results for the same Cartesian grid for the neighbourhood tracking are available for the naïve and the new algorithm. Again, we show results for several different computing platforms. Here, the exploitation of the symmetry in the SPH equations results in a lower execution time in all cases. On the other hand,

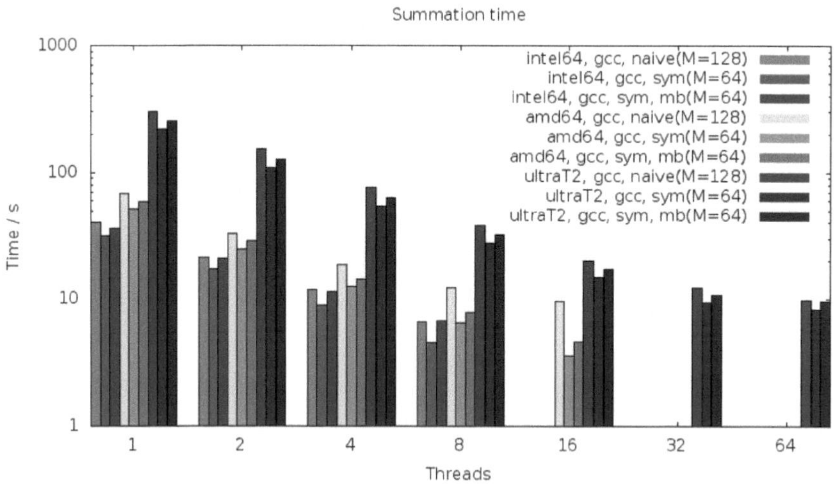

Fig. 5. Total run times of naïve summation with $M = 128$, symmetric summation with $M = 64$, and the multi-buffer summation with $M = 64$ on three computing platforms.

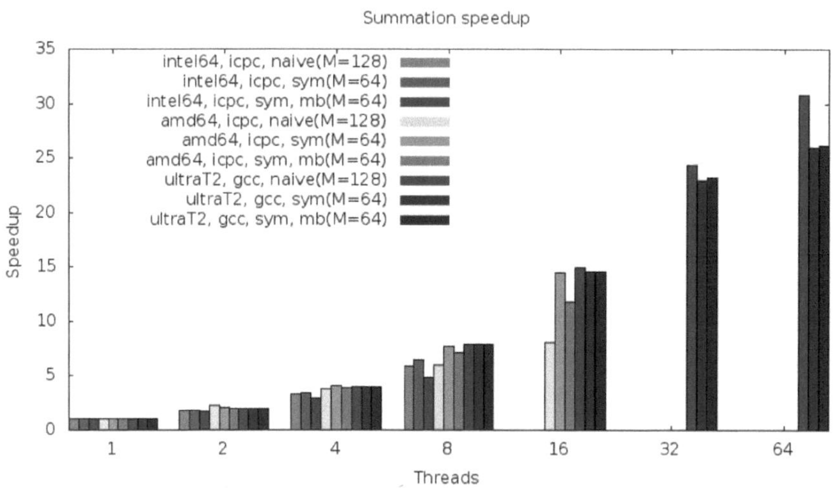

Fig. 6. Run time speedups of naïve summation with $M = 128$, symmetric summation with $M = 64$, and the multi-buffer summation with $M = 64$ on three computing platforms.

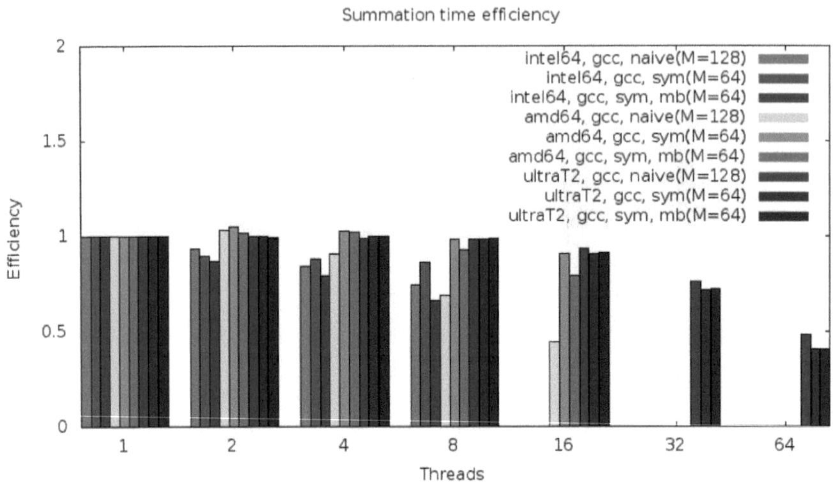

Fig. 7. Run time efficiency of naïve summation with $M = 128$, symmetric summation with $M = 64$, and the multi-buffer summation with $M = 64$ on three computing platforms.

the factor of two which could be expected by exploiting symmetry is in most cases not achieved.

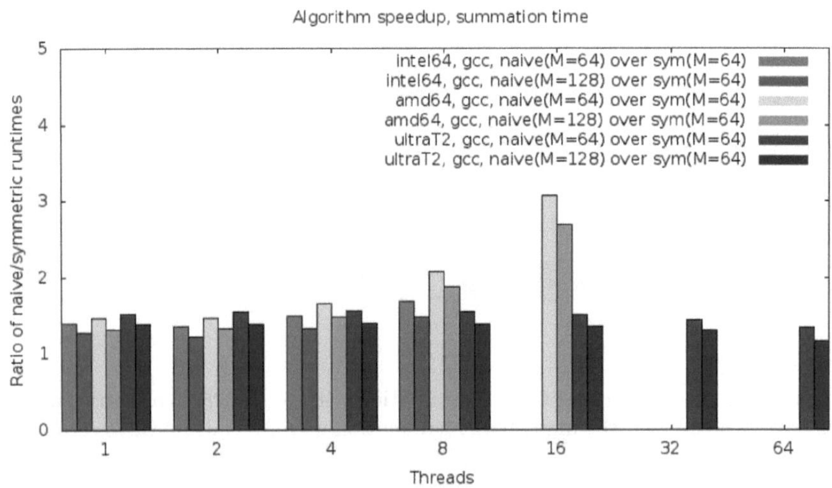

Fig. 8. Total speedup of the force summation with the symmetric algorithm over the naïve version of the force summation on three computing platforms.

6 Conclusion

One of the key ingredients to an efficient parallel implementation of the smoothed particle hydrodynamics (SPH) method is the management of particle neighbourhoods. This management is crucial for the summation of contributions from neighbouring particles rather than from all particles. We introduced a new algorithm for the parallel summation of particle-particle interactions exploiting symmetry in the computations. The overall idea is to track the neighbourhood of particles on a Cartesian grid of cells in which the particles are allowed to move freely. The approach is generalised to an arbitrary number of spatial dimensions. We implemented the new algorithm on three different shared-memory computers. Performance results using up to 64 threads demonstrate that the implementation of the new algorithm exhibits lower run times than a corresponding naïve implementation that does not exploit symmetry. However, the scalability of the naïve algorithm is somewhat better than for the new algorithm if the parameter specifying the number of cells in a single dimension of the grid is chosen appropriately.

Acknowledgement

The collaboration with the Institute for Hydraulic Engineering and Water Resources Management is carried out within the FLOWRUN project funded by the section Simulation Sciences of the Jülich-Aachen Research Alliance (JARA-SIM).

References

1. B. Chapman, G. Jost, and R. van der Pas, *Using OpenMP: Portable Shared Memory Parallel Programming*, The MIT Press, 2007.
2. J. A. Faber and F. A. Rasio, *Post-Newtonian SPH calculations of binary neutron star coalescence: Method and first results*, Phys. Rev. D **62** (2000), no. 6, 064012.
3. A. Ferrari, M. Dumbser, E. F. Toro, and A. Armanini, *A new 3D parallel SPH scheme for free surface flows*, Computers & Fluids **38** (2009), no. 6, 1203–1217.
4. R. A. Gingold and J. J. Monaghan, *Smoothed particle hydrodynamics — Theory and application to non-spherical stars*, Mon. Not. Roy. Astron. Soc. **181** (1977), 375–389.
5. M. Gómez-Gesteira, B. D. Rogers, R. A. Dalrymple, A. J. C. Crespo, and M. Narayanaswamy, *User guide for the SPHysics code v1.2*, 2008.
6. G. R. Liu and M. B. Liu, *Smoothed Particle Hydrodynamics*, Singapore, World Scientific Publishing Co. Pte. Ltd, 2003.
7. L. B. Lucy, *A numerical approach to the testing of the fission hypothesis*, Astronomical Journal **82** (1977), 1013–1024.
8. P. Maruzewski, D. Le Touzé, G. Oger, and F. Avellan, *SPH high-performance computing simulations of rigid solids impacting the free-surface of water*, Journal of Hydraulic Research **48** (2010), in Press.

9. J. J. Monaghan, *Simulating free surface flows with SPH*, J. Comput. Phys. **110** (1994), no. 2, 399–406.
10. I. F. Sbalzarini, J. H. Walther, M. Bergdorf, S. E. Hieber, E. M. Kotsalis, and P. Koumoutsakos, *PPM: a highly efficient parallel particle-mesh library for the simulation of continuum systems*, J. Comput. Phys. **215** (2006), no. 2, 566–588.
11. R. Speith, E. Schnetter, S. Kunze, and H. Riffert, *Distributed implementation of SPH for simulations of accretion disks*, Molecular Dynamics On Parallel Computers, Proceedings of the NIC-Workshop, Jülich, February 8–10, 1999 (Singapore), World Scientific Publishing Co., 2000, pp. 276–285.
12. V. Springel, *The cosmological simulation code GADGET-2*, Monthly Notices of the Royal Astronomical Society **364** (2005), no. 4, 1105–1134.
13. S. Vanaverbeke, R. Keppens, S. Poedts, and H. Boffin, *GRADSPH: A parallel smoothed particle hydrodynamics code for self-gravitating astrophysical fluid dynamics*, Computer Physics Communications **180** (2009), no. 7, 1164–1182.
14. D. W. Walker and J. J. Dongarra, *MPI: a standard Message Passing Interface*, Supercomputer **12** (1996), no. 1, 56–68.
15. J. Willkomm and H. M. Bücker, *A shared-memory parallel smoothed particle hydrodynamics simulation*, Proceedings of the 28th International Conference of the Chilean Computing Science Society, Santiago de Chile, Chile, November 9–14, 2009, 2009, pp. 41–48.

Meshfree Wavelet-Galerkin Method for Steady-State Analysis of Nonlinear Microwave Circuits

Alla Brunner

42349, Wuppertal, Germany ja-brunner@web.de

Summary. The paper presents a Wavelet-Galerkin method to compute the non-periodic steady-state response of a nonlinear dynamic microwave circuit with m pulsed input signals and n outputs. The method is based on both time-domain and frequency domain approaches. In the formulation of the equations describing the network, the elements of the circuit are assumed to be regrouped into linear and nonlinear subcircuits by loop analysis concepts. The linear part is transferred into the frequency domain by means of the Laplace transformation where the solution can by computed analytically. For the nonlinear subcircuit, the Bubnov-Galerkin method is applied in the time-domain. The discretization of the steady-state response in the Haar-wavelet basis results in a nonlinear system of algebraic equations for the expansion coefficients. The system is solved by means of the Newton-Raphson method for which the linear part of the multidimensional Taylor series expansion with respect to the expansion coefficients serves as start value. The performance of this new approach is illustrated by two examples.

Key words: Meshfree Wavelet-Galerkin Method; Nonlinear Dynamic Circuits.

1 Introduction

Time-domain circuit simulation plays an important role in the design and analysis of nonlinear dynamic radio frequency (RF), microwave and pulse signal circuits, such as pulse amplifiers, high-speed switches, modulators, multivibrators, blocking oscillators, triggers, functional pulse and digital circuits etc. These circuits, working with pulse input- or output signals, have strongly nonlinear characteristics and widely separated time constants.

The nonlinear distortions of pulse waveforms of these devices are shown in distortions of the pulse output waveform: the rise time, the fall time, the settling time, the pulse duration and the pulse amplitude. All key parameters and characteristics of the nonlinear dynamic circuit can be defined from the steady-state response of the circuit in the time-domain [2–4].

M. Griebel, M.A. Schweitzer (eds.), *Meshfree Methods for Partial Differential Equations V*, Lecture Notes in Computational Science and Engineering 79,
DOI 10.1007/978-3-642-16229-9_16,

To describe such nonlinear dynamic circuits, a multidimensional model is used in the time-domain which transfers the circuit's differential algebraic equations (DAEs) into a system of multivariate partial differential algebraic equations (MPDAEs).

Traditionally, such equations are solved by a discretization using a mesh and numerical schemes, such as the finite difference method. The steady-state response is calculated through time-consuming transient analyses with step-by-step procedure. For the microwave and RF circuits with largely different time scales the numerical methods have multi-scale problems and very long transient simulation (integration) intervals with rather small step sizes.

On the other hand, these exist many frequency domain approaches among these, mention should be made of the harmonic balance method, as a widely used equivalent version of the Galerkin projection method [7]. However, approximation of the steady-state response by truncated Fourier series should use many harmonics for a pulse input signal. The time-consuming Fourier-transform and Inverse-Fourier-transform are calculated in each iteration step.

A number of meshfree methods for partial differential equations have been proposed and used successfully to solve some engineering problems [1, 6, 8, 9].

In this paper, a numeric-analytical meshfree method is presented to compute the non-periodic steady-state (NPSS) response of a nonlinear dynamic microwave circuit with m pulsed input signals and n outputs. Considering non-periodic signals, this approach based on a Bubnov-Galerkin method with Haar-wavelets as orthogonal basis applies efficiently the techniques of Harmonic Balance to these non-trigonometric basis functions.

2 Wavelet-Galerkin method

We apply the Bubnov-Galerkin projection method with Haar-wavelets as basis functions and the techniques of Harmonic Balance to the set of MPDAEs which model the multidimensional NPSS response of nonlinear dynamic circuit with m input signals, including pulsed inputs, and n outputs.

2.1 Bubnov-Galerkin projection method

Let $\boldsymbol{\mathcal{F}}$ be a nonlinear operator with domain of definition and range of values in a Hilbert space $\boldsymbol{X}$. In this paper, the problem of determining the NPSS response of nonlinear dynamic circuit is handled as a nonlinear non-periodic problem of the operator equation

$$\boldsymbol{\mathcal{F}}(\boldsymbol{u}(t)) = 0\,, \quad \forall \quad t \in [0, T) \tag{1}$$

with initial condition $\boldsymbol{u}(0) = \boldsymbol{u}_0$, where $\boldsymbol{u}(t) \in \boldsymbol{X}$, $\boldsymbol{\mathcal{F}} : \boldsymbol{X_k} \subseteq \boldsymbol{X} \to \boldsymbol{X}$ is defined on a dense subspace $\boldsymbol{X_k}$ of a Hilbert space $\boldsymbol{X}$ with scalar product $\langle .,. \rangle$.

Let $\{\phi_1(t), \phi_2(t), \ldots\} \subseteq \boldsymbol{X_k}$ be a denumerable orthonormal basis of $\boldsymbol{X}$ with interval of orthogonality T. The Bubnov-Galerkin projection method computes an approximate solution of (1) in the form of a linear combination of finitely many basis test functions:

$$\boldsymbol{u}_k(t) = \sum_{i=1}^{K} c_i \cdot \phi_i(t) \,, \qquad \boldsymbol{u}_k(t) \in \boldsymbol{X}_k \,. \tag{2}$$

The expansion coefficients $c_1, c_2, \ldots, c_K$ are found from the Galerkin orthogonality relation for the error:

$$\left\langle \mathcal{F}\left(\sum_{i=1}^{K} c_i \cdot \phi_i(t)\right), \phi_j(t) \right\rangle = 0 \,, \qquad (j = 1, 2, \ldots K) \,. \tag{3}$$

2.2 Haar-wavelets basis

The Haar-wavelets are non-harmonious (rectangular) functions and therefore pulse signals are described effectively by these functions. A wavelet transform has advantages over traditional Fourier transform for accurately deconstructing and reconstructing finite or non-periodic signals [10]. The set of Haar-wavelets forms a complete orthonormal basis of the square-integrable functions on the unit interval $x \in [0, 1)$.

Denote by $\boldsymbol{L}^2(\boldsymbol{R})$ the space of finite energy signals, i.e.

$$\boldsymbol{L}^2(\boldsymbol{R}) = \left\{ f(t) : \int_{-\infty}^{\infty} |f(t)|^2 dt < \infty \right\} .$$

The orthonormal basis $\{\psi_{l,n}(x)\}$ of the wavelets is formed by dilations and translations of the mother wavelet $\psi(x) \in \boldsymbol{L}^2(\boldsymbol{R})$:

$$\psi_{l,n}(x) = 2^{l/2}\psi(2^l x - n), \quad l, n \in \boldsymbol{Z}; \quad x \in \boldsymbol{R} \,.$$

The orthonormal Haar-wavelet basis can be constructed using a scaling function $\varphi(x)$ which generates a multi-resolution scaling or multi-resolution analysis. The function $\psi(x)$ and scaling function $\varphi(x)$ satisfy the following relations:

$$\varphi(x) = \sum_{i \in \boldsymbol{Z}} \alpha_i \cdot \varphi(2x - i) \,,$$

$$\psi(x) = \sum_{i \in \boldsymbol{Z}} (-1)^i \cdot \alpha_{1-i} \cdot \varphi(2x - i) \,,$$

with coefficients $\alpha_i (i = 0, 1, \ldots K - 1)$ such that

$$\int_{-\infty}^{\infty} \varphi(x) \cdot \psi(x - n) dx = 0$$

for any integer n.

The scaling function for Haar-wavelets is described by

$$\varphi(x) = \begin{cases} 1, & 0 \leq x < 1 \\ 0, & otherwise \end{cases},$$

the mother wavelet $\psi(x)$ is defined by

$$\psi(x) = \begin{cases} 1, & 0 \leq x < 1/2 \\ -1, & 1/2 \leq x < 1 \\ 0 & otherwise \end{cases}.$$

Every function $\boldsymbol{f}(x) \in \boldsymbol{L}^2(\boldsymbol{R})$ may be expanded in the orthonormal wavelet basis as

$$\boldsymbol{f}(x) = \sum_{j,k=-\infty}^{\infty} c_{j,k} \cdot \psi_{j,k}(x) , \tag{4}$$

with wavelet coefficients

$$c_{j,k} := \langle \boldsymbol{f}(x), \psi_{j,k}(x) \rangle \quad \forall \quad j,k \in \boldsymbol{Z}.$$

The coefficients $c_{j,k}(\boldsymbol{f}(x))$ are called the wavelet spectrum of the signal $\boldsymbol{f}(x)$ with Kronecker delta property $\langle \psi_{j,k}, \psi_{l,m} \rangle = \delta_{j,l} \cdot \delta_{k,m} \quad \forall \quad j,k,l,m \in \boldsymbol{Z}$.

The Haar-wavelet transformation provides two-dimensional development of an one-dimensional signal. The set of wavelet coefficients $c_{j,k}(\boldsymbol{f}(x))$ captures both time and frequency behavior of the signal $\boldsymbol{f}(x)$.

The orthonormal Haar-wavelets basis can be applied to construct the space of trial functions of a coordinate system and the space of test functions of a projection system for the Bubnov-Galerkin approximation in (2) and (3). Both spaces are K-dimensional subspaces of a Hilbert space $\boldsymbol{X}$.

3 Meshfree Wavelet-Galerkin Method

The multidimensional NPSS response at all n outputs of the nonlinear dynamic circuits with m pulsed inputs is calculated according to the operator equation (1). The elements of the circuit are regrouped into linear and nonlinear subcircuits by loop analysis concepts. The linear part is transferred into the frequency domain by means of the Laplace transformation where the solution can by computed analytically. The inverse Laplace transform makes it possible to calculate the vector of Haar-wavelet coefficients for output currents of the linear two-terminals in the time-domain. For the nonlinear subcircuit, the Bubnov-Galerkin method is applied in the time-domain. The discretization of the steady-state response in the Haar-wavelet basis results in a nonlinear system of algebraic equations for the expansion coefficients. The system is solved by means of the Newton-Raphson method for which the linear part of the multidimensional Taylor series expansion with respect to the expansion coefficients serves as start value.

3.1 The network equations formulation

Usually, the electric circuit consists of linear elements (resistors, capacitors, inductors), voltage and current sources, as well as of nonlinear semiconductor devices, such as transistors and diodes (e.g. Fig. 1(a), Fig. 2(a)). The linear and nonlinear elements can be replaced by equivalent circuits that include models of linear and nonlinear elements and controlled sources (e.g. Fig. 1(b), Fig. 2(b)).

All independent voltage and current sources and all linear passive (R) and dynamic (L, C) circuit elements are converted in impedance form of the linear two-terminals $Z(\omega)$ in the frequency domain into the Thevenin equivalent circuits. For example, the impedance of the linear two-terminal $Z_2(\omega)$ (Fig. 1(b)) is defined by

$$Z_2(\omega) = R_e + R_4 + 1/\left(\frac{1}{R_5} + i \cdot \omega \cdot C_2\right), \quad i = \sqrt{-1}\,.$$

The nonlinear two-terminals are described by nonlinear functions. The current-voltage characteristics of resistive nonlinear two-terminals are described by $i_\gamma(u_\gamma(t)), \quad \gamma = 1, \ldots, \Gamma$. We denote by $i_\gamma : [0, T) \to \boldsymbol{R}$ the current function and by $u_\gamma(t) \in \boldsymbol{R}$ the voltage of γ-th resistive nonlinear two-terminal, Γ is the number of such two-terminals.

The nonlinear capacitors are described by electrical charges $q_\lambda(u_\lambda(t)), \quad \lambda = 1, \ldots, \Lambda$. We denote by $q_\lambda : [0, T) \to \boldsymbol{R}$ the charge function and by $u_\lambda(t) \in \boldsymbol{R}$ the voltage of λ-th nonlinear capacitor, Λ is the number of such two-terminals.

The inductive nonlinear two-terminals are described by flux-current characteristics $\phi_j(i_j(t)), \quad j = 1, \ldots, J$. We denote by $\phi_j : [0, T) \to \boldsymbol{R}$ the flux function and by $i_j(t) \in \boldsymbol{R}$ the current of j-th nonlinear inductance, J is the number of such two-terminals.

The Thevenin simplifications should be such that $n_1 \geq n_2$. By n_1 we denote the number of linear two-terminals, n_2 is the number of nonlinear two-terminals.

Hence, we have a nonlinear dynamic circuit with linear and nonlinear two-terminals, m input signal sources including pulsed inputs and other independent voltage sources, and n outputs.

Let us construct a network graph and the fundamental tree is chosen in such a manner that the linear two-terminals should be located in branches of the fundamental tree and all the nonlinear two-terminals should be located in the chords [5]. This network graph contains p nodes except ground (and hence p branches of the fundamental tree), b branches and l fundamental loops. Each fundamental loop contains one and only one chord with a nonlinear two-terminal. The fundamental loops can contain also some linear two-terminals, we denote by n_μ the number of such two-terminals.

We denote by subscript "L" the linear subnetwork and by subscript "N" the nonlinear subnetwork. Thus, the linear subnetwork contains $p = n_1 - n_\mu$

linear two-terminals, and the nonlinear subnetwork contains $l = n_2 + n_\mu$ two-terminals: n_2 is the number of nonlinear- and n_μ is the number of linear two-terminals.

For this network graph, the Kirchhoff's Voltage Law (KVL) can be written as

$$\boldsymbol{u}_N(t) = -\boldsymbol{B}_T \cdot \boldsymbol{v}_L(t) . \tag{5}$$

We denote by $\boldsymbol{u}_N(t) \in \boldsymbol{R}^l$ the vector of voltages at the elements of the nonlinear subnetwork, $\boldsymbol{v}_L(t) \in \boldsymbol{R}^p$ is the vector of voltages at the branches of fundamental tree in the linear subnetwork. By $\boldsymbol{B} \in \{1, -1, 0\}^{l \times b}$ we denote the fundamental branch-loop matrix. The matrix $\boldsymbol{B}$ can be regrouped and expressed in the form $\boldsymbol{B} = [\boldsymbol{B}_T \mid \mathbf{1}_l]$, where $\mathbf{1}_l$ is a unit matrix of dimension l and $\boldsymbol{B_T} \in \{1, -1, 0\}^{l \times p}$.

Each branch of fundamental tree is the Thevenin equivalent circuit with single voltage source $e^{(r)}(t), (r = 1, \ldots, p)$ and a single linear two-terminal. Therefore $\boldsymbol{v}_L(t) = \boldsymbol{e}(t) - \boldsymbol{u}_L(t)$, where $\boldsymbol{e}(t) \in \boldsymbol{R}^p$ with m non-zero components is the vector of input signal sources including pulsed inputs and other independent voltage sources. We denote by $\boldsymbol{u}_L(t) \in \boldsymbol{R}^p$ the vector of voltages at linear two-terminals in the branches of fundamental tree. Each component of the vector $\boldsymbol{v}_L(t)$ is the voltage between two nodes in the branch of fundamental tree.

The Kirchhoff's Current Law (KCL) is described by

$$\boldsymbol{i}_L(t) = \boldsymbol{B}_T^* \cdot \boldsymbol{i}_N(t) . \tag{6}$$

We denote by $\boldsymbol{i}_L(t) \in \boldsymbol{R}^p$ the vector of currents in the branches of fundamental tree (i.e. at the two-terminals in the linear subnetwork), $\boldsymbol{i}_N(t) \in \boldsymbol{R}^l$ is the vector of currents in the chords (i.e. at the two-terminals in the nonlinear subnetwork).

Using the orthonormal system of Haar-wavelet functions we represent the each element of the vector of input signals $\boldsymbol{e}(t)$ in the form

$$e^{(r)}(t) = \sum_{j,k \in \boldsymbol{Z}} E_{j,k}^{(r)} \cdot \psi_{j,k}(\theta) , \tag{7}$$

where $e^{(r)}(t)$ is the r-th component of the vector $\boldsymbol{e}(t) \in \boldsymbol{R}^p, (r = 1, \ldots, p)$ $\psi_{j,k}(\theta)$ are the Haar-wavelet functions of the variable $\theta \in [0, 1)$. The numerical coefficients $E_{j,k}^{(r)}$ are found from the system of equations

$$E_{j,k}^{(r)} = \int_0^1 e^{(r)}(t) \cdot \psi_{j,k}(\theta) d\theta, \quad t \in [0, T), \quad \theta = t/T .$$

By $E^{(r)} \in \boldsymbol{R}^{\boldsymbol{K}}$ we denote the vector of Haar-wavelet coefficients for the r-th component of the vector $\boldsymbol{E}$. The vector $\boldsymbol{E} \in \boldsymbol{R}^{p \times K}$ consists of the Haar-wavelet coefficients of the multidimensional input signal.

Each element of the vector of voltages at the linear two-terminals $\boldsymbol{u}_L(t)$ and each element of the vector of node voltages $\boldsymbol{v}_L(t)$ can be also represented in the Haar-wavelets orthonormal basis. Then

$$u_L^{(r)}(t) = \sum_{j,k\in \boldsymbol{Z}} \left(U_L^{(r)}\right)_{j,k} \cdot \psi_{j,k}(\theta) \,, \tag{8a}$$

$$v_L^{(r)}(t) = \sum_{j,k\in \boldsymbol{Z}} \left(V_L^{(r)}\right)_{j,k} \cdot \psi_{j,k}(\theta) = \sum_{j,k\in \boldsymbol{Z}} \left[E_{j,k}^{(r)} - \left(U_L^{(r)}\right)_{j,k}\right] \cdot \psi_{j,k}(\theta) \,, \tag{8b}$$

where $u_L^{(r)}(t) \in \boldsymbol{R}$ is the r-th component of the vector $\boldsymbol{u}_L(t)$, $v_L^{(r)}(t) \in \boldsymbol{R}$ is the r-th component of the vector $\boldsymbol{v}_L(t)$, $(r = 1, \ldots, p)$, and $E_{j,k}^{(r)}$ are the coefficients from (7). By $\boldsymbol{U}_L^{(r)} \in \boldsymbol{R^K}$ we denote the vector of Haar-wavelet coefficients for the r-th component of the vector $\boldsymbol{u}_L$. By $\boldsymbol{V}_L^{(r)} \in \boldsymbol{R^K}$ we denote the vector of Haar-wavelet coefficients for $v_L^{(r)}(t)$. We denote by $\boldsymbol{U}_L \in \boldsymbol{R}^{p\times K}$ the vector of unknown Haar-wavelet coefficients of the multidimensional NPSS response at all p linear two-terminals, including m inputs and n outputs, $\boldsymbol{V_L} \in \boldsymbol{R}^{p\times K}$ is the vector of the Haar-wavelet coefficients of the vector network node voltages.

We shall assume that, the circuit has a unique steady-state solution described by (8a). This solution for the vector of Haar-wavelet coefficients $\boldsymbol{U}_L$ can be found from the Galerkin orthogonality relation for the error (3) in the form:

$$\left\langle \boldsymbol{\mathcal{F}} \left(\sum_{j,k\in \boldsymbol{Z}} \boldsymbol{U}_L^{(j,k)}(t) \cdot \psi_{j,k}(t) \right), \psi_{l,m}(t) \right\rangle = 0 \,, \qquad \forall \qquad l,m \in Z \,. \tag{9}$$

3.2 Solution of the linear subnetwork

The vector $\boldsymbol{i}_L(t)$ of currents at the linear two-terminals in the linear subnetwork is described by the linear differential equations with constant coefficients in the operational form as follows:

$$\boldsymbol{i}_L(t) = \boldsymbol{M}(P) \cdot \boldsymbol{u}_L(t) \,. \tag{10}$$

$\boldsymbol{M}(P)$ is the diagonal matrix

$$\boldsymbol{M}(P) = diag[M_1(P), M_2(P), \ldots, M_p(P)] \,,$$

where $M_i(P)$ is a polinomial of $P = d/dt$ for the i-th two-terminal, $i = 1, \ldots, p$.

Hence, from (8a) and differential equation (10) we transfer the linear network into the frequency domain by means of the Laplace transformation

$$\boldsymbol{I}_L(s) = \boldsymbol{\mathcal{L}}\left\{\boldsymbol{i}_L(t)\right\} = \boldsymbol{M(s)} \sum_{j,k\in \boldsymbol{Z}} \boldsymbol{U}_L^{(j,k)} \cdot \Phi_{j,k}(s) \,, \tag{11}$$

where $\boldsymbol{I}_L(s)$ is the vector of Laplace transformation of the currents at the elements of the linear subnetwork; matrix $\boldsymbol{M}(\boldsymbol{s}) = \boldsymbol{\mathcal{L}}\{\boldsymbol{M}(\boldsymbol{t})\}$; the parameter s is in general complex, $s = \sigma + i\omega$; $\Phi_{j,k}(s)$ is the Laplace transformation of the Haar-wavelets basis functions.

Inverse Laplace transform of (11) makes it possible to find the vector of currents of the linear two-terminals

$$\boldsymbol{i}_L(t) = \boldsymbol{\mathcal{L}}^{-1}\{\boldsymbol{I}_L(s)\} = \sum_{j,k\in Z} \boldsymbol{U}_L^{(j,k)} \cdot \beta_{j,k}(t)\,. \tag{12a}$$

where

$$\beta_{j,k}(t) = \frac{1}{2\pi i}\int_{c-i\cdot\infty}^{c+i\cdot\infty} \boldsymbol{M}(s)\cdot\Phi_{j,k}(s)\cdot e^{st}ds \tag{12b}$$

is a diagonal matrix, $\beta_{j,k}(t) \in \boldsymbol{R}^{p\times K}$. In the integrand (12b) $\Phi_{j,k}(s)$ is the Laplace transform of the Haar-wavelet functions, each component of the matrix $\boldsymbol{M}(s)$ is the admittance of the linear two-terminals in the frequency domain, each component of the diagonal matrix $\boldsymbol{\beta}_{j,k}(t)$ is the current at this two-terminal by the voltage waveform of Haar-wavelet function $\psi_{j,k}(\theta)$. The functions $\Phi_{j,k}(s)$ and $\psi_{j,k}(\theta)$ can be defined as a sequence of the Heaviside step functions $H(t)$ with the corresponding coefficients. Hence, each component of the matrix $\beta_{j,k}(t)$ is the step response of the linear two-terminal to the Heaviside step function. The integral (12b) is determined by application of the residues theory.

3.3 Solution of the nonlinear subnetwork

For the nonlinear subnetwork we have from (8b) and (5) the vector of chord voltages as a Galerkin approximation in the Haar-wavelets basis in the following form:

$$\boldsymbol{u}_N(t) = -\sum_{j,k\in\boldsymbol{Z}} \boldsymbol{B}_T[\boldsymbol{E}_{j,k} - \boldsymbol{U}_L^{(j,k)}]\cdot\psi_{j,k}(\theta)\,. \tag{13}$$

Since the characteristics of nonlinear two-terminals are known, the vector of currents in the chords $\boldsymbol{i}_N$ is described as follows:

$$\boldsymbol{i}_N(t) = [\ldots i_\gamma(u_\gamma(t)),\ldots,q'_\lambda(u_\lambda(t)),\ldots,i_j(\phi_j(t)),\ldots,i_\mu(t),\ldots]^*\,, \tag{14}$$

where $i_\mu(t)$ are currents of the linear two-terminals found in the fundamental loops, $\mu = 1,\ldots,n_\mu$, n_μ is the number of such two-terminals.

The vector of currents of the linear subnetwork is described from (13), (14) and (6) as

$$\boldsymbol{i}_L(t) = \boldsymbol{B}_T^*\boldsymbol{i}_N\left[-\sum_{j,k\in\boldsymbol{Z}} \boldsymbol{B}_T\cdot\left(\boldsymbol{E}_{j,k} - \boldsymbol{U}_L^{(j,k)}\right)\cdot\psi_{j,k}(\theta)\right]\,. \tag{15}$$

Thus, we have from (15) and (12a) the following system of equations for the error in Galerkin approach:

$$\sum_{j,k\in \boldsymbol{Z}} \boldsymbol{U}_L^{(j,k)} \cdot \boldsymbol{\beta}_{j,k}(t) = \boldsymbol{B}_T^* \boldsymbol{i}_N \left[- \sum_{j,k\in \boldsymbol{Z}} \boldsymbol{B}_T \cdot \left(\boldsymbol{E}_{j,k} - \boldsymbol{U}_L^{(j,k)} \right) \cdot \psi_{j,k}(\theta) \right] . \quad (16)$$

The Haar-wavelet coefficients $\boldsymbol{U}_L^{(j,k)}$ of the NPSS response in (16) are found from the Galerkin orthogonality relation for the error (9). Equating the coefficients of identical Haar-wavelet functions, we obtain the Harmonic Balance equations for non-trigonometric basis functions:

$$\sum_{j,k\in \boldsymbol{Z}} \boldsymbol{U}_L^{(j,k)} \int_0^1 \boldsymbol{\beta}_{j,k}(t) \cdot \psi_{l,m}(\theta) d\theta =$$

$$\boldsymbol{B}_T^* \int_0^1 \boldsymbol{i}_N \left[- \sum_{j,k\in \boldsymbol{Z}} \boldsymbol{B}_T \cdot \left(\boldsymbol{E}_{j,k} - \boldsymbol{U}_L^{(j,k)} \right) \cdot \psi_{j,k}(\theta) \right] \cdot \psi_{l,m}(\theta) d\theta, (l,m) \in \boldsymbol{Z} ,$$

or in the matrix form:

$$\boldsymbol{Y} \cdot \boldsymbol{U}_L = \boldsymbol{B}_T^* \cdot \boldsymbol{I}_N(\boldsymbol{U}_L) . \quad (17)$$

This is a system of equations for the nonlinear dynamic circuit to compute the vector of the Haar-wavelet coefficients $\boldsymbol{U}_L$ of the NPSS response. By $\boldsymbol{Y}$ we denote the block-diagonal Nodal Admittance matrix, $\boldsymbol{Y} = diag[\boldsymbol{Y}^{(1)}, \dots \boldsymbol{Y}^{(p)}]$ obtained from the linear subcircuit. Each component of this matrix has the form:

$$\boldsymbol{Y}_{j,k,l,m}^{(\alpha)} = \int_0^1 \boldsymbol{\beta}_{j,k}^{(\alpha)}(t) \cdot \psi_{l,m}(\theta) d\theta, \quad \alpha = (1, 2, \dots p) .$$

$\boldsymbol{I}_N(\boldsymbol{U}_L)$ is the vector of coefficients of Galerkin approximation for the nonlinear currents $\boldsymbol{i}_N$ (14) in the Haar-wavelets basis:

$$\boldsymbol{I}_N(\boldsymbol{U}_L) = \int_0^1 \boldsymbol{i}_N \left[- \sum_{j,k\in \boldsymbol{Z}} \boldsymbol{B}_T \cdot \left(\boldsymbol{E}_{j,k} - \boldsymbol{U}_L^{(j,k)} \right) \cdot \psi_{j,k}(\theta) \right] \cdot \psi_{l,m}(\theta) d\theta . \quad (18)$$

The system of equations (17) is a set of nonlinear algebraic equations in the time-domain which contains $p \cdot K$ equations, where K is number of Haar-wavelet functions in the basis.

4 Solution of the network equations

The solution of system (17) to compute the vector of Haar-wavelet coefficients $\boldsymbol{U}_L$ of the multidimensional NPSS response can be found by the Newton-Raphson method and an initial approximation to the solution, or in the form of a multidimensional Taylor series expansion with respect to the Haar-wavelet coefficients in the input signal approximation:

$$\boldsymbol{U}_L = \sum_{k=0}^{\infty} \frac{1}{k_1! \cdots k_d!} \left(\frac{\partial^{k_1+\cdots+k_d} U_L}{\partial E_1^{k_1} \cdots \partial E_d^{k_d}} \right)_0 E_1^{k_1} \cdots E_d^{k_d} , \tag{19}$$

where $k = k_1 + \cdots + k_d$; E_j are non-zero Haar-wavelets coefficients in the input signal approximation (7), $(j = 1 \cdots d)$.

The linear part of the multidimensional Taylor expansion around fixed point $\boldsymbol{u_L(0)} = \boldsymbol{u_0}$ in the form of Taylor series (19) truncated after the first-order terms can be used as start value in the Newton-Raphson method:

$$\boldsymbol{U}_L^0 = \sum_{j=0}^{d} \left(\frac{\partial U_L}{\partial E_j} \right)_0 \cdot E_j . \tag{20}$$

The coefficients of this expansion are determined from system (17) as follows:

$$\boldsymbol{Y} \frac{\partial \boldsymbol{U}_L}{\partial E_j} = \boldsymbol{B}_T^* \frac{\partial \boldsymbol{I}_N}{\partial \boldsymbol{U}_N} \frac{\partial \boldsymbol{U}_N}{\partial \boldsymbol{E}_j} . \tag{21}$$

The Jacobian matrix $\frac{\partial \boldsymbol{I}_N}{\partial \boldsymbol{U}_N}$ is described by

$$\left(\frac{\partial I_N^{(k)}}{\partial U_N^{(j)}} \right)_0 = \begin{cases} i'_N(0) & for \quad k+j \equiv 0 \pmod{2} \\ 0 & for \quad k+j \not\equiv 0 \pmod{2} \end{cases}$$

Jacobian is a block-diagonal $K \cdot l \times K \cdot l$ matrix, l is the number of two-terminals in the chords of the network graph. Thus, we have from (18) and (21) the following equation:

$$\boldsymbol{Y} \frac{\partial \boldsymbol{U}_L^{(j)}}{\partial E_j} = -\boldsymbol{B}_T^* \frac{\partial \boldsymbol{I}_N^{(j)}}{\partial \boldsymbol{U}_N^{(j)}} \cdot \boldsymbol{B}_T \cdot \sigma_j + \boldsymbol{B}_T^* \cdot \frac{\partial \boldsymbol{I}_N^{(j)}}{\partial \boldsymbol{U}_N^{(j)}} \cdot \boldsymbol{B}_T \cdot \frac{\partial \boldsymbol{U}_L^{(j)}}{\partial \boldsymbol{E}_j} , \tag{22}$$

where σ_j is the column-matrix which j–th non-zero element equal to 1 according to the j–th component in the Haar-wavelets approximation of the input signals.

The decision of system (22) yields the set of equations for the initial approximation of the Taylor series expansion (20) at the fixed point:

$$\left(\frac{\partial \boldsymbol{U}_L^{(j)}}{\partial E_j} \right)_0 = \left[\boldsymbol{B}_T^* \frac{\partial \boldsymbol{I}_N^{(j)}}{\partial \boldsymbol{U}_N^{(j)}} \cdot \boldsymbol{B}_T - \boldsymbol{Y} \right]^{-1} \cdot \boldsymbol{B}_T^* \cdot \frac{\partial \boldsymbol{I}_N^{(j)}}{\partial \boldsymbol{U}_N^{(j)}} \cdot \boldsymbol{B}_T \cdot \sigma_j .$$

Thus, we can calculate the multidimensional NPSS response by a time-domain approach by means of the Newton-Raphson method in the following form:

$$\boldsymbol{U}_L^{(k+1)} = \boldsymbol{U}_L^{(k)} - \left[\boldsymbol{B}_T^* \frac{\partial \boldsymbol{I}_N}{\partial \boldsymbol{U}_N} \boldsymbol{B}_T + \boldsymbol{Y}\right]_{(k)}^{-1} \cdot [\boldsymbol{Y}\boldsymbol{U}_L - \boldsymbol{B}_T^* \boldsymbol{I}_N]_{(k)} \ .$$

This circuit solution is used as the next "guess" and the process is repeated according to the given tolerance.

The time-domain voltage waveforms of the multidimensional NPSS response at all p linear two-terminals of the network are calculated from the vector of Haar-wavelets coefficients $\boldsymbol{U}_L$ according to (8a).

5 Illustrative examples

The presented method was applied to the transistor broadband amplifier and Schmitt trigger which have found many applications in numerous circuits, both analog and digital.

5.1 Simulation results of the Broadband amplifier

The analog broadband pulse amplifier shown in Fig. 1(a) is used in the digital electronics as an inverter. Fig. 1(b) shows an equivalent circuit of the amplifier. The bipolar transistor is modeled by the Ebers-Moll model.

The parameters associated with this model are described from the input and output characteristics of the transistor as follows: emitter junction current $I_e = I_{0e}\left(\mathrm{e}^{b_0 \cdot u_{be}} - 1\right)$; current of collector current-controlled source $I_c = \alpha \cdot I_e$; diffusion capacitance of emitter-base junction $C_{de} = C_{0e}\left(\mathrm{e}^{b_0 \cdot u_{be}} - 1\right)$; barrier capacitance of collector-base junction $C_{tc} = \quad C_{0c}\left(1 - u_{bc}/\varphi_T\right)^{-n}$.

For the model of the amplifier with 3 node and 4 branches, the branch-loop matrix has the following form:

$$B_T = \begin{bmatrix} 1 & 1 & 0 \\ 1 & 1 & 0 \\ 1 & 0 & -1 \\ -1 & 0 & 1 \end{bmatrix} .$$

The relation (6) of currents in the branches $\boldsymbol{i}_L(t)$ with the currents in the chords $\boldsymbol{i}_N(t) = [i_4(t), i_5(t), i_6(t), i_7(t)]^*$ is described for this equivalent circuit of the amplifier by

$$\boldsymbol{i}_L(t) = \begin{bmatrix} i_1 \\ i_2 \\ i_3 \end{bmatrix} = \boldsymbol{B}_T^* \cdot \boldsymbol{i}_N(t) = \begin{bmatrix} i_4 + i_5 + i_6 - i_7 \\ i_4 + i_5 \\ -i_6 + i_7 \end{bmatrix} .$$

The Fig. 1(c) and Fig. 1(d) show the steady-state wave forms of $V_1(t)$, $V_2(t)$, $V_3(t)$ at three nodes of amplifier for a two pulse input signals $e(t) = e_1(t)$ and $e(t) = e_2(t)$ respectively with amplitudes $\pm E_0$. The vector of

Haar-wavelet coefficients for the input signal Fig. 1(c) according to (7) is $(0,0,0,0,0,\frac{1}{2}E_0,0,0)$ in basis with $K = 8$ functions. For the input signal from Fig. 1(d) these coefficients are $(0,E_0,0,0,0,0,0,0)$, and for the collector voltage source E_c, the Haar-wavelets spectrum is $(E_c,0,0,0,0,0,0,0)$.

5.2 Simulation results of the Schmitt-Trigger circuit

The Schmitt-Trigger circuit with two n-p-n transistors is shown in Fig. 2(a). The model is shown in a Fig. 2(b). This is a comparator circuit that incorporates feedback. The ideal feedback model shown in Fig. 2(c) and includes an amplifier (active element) and feedback loop (passive two-terminal). We will apply the following algorithm: we represent this circuit as is shown in Fig. 2(d). We break off a feedback, the NPSS response of this new circuit can be calculated, as shown in the first example. Thus it is necessary to use a voltage controlled voltage source $E_r(t)$ in emitter of the transistor T_1 as shown in the trigger model Fig. 2(b). The voltage of this source should be equal to the voltage $U_{R_3}(t)$ in emitter of transistor T_2 on each step of the iteration process.

The $n-p-n$ bipolar transistors T1 and T2 are modeled by an Ebers-Moll model. The parameters associated with this model are calculated, as shown in the first example. A triangle pulse input signal of the trigger is shown in Fig. 2(e). The vector of Haar-wavelet coefficients is described for this signal in the basis with $K = 8$ Haar-wavelet functions according to (7) by

$$(\frac{15}{32}E_0, 0, -\frac{11\sqrt{2}}{64}E_0, \frac{9\sqrt{2}}{64}E_0, -\frac{1}{16}E_0, -\frac{3}{32}E_0, \frac{1}{8}E_0, \frac{1}{32}E_0)\ .$$

For the model of the Schmitt-Trigger which contains 6 nodes and 5 branches the branch-loop matrix has the following form:

$$B_T = \begin{bmatrix} -1 & 0 & 1 & 0 & 0 & 0 \\ 1 & -1 & 0 & 0 & 0 & 0 \\ 0 & -1 & 0 & 1 & 0 & 0 \\ 0 & 0 & 0 & -1 & 0 & 1 \\ 0 & 0 & 0 & 1 & -1 & 0 \end{bmatrix}\ .$$

The relation (6) of currents in the branches $\boldsymbol{i}_L(t)$ with the currents in the chords according (14) $\boldsymbol{i}_N(t) = [i_7(t), i_8(t), i_9(t), i_{10}(t), i_{11}(t)]^*$ is described for this equivalent circuit of the trigger by

$$\boldsymbol{i}_L(t) = \begin{bmatrix} i_1 \\ i_2 \\ i_3 \\ i_4 \\ i_5 \\ i_6 \end{bmatrix} = \boldsymbol{B}_T^* \cdot \boldsymbol{i}_N(t) = \begin{bmatrix} -i_7 + i_8 \\ -i_8 - i_9 \\ i_7 \\ i_9 - i_{10} + i_{11} \\ -i_{11} \\ i_{10} \end{bmatrix}\ .$$

where $i_9(t)$ is the current of the linear two-terminal Z_7 in the chord. Fig. 2(f) shows the steady-state waveform $V_{out}(t)$ in the output of the Schmitt-Trigger.

6 Conclusions

The paper presents a meshfree numeric-analytical Wavelet-Galerkin method to compute the non-periodic steady-state response of a nonlinear dynamic microwave circuit with m pulsed input signals and n outputs in the time-domain. The method is free from time-consuming step-by-step algorithms, which are used for handling the transient evolution and the numerical integration.

In this paper, the NPSS response of nonlinear dynamic circuit has been calculated by Bubnov-Galerkin method with Haar-wavelets as basis functions in the time-domain without the transient analysis and without numerical integration, which are based on multi-step schemes. The presented method uses the Laplace transformation for analytical solution of the linear subnetwork in the frequency domain. For the nonlinear subcircuit, the Bubnov-Galerkin method is applied in the time-domain. The discretization of the steady-state response in the Haar-wavelet basis results in a nonlinear system of algebraic equations for the expansion coefficients. The system is solved for the whole problem domain by means of the Newton-Raphson iteration method.

The presented meshfree method can be applied efficiently for calculating the steady-state response of nonlinear RF and mixed-signal circuits with very different time constants and particularly for circuits with pulsed inputs.

References

1. T. Belytschko, Y. Y. Lu, and L. Gu, *Element-free Galerkin methods*, Int. J. Numer. Meth. Engrg. **37** (1994), 229–256.
2. M. Liu, C. K. Tse, and J. Wu, *A wavelet approach to fast approximation of steady-state waveforms of power electronics circuits*, Int. J. Circuit Theory Appl. **37** (2003), 591–610.
3. A. Ushida, T. Adachi, and L. O. Chua, *Steady-state analysis of nonlinear circuits based on hybrid methods*, IEEE Trans. Circuits Syst. I. **39** (1992), 649–661.
4. E. A. Volkov and A. N. Kapitanova, *Analysis of nonlinear lag circuits under pulse actions in the Walsh functions basis*, Engineering Simulation. **11** (1993), 74–85.
5. L. O. Chua and P. M. Lin, *Computer-Aided Analysis of Electronic Circuits*, Prentice-Hall, 1975.
6. S. Li and W. K. Liu, *Meshfree Particle Methods*, Springer, 2004.
7. S. A. Maas, *Nonlinear microwave circuits*, IEEE, Inc., 1997.
8. M. Griebel and M. A. Schweitzer (eds.), *Meshfree Methods for Partial Differential Equations*, Lecture Notes in Computational Science and Engineering, vol. 26, Springer, 2002.
9. M. Griebel and M. A. Schweitzer (eds.), *Meshfree Methods for Partial Differential Equations II*, Lecture Notes in Computational Science and Engineering, vol. 43, Springer, 2005.
10. K. Urban, *Wavelets in Numerical Simulation: Problem Adapted Construction and Applications*, Lecture Notes in Computational Science and Engineering, vol. 22, Springer, 2002.

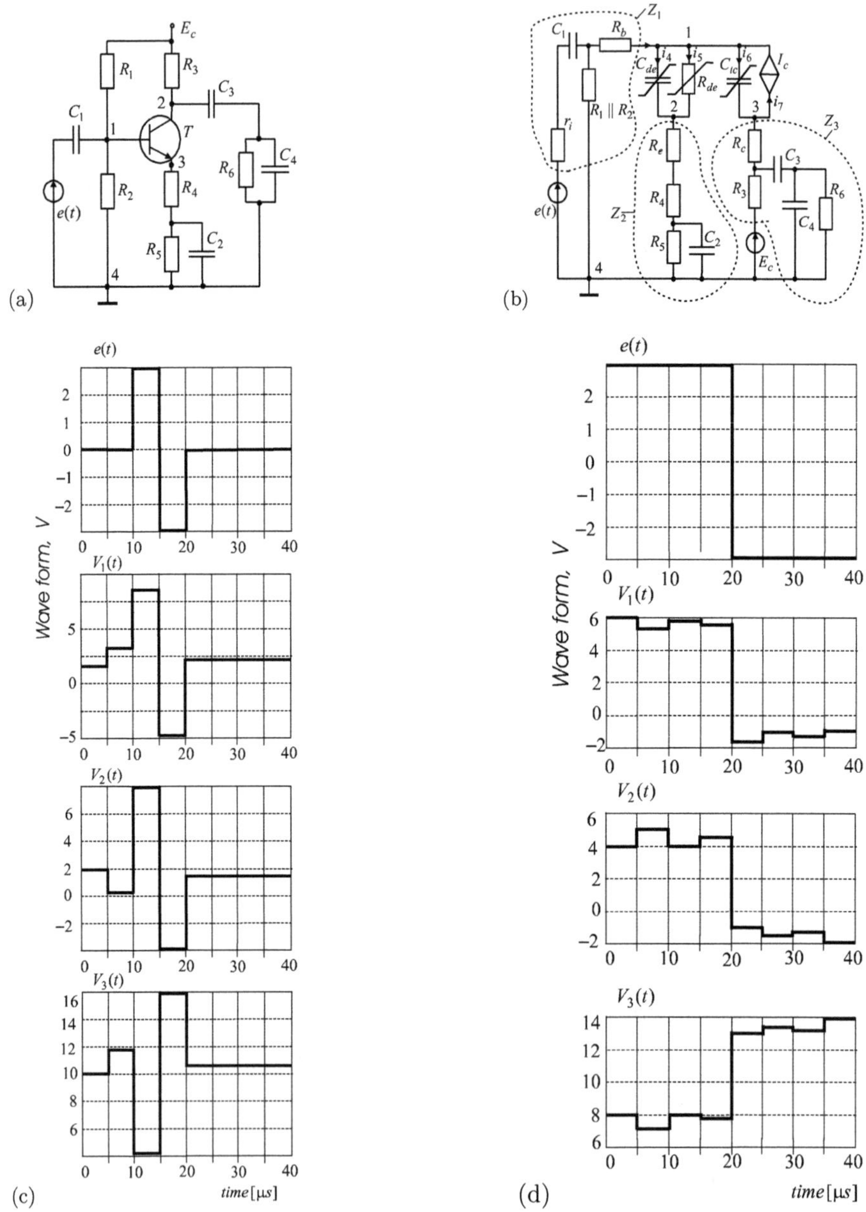

Fig. 1. A pulse amplifier: (a) the electrical circuit with $R_1 = 62[k\Omega]$, $R_2 = 7,5[k\Omega]$, $R_3 = 910[\Omega]$, $R_4 = 120[\Omega]$, $R_5 = 680[\Omega]$, $R_6 = 10[k\Omega]$, $C_1 = 68[\mu F]$, $C_2 = 68[\mu F]$, $C_3 = 62[\mu F]$, $C_4 = 100[pF]$, $E_c = 27[V]$, $E_0 = 3[V]$; (b) the equivalent circuit, where resistances of the base, emitter and collector are $R_b = R_e = R_c = 10[\Omega]$, $r_i = 10[\Omega]$, $I_{0e} = 8.4 \cdot 10^{-5}[A]$, $b_0 = 28.6$, $\alpha = 0.98$, $C_{0e} = 2.1 \cdot 10^{-11}[F]$, $C_{0c} = 7.5 \cdot 10^{-12}[F]$, $n = 0.362$, $\varphi_T = 0.65$; (c) the steady-state wave form of $V_1(t), V_2(t), V_3(t)$ in three nodes of amplifier for a input signal $e(t) = e_1(t)$, $T = 40[\mu s]$; (d) the steady-state wave form of $V_1(t), V_2(t), V_3(t)$ for a input signal $e(t) = e_2(t)$.

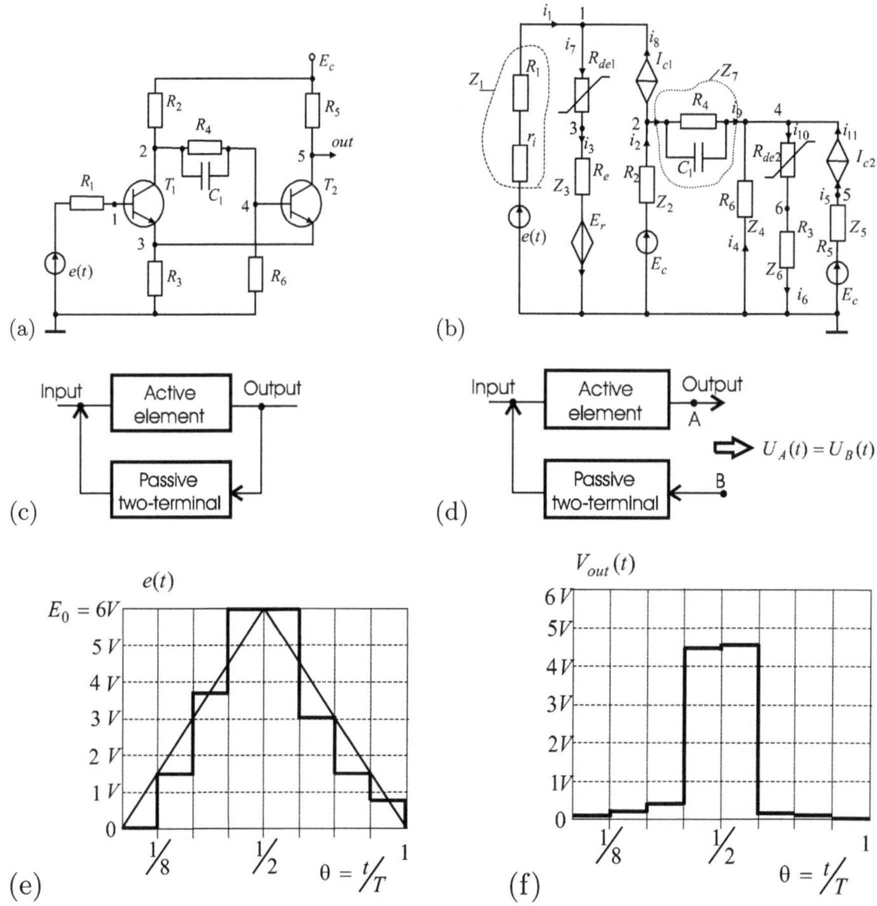

Fig. 2. A Schmitt-Trigger circuit: (a) the electrical trigger circuit with $R_1 = 10[k\Omega]$, $R_2 = 1[k\Omega]$, $R_3 = 100[\Omega]$, $R_4 = 4.7[k\Omega]$, $R_5 = 220[\Omega]$, $R_6 = 3.3[k\Omega]$, $C_1 = 470[pF]$, $E_c = 12[V]$, $E_0 = 6[V]$; (b) the equivalent circuit of the Schmitt-Trigger with resistance of the emitter $R_e = 1[\Omega]$, output impedance of the input voltage source $r_i = 10[\Omega]$, the voltage controlled voltage source E_r; (c) the ideal feedback model; (d) the feedback model with a voltage controlled voltage source; (e) triangle pulse input signal and its approximation in the basis with $K = 8$ Haar-wavelet functions, $T = 2[ms]$; (f) the steady-state output waveform of $V_{out}(t)$.

Editorial Policy

1. Volumes in the following three categories will be published in LNCSE:

i) Research monographs
ii) Tutorials
iii) Conference proceedings

Those considering a book which might be suitable for the series are strongly advised to contact the publisher or the series editors at an early stage.

2. Categories i) and ii). Tutorials are lecture notes typically arising via summer schools or similar events, which are used to teach graduate students. These categories will be emphasized by Lecture Notes in Computational Science and Engineering. **Submissions by interdisciplinary teams of authors are encouraged.** The goal is to report new developments – quickly, informally, and in a way that will make them accessible to non-specialists. In the evaluation of submissions timeliness of the work is an important criterion. Texts should be well-rounded, well-written and reasonably self-contained. In most cases the work will contain results of others as well as those of the author(s). In each case the author(s) should provide sufficient motivation, examples, and applications. In this respect, Ph.D. theses will usually be deemed unsuitable for the Lecture Notes series. Proposals for volumes in these categories should be submitted either to one of the series editors or to Springer-Verlag, Heidelberg, and will be refereed. A provisional judgement on the acceptability of a project can be based on partial information about the work: a detailed outline describing the contents of each chapter, the estimated length, a bibliography, and one or two sample chapters – or a first draft. A final decision whether to accept will rest on an evaluation of the completed work which should include

- at least 100 pages of text;
- a table of contents;
- an informative introduction perhaps with some historical remarks which should be accessible to readers unfamiliar with the topic treated;
- a subject index.

3. Category iii). Conference proceedings will be considered for publication provided that they are both of exceptional interest and devoted to a single topic. One (or more) expert participants will act as the scientific editor(s) of the volume. They select the papers which are suitable for inclusion and have them individually refereed as for a journal. Papers not closely related to the central topic are to be excluded. Organizers should contact the Editor for CSE at Springer at the planning stage, see *Addresses* below.

In exceptional cases some other multi-author-volumes may be considered in this category.

4. Only works in English will be considered. For evaluation purposes, manuscripts may be submitted in print or electronic form, in the latter case, preferably as pdf- or zipped ps-files. Authors are requested to use the LaTeX style files available from Springer at http:// www. springer.com/authors/book+authors?SGWID=0-154102-12-417900-0.

For categories ii) and iii) we strongly recommend that all contributions in a volume be written in the same LaTeX version, preferably LaTeX2e. Electronic material can be included if appropriate. Please contact the publisher.

Careful preparation of the manuscripts will help keep production time short besides ensuring satisfactory appearance of the finished book in print and online.

5. The following terms and conditions hold. Categories i), ii) and iii):

Authors receive 50 free copies of their book. No royalty is paid.
Volume editors receive a total of 50 free copies of their volume to be shared with authors, but no royalties.

Authors and volume editors are entitled to a discount of 33.3 % on the price of Springer books purchased for their personal use, if ordering directly from Springer.

6. Commitment to publish is made by letter of intent rather than by signing a formal contract. Springer-Verlag secures the copyright for each volume.

Addresses:

Timothy J. Barth
NASA Ames Research Center
NAS Division
Moffett Field, CA 94035, USA
barth@nas.nasa.gov

Michael Griebel
Institut für Numerische Simulation
der Universität Bonn
Wegelerstr. 6
53115 Bonn, Germany
griebel@ins.uni-bonn.de

David E. Keyes
Mathematical and Computer Sciences
and Engineering
King Abdullah University of Science
and Technology
P.O. Box 55455
Jeddah 21534, Saudi Arabia
david.keyes@kaust.edu.sa

and

Department of Applied Physics
and Applied Mathematics
Columbia University
500 W. 120 th Street
New York, NY 10027, USA
kd2112@columbia.edu

Risto M. Nieminen
Department of Applied Physics
Aalto University School of Science
and Technology
00076 Aalto, Finland
risto.nieminen@tkk.fi

Dirk Roose
Department of Computer Science
Katholieke Universiteit Leuven
Celestijnenlaan 200A
3001 Leuven-Heverlee, Belgium
dirk.roose@cs.kuleuven.be

Tamar Schlick
Department of Chemistry
and Courant Institute
of Mathematical Sciences
New York University
251 Mercer Street
New York, NY 10012, USA
schlick@nyu.edu

Editor for Computational Science
and Engineering at Springer:
Martin Peters
Springer-Verlag
Mathematics Editorial IV
Tiergartenstrasse 17
69121 Heidelberg, Germany
martin.peters@springer.com

Lecture Notes in Computational Science and Engineering

1. D. Funaro, *Spectral Elements for Transport-Dominated Equations.*
2. H.P. Langtangen, *Computational Partial Differential Equations.* Numerical Methods and Diffpack Programming.
3. W. Hackbusch, G. Wittum (eds.), *Multigrid Methods V.*
4. P. Deuflhard, J. Hermans, B. Leimkuhler, A.E. Mark, S. Reich, R.D. Skeel (eds.), *Computational Molecular Dynamics: Challenges, Methods, Ideas.*
5. D. Kröner, M. Ohlberger, C. Rohde (eds.), *An Introduction to Recent Developments in Theory and Numerics for Conservation Laws.*
6. S. Turek, *Efficient Solvers for Incompressible Flow Problems.* An Algorithmic and Computational Approach.
7. R. von Schwerin, *Multi Body System SIMulation.* Numerical Methods, Algorithms, and Software.
8. H.-J. Bungartz, F. Durst, C. Zenger (eds.), *High Performance Scientific and Engineering Computing.*
9. T.J. Barth, H. Deconinck (eds.), *High-Order Methods for Computational Physics.*
10. H.P. Langtangen, A.M. Bruaset, E. Quak (eds.), *Advances in Software Tools for Scientific Computing.*
11. B. Cockburn, G.E. Karniadakis, C.-W. Shu (eds.), *Discontinuous Galerkin Methods.* Theory, Computation and Applications.
12. U. van Rienen, *Numerical Methods in Computational Electrodynamics.* Linear Systems in Practical Applications.
13. B. Engquist, L. Johnsson, M. Hammill, F. Short (eds.), *Simulation and Visualization on the Grid.*
14. E. Dick, K. Riemslagh, J. Vierendeels (eds.), *Multigrid Methods VI.*
15. A. Frommer, T. Lippert, B. Medeke, K. Schilling (eds.), *Numerical Challenges in Lattice Quantum Chromodynamics.*
16. J. Lang, *Adaptive Multilevel Solution of Nonlinear Parabolic PDE Systems.* Theory, Algorithm, and Applications.
17. B.I. Wohlmuth, *Discretization Methods and Iterative Solvers Based on Domain Decomposition.*
18. U. van Rienen, M. Günther, D. Hecht (eds.), *Scientific Computing in Electrical Engineering.*
19. I. Babuška, P.G. Ciarlet, T. Miyoshi (eds.), *Mathematical Modeling and Numerical Simulation in Continuum Mechanics.*
20. T.J. Barth, T. Chan, R. Haimes (eds.), *Multiscale and Multiresolution Methods.* Theory and Applications.
21. M. Breuer, F. Durst, C. Zenger (eds.), *High Performance Scientific and Engineering Computing.*
22. K. Urban, *Wavelets in Numerical Simulation.* Problem Adapted Construction and Applications.

23. L.F. Pavarino, A. Toselli (eds.), *Recent Developments in Domain Decomposition Methods.*

24. T. Schlick, H.H. Gan (eds.), *Computational Methods for Macromolecules: Challenges and Applications.*

25. T.J. Barth, H. Deconinck (eds.), *Error Estimation and Adaptive Discretization Methods in Computational Fluid Dynamics.*

26. M. Griebel, M.A. Schweitzer (eds.), *Meshfree Methods for Partial Differential Equations.*

27. S. Müller, *Adaptive Multiscale Schemes for Conservation Laws.*

28. C. Carstensen, S. Funken, W. Hackbusch, R.H.W. Hoppe, P. Monk (eds.), *Computational Electromagnetics.*

29. M.A. Schweitzer, *A Parallel Multilevel Partition of Unity Method for Elliptic Partial Differential Equations.*

30. T. Biegler, O. Ghattas, M. Heinkenschloss, B. van Bloemen Waanders (eds.), *Large-Scale PDE-Constrained Optimization.*

31. M. Ainsworth, P. Davies, D. Duncan, P. Martin, B. Rynne (eds.), *Topics in Computational Wave Propagation.* Direct and Inverse Problems.

32. H. Emmerich, B. Nestler, M. Schreckenberg (eds.), *Interface and Transport Dynamics.* Computational Modelling.

33. H.P. Langtangen, A. Tveito (eds.), *Advanced Topics in Computational Partial Differential Equations.* Numerical Methods and Diffpack Programming.

34. V. John, *Large Eddy Simulation of Turbulent Incompressible Flows.* Analytical and Numerical Results for a Class of LES Models.

35. E. Bänsch (ed.), *Challenges in Scientific Computing - CISC 2002.*

36. B.N. Khoromskij, G. Wittum, *Numerical Solution of Elliptic Differential Equations by Reduction to the Interface.*

37. A. Iske, *Multiresolution Methods in Scattered Data Modelling.*

38. S.-I. Niculescu, K. Gu (eds.), *Advances in Time-Delay Systems.*

39. S. Attinger, P. Koumoutsakos (eds.), *Multiscale Modelling and Simulation.*

40. R. Kornhuber, R. Hoppe, J. Périaux, O. Pironneau, O. Wildlund, J. Xu (eds.), *Domain Decomposition Methods in Science and Engineering.*

41. T. Plewa, T. Linde, V.G. Weirs (eds.), *Adaptive Mesh Refinement – Theory and Applications.*

42. A. Schmidt, K.G. Siebert, *Design of Adaptive Finite Element Software.* The Finite Element Toolbox ALBERTA.

43. M. Griebel, M.A. Schweitzer (eds.), *Meshfree Methods for Partial Differential Equations II.*

44. B. Engquist, P. Lötstedt, O. Runborg (eds.), *Multiscale Methods in Science and Engineering.*

45. P. Benner, V. Mehrmann, D.C. Sorensen (eds.), *Dimension Reduction of Large-Scale Systems.*

46. D. Kressner, *Numerical Methods for General and Structured Eigenvalue Problems.*

47. A. Boriçi, A. Frommer, B. Joó, A. Kennedy, B. Pendleton (eds.), *QCD and Numerical Analysis III.*

48. F. Graziani (ed.), *Computational Methods in Transport.*

49. B. Leimkuhler, C. Chipot, R. Elber, A. Laaksonen, A. Mark, T. Schlick, C. Schütte, R. Skeel (eds.), *New Algorithms for Macromolecular Simulation.*

50. M. Bücker, G. Corliss, P. Hovland, U. Naumann, B. Norris (eds.), *Automatic Differentiation: Applications, Theory, and Implementations.*

51. A.M. Bruaset, A. Tveito (eds.), *Numerical Solution of Partial Differential Equations on Parallel Computers.*

52. K.H. Hoffmann, A. Meyer (eds.), *Parallel Algorithms and Cluster Computing.*

53. H.-J. Bungartz, M. Schäfer (eds.), *Fluid-Structure Interaction.*

54. J. Behrens, *Adaptive Atmospheric Modeling.*

55. O. Widlund, D. Keyes (eds.), *Domain Decomposition Methods in Science and Engineering XVI.*

56. S. Kassinos, C. Langer, G. Iaccarino, P. Moin (eds.), *Complex Effects in Large Eddy Simulations.*

57. M. Griebel, M.A Schweitzer (eds.), *Meshfree Methods for Partial Differential Equations III.*

58. A.N. Gorban, B. Kégl, D.C. Wunsch, A. Zinovyev (eds.), *Principal Manifolds for Data Visualization and Dimension Reduction.*

59. H. Ammari (ed.), *Modeling and Computations in Electromagnetics: A Volume Dedicated to Jean-Claude Nédélec.*

60. U. Langer, M. Discacciati, D. Keyes, O. Widlund, W. Zulehner (eds.), *Domain Decomposition Methods in Science and Engineering XVII.*

61. T. Mathew, *Domain Decomposition Methods for the Numerical Solution of Partial Differential Equations.*

62. F. Graziani (ed.), *Computational Methods in Transport: Verification and Validation.*

63. M. Bebendorf, *Hierarchical Matrices.* A Means to Efficiently Solve Elliptic Boundary Value Problems.

64. C.H. Bischof, H.M. Bücker, P. Hovland, U. Naumann, J. Utke (eds.), *Advances in Automatic Differentiation.*

65. M. Griebel, M.A. Schweitzer (eds.), *Meshfree Methods for Partial Differential Equations IV.*

66. B. Engquist, P. Lötstedt, O. Runborg (eds.), *Multiscale Modeling and Simulation in Science.*

67. I.H. Tuncer, Ü. Gülcat, D.R. Emerson, K. Matsuno (eds.), *Parallel Computational Fluid Dynamics 2007.*

68. S. Yip, T. Diaz de la Rubia (eds.), *Scientific Modeling and Simulations.*

69. A. Hegarty, N. Kopteva, E. O'Riordan, M. Stynes (eds.), *BAIL* 2008 – *Boundary and Interior Layers.*

70. M. Bercovier, M.J. Gander, R. Kornhuber, O. Widlund (eds.), *Domain Decomposition Methods in Science and Engineering XVIII.*

71. B. Koren, C. Vuik (eds.), *Advanced Computational Methods in Science and Engineering.*

72. M. Peters (ed.), *Computational Fluid Dynamics for Sport Simulation.*

73. H.-J. Bungartz, M. Mehl, M. Schäfer (eds.), *Fluid Structure Interaction II - Modelling, Simulation, Optimization.*

74. D. Tromeur-Dervout, G. Brenner, D.R. Emerson, J. Erhel (eds.), *Parallel Computational Fluid Dynamics 2008.*

75. A.N. Gorban, D. Roose (eds.), *Coping with Complexity: Model Reduction and Data Analysis.*

76. J.S. Hesthaven, E.M. Rønquist (eds.), *Spectral and High Order Methods for Partial Differential Equations.*

77. M. Holtz, *Sparse Grid Quadrature in High Dimensions with Applications in Finance and Insurance.*

78. Y. Huang, R. Kornhuber, O.Widlund, J. Xu (eds.), *Domain Decomposition Methods in Science and Engineering XIX.*

79. M. Griebel, M. A. Schweitzer (eds.), *Meshfree Methods for Partial Differential Equations V.*

For further information on these books please have a look at our mathematics catalogue at the following URL: `www.springer.com/series/3527`

Monographs in Computational Science and Engineering

1. J. Sundnes, G.T. Lines, X. Cai, B.F. Nielsen, K.-A. Mardal, A. Tveito, *Computing the Electrical Activity in the Heart.*

For further information on this book, please have a look at our mathematics catalogue at the following URL: `www.springer.com/series/7417`

Texts in Computational Science and Engineering

1. H. P. Langtangen, *Computational Partial Differential Equations.* Numerical Methods and Diffpack Programming. 2nd Edition
2. A. Quarteroni, F. Saleri, P. Gervasio, *Scientific Computing with MATLAB and Octave.* 3rd Edition
3. H. P. Langtangen, *Python Scripting for Computational Science.* 3rd Edition
4. H. Gardner, G. Manduchi, *Design Patterns for e-Science.*
5. M. Griebel, S. Knapek, G. Zumbusch, *Numerical Simulation in Molecular Dynamics.*
6. H. P. Langtangen, *A Primer on Scientific Programming with Python.*
7. A. Tveito, H. P. Langtangen, B. F. Nielsen, X. Cai, *Elements of Scientific Computing.*

For further information on these books please have a look at our mathematics catalogue at the following URL: `www.springer.com/series/5151`

MIX
Papier aus verantwortungsvollen Quellen
Paper from responsible sources
FSC® C105338

If you have any concerns about our products,
you can contact us on
ProductSafety@springernature.com

In case Publisher is established outside the EU,
the EU authorized representative is:
Springer Nature Customer Service Center GmbH
Europaplatz 3, 69115 Heidelberg, Germany

Printed by Libri Plureos GmbH
in Hamburg, Germany